Time and History

MAKING SENSE OF HISTORY

Studies in Historical Cultures

General Editor: Jörn Rüsen, Allan Megill and Alan Confino in Association with Angelika Wülff

Volume 1
Western Historical Thinking: An Intercultural Debate
Edited by Jörn Rüsen

Volume 2
Identities: Time, Difference, and Boundaries
Edited by Heidrun Friese

Volume 3
Narration, Identity, and Historical Consciousness
Edited by Jürgen Straub

Volume 4
Thinking Utopia: Steps into Other Worlds
Edited by Jörn Rüsen, Michael Fehr, and Thomas W. Rieger

Volume 5
History: Narration, Interpretation, Orientation
Jörn Rüsen

Volume 6
The Dynamics of German Industry: Germany's Path toward the New Economy and the American Challenge
Werner Abelshauser

Volume 7
Meaning and Representation in History
Edited by Jörn Rüsen

Volume 8
Remapping Knowledge: Intercultural Studies for a Global Age
Mihai Spariosu

Volume 9
Cultures of Technology and the Quest for Innovation
Edited by Helga Nowotny

Volume 10
Time and History: The Variety of Cultures
Edited by Jörn Rüsen

TIME AND HISTORY: THE VARIETY OF CULTURES

Edited by
Jörn Rüsen

Berghahn Books
New York • Oxford

Published in 2007 by

Berghahn Books
www.berghahnbooks.com

Library of Congress Cataloguing-in-Publication Data

Time and history : the variety of cultures / edited by Jörn Rüsen.
p. cm. — (Making sense of history)
Includes bibliographical references and index.
ISBN 978-1-84545-349-7 (hardback : alk. paper)
1. Time. 2. History. 3. Civilization, Oriental. 4. India—Civilization.
I. Rüsen, Jörn.

BD638.T5435 2007
115—dc22

2007023971

British Library Cataloguing in Publication Data

A catalogue record for this book is available from the British Library.

Printed in the United States on acid-free paper.

ISBN 978-1-84545-349-7 hardback

Contents

Part II: History

Preface to the Series

Jörn Rüsen

At the turn of the twenty-first century the term "history" brings extremely ambivalent associations to mind. On the one hand, the last decade has witnessed numerous declarations of history's end. Whether in reference to the fundamental change of the global political situation around 1989/90, or to so-called postmodernism, or to the challenge to Western dominance by decolonization and multiculturalism, "history" as we know it has been declared to be dead, outdated, overcome, and at its end. On the other hand, there has been a global wave of intellectual explorations into fields that are "historical" by their nature: the building of personal and collective identity through "memory"; the cultural, social, and political use and function of "narrating the past"; and the psychological structures of remembering, repressing, and recalling. Even the subjects that seemed to call for an "end of history" (globalization, postmodernism, multiculturalism) quickly turned out to be intrinsically "historical" phenomena. Moreover, "history" and "historical memory" have entered the sphere of popular culture, from history channels to Hollywood movies. They have also become an ever more important factor in public debates and political negotiations (the discussions about the aftermath of the wars in the former Yugoslavia, European unification, or the various heritages of totalitarian systems, to name a few). In other words, after "history" was declared to be at its end, "historical matters" have come back with a vengeance.

This paradox calls for a new orientation or at least a new theoretical reflection. Indeed, it calls for a new theory of history. Such a theory should serve neither as a subdiscipline reserved for historians nor as a systematic collection of definitions, "laws," and rules claiming universal validity. What is needed is an interdisciplinary and intercultural field of study. Hayden White's deconstruction of the narrative strategies of the nineteenth-century historicist paradigm somehow came to be regarded by many as historical theory's final word, as if the critique of the discipline's claim to rationality could put an end to the rational self-reflection of that discipline—as if this critique were not a rational self-reflection in itself.

In the late 1980s the "critical study of historical memory" began to substitute for historical theory. Overlooked in this substitution is the fact that any exploration into the ways of historical memory in different cultural contexts not only crosses into the field of critical studies, but also contains the keystones for a more general theory of history. Analysis of even a simple instance of historical memory cannot avoid questions of the theory and philosophy of history. And vice versa: the most abstract thoughts of philosophers of history have an intrinsic counterpart in the most secular procedures of memory (for example, when parents narrate past experiences to their children, or when an African community remembers its own colonial subordination and its liberation from it). As long as we fail to acknowledge the intrinsic connection between the most sophisticated historical theory and the procedures of historical memory most deeply imbedded in the culture and the everyday life of people, we remain caught in an ideology of linear progress that regards cultural forms of memory simply as interesting objects of study instead of recognizing them as examples of "how to make sense of history."

The series "Making Sense of History" aims at bridging the gap between historical theory and the study of historical memory. It is not exclusively related to historical studies as one academic discipline among others in the humanities and the social sciences. Its contributions, from virtually all fields of cultural and social studies, explore a wide range of phenomena that can be labeled "making historical sense" (*Historische Sinnbildung*). The series crosses the boundaries between academic disciplines as well as those between cultural, social, political, and historical contexts. Instead of reducing historical memory to just another form of the social or cultural "construction of reality," its contributions deal with concrete phenomena of historical memory: it seeks to interpret them as case studies in the emerging empirical and theoretical field of "making historical sense." Along the same line, it includes rather theoretical essays that aim not only to establish new methods and theories for historical research but also to provide perspectives for a comparative, interdisciplinary, and intercultural understanding of what could be called the "global work of historical memory" or the "cultural strategy to orient human life in the course of time."

This does not imply the exclusion of critical evaluations of the ideological functions of historical memory. But it is not the major aim of the series to find an ideal, politically correct, ideology-free mode or method of how to make sense of history. The aim is rather to explore the cultural practices involved in generating historical sense as an extremely important realm of human thought and action, the study of which may contribute to new forms of mutual understanding. In an age of rapid globalization that manifests itself primarily on an economic and political—and, much less so, on a cultural level—finding such forms is an urgent task.

This is why this series starts out with a volume documenting an intercultural debate. This volume questions whether or not the academic discipline of "history"—as developed at Western universities over the course of the last two hundred years—represents a specific mode or type of historical thinking that can be differentiated from other forms and practices of historical consciousness. Subsequent volumes present history as a genuinely interdisciplinary field of research. Historians, anthropologists, philosophers, sociologists, psychologists, and literary theorists, as well as specialists in fields such as media and cultural studies, explore such questions as: What constitutes a specifically historical "sense" and meaning? What are the concepts of time underlying different historical cultures? Which specific forms of "perception" inform these concepts, and which general problems are connected with them? What are the dominant strategies used to represent historical meaning? What function does the generating of historical sense fulfill in practical life?

Ranging from general overviews and theoretical reflections to case studies, the essays cover a wide range of contexts related to the question of "historical sense," among them topics such as collective identity, the psychology and psychoanalysis of historical memory, and the intercultural dimension of historical thinking. Additionally, the books of this series address the place of history in the humanities, and the humanities in general as an essential place for sense generation in modern societies. Even modes of sense generation that are not specifically historical can be dealt with, as long as they share with history the concern for coming to terms with time in human life. For the most part, historical memory is not an arbitrary function of the cultural practices used by human beings to orient themselves in the world in which they are born, but covers, rather, those domains of human life that seek to orient life temporally. These domains demand mental procedures for connecting past, present, and future that became generalized and institutionalized in the West as that specific field of culture we call "history." The areas of human thought, action, and suffering that call for a specifically "historical thinking" include (1) the construction and perpetuation of collective identity, (2) the reconstruction of patterns of orientation after catastrophes and events of massive destruction, (3) the challenge of given patterns of orientation presented by and through the confrontation with radical otherness, and (4) the general experience of change and contingency.

In accordance with the general aim of the series "Making Sense of History" to outline a new field of interdisciplinary research (rather than to offer a single theory), the volumes in the series are not designed to establish a new historiographical approach. Rather, they seek to participate in an interdisciplinary study of historical cultures and related subjects. One focus, for instance, is on the notion of collective identity. General theoretical aspects and problems of this field are considered, most importantly the interrelationships among identity, otherness, and representation. But case studies of the construction of gender identities (especially of women), of ethnic identities, and of different forms and politics of national identity are also included. The essays on this subject point out that any concept of identity as being disconnected from historical change not only leads to theoretical problems, but also covers over the fact that most modern forms of collective identity take into account the possibility of their own historical transformation. Thus the essays in this series that are concerned with identity suggest that identity ought to be regarded not as a function of difference, but as a concrete cultural and ongoing *practice* of difference. They show that the production of "sense" is an epistemological starting point, as well as a theoretical and empirical research-field in and of itself.

Another volume focuses on the psychological construction of time and history, analyzing the interrelation between memory, morality, and authenticity in different forms of historical or biographical narration. The findings of empirical psychological studies (on the development of temporal and historical consciousness in children, or on the psychological mechanisms of reconstructing past experiences) are discussed in the light of attempts to outline a psychological concept of historical consciousness around the notions of "narration" and the "narrative structure of historical time."

A special volume is dedicated to psychoanalytical approaches to the study of historical memory. It reconsiders older debates on the relation between psychoanalysis and history and introduces more recent research projects. Instead of simply pointing out some psychoanalytical insights that can be adopted and applied in certain areas of historical studies, this volume aims at combining psychoanalytical and historical perspectives, thus exploring the history of psychoanalysis itself, as well as the "unconscious" dimensions underlying and informing academic and nonacademic forms of historical memory. Moreover, it puts special emphasis on transgenerational forms of remembrance, on the notion of trauma as a key concept in this field, and on case studies that may indicate directions for further research.

Cultural differences in historical thinking that arise from different time concepts are the subject of another volume. With a view to encouraging comparative research, this volume offers general essays and case studies written with the intention of providing comparative interpretations of concrete material, as well as possible paradigmatic research-questions for further comparisons. In the light of the recent resurgence of ethnocentric world-views, this volume focuses on the question of how cultural and social studies should react to this challenge. It aims at counteracting ethnocentrism by bridging the current gap between a rapid globalization manifesting itself in ever increasing economic and political interdependencies of states and continents, and the corresponding lack of mutual understanding in the realm of culture. Its essays illustrate the necessity of intercultural communication concerning the common grounds of the various historical cultures as well as concerning the differences between them. Such communication seems not only a possible, but indeed a necessary presupposition of any attempt to negotiate cultural differences on a political level, whether between states or within the increasingly multicultural societies in which we live.

The special emphasis the series puts on the problem of cultural differences and intercultural communication shows the editors' intentions to aim beyond the realm of merely academic interest. Intercultural communication represents a great challenge, as well as a great hope, to a project committed to general theoretical reflection on the universal phenomenon of "remembering the past." Despite the fact that cultural difference has become something like a master phrase of the 1990s, this topic is characterized by a paradox quite similar to that underlying the current fate of the notion of history.

The past fifteen years have witnessed intensified interventions by the industrialized states in the political and economic affairs of the rest of the world, as well as an increased (if sometimes peculiar) appropriation of modern economic and political structures in developing countries, and in the former or still officially "communist" states. But this process of mutual rapprochement on the political and economic level is characterized by a remarkable lack of knowledge of, or even interest in, the cultural and historical backgrounds of the respective nations. Thus, the existing official forms of intercultural communication lack an adequate cultural dimension, leaving the themes and problems analyzed in this book series (identity, memory, cultural practices, history, religion, philosophy, literature) outside of what is explicitly communicated, as if such matters would not strongly affect political as well as economic agendas.

On the other hand, the currently dominant approaches of cultural theorists and critical thinkers in the West either claim that an intercultural communication concerning the common grounds of cultural identities is impossible—based on the assumption that there are no common grounds (the hypostatization of difference)—or politicize cultural differences in such a way that they are relegated to mere material for the construction of cultural subject-positions. Despite their self-understanding as "critique," these approaches amount to the exclusion of culture on the level of state politics and economic exchange alike. Thus, cultural theory seems to react to the marginalization of culture by way of its own self-marginalization.

The series "Making Sense of History" intends to challenge this marginalization by introducing a form of cultural studies that takes the term *culture* seriously again, without dissolving it into identity politics or into a hypostatized concept of unbridgeable difference. At the same time it wants to reintroduce a notion of "historical theory" that no longer disconnects itself from historical memory and remembrance as concrete cultural practices, but seeks to explore those practices, interpreting them as different articulations of the universal (if heterogeneous) effort to make sense of history. Thus, the series relies on the idea that an academic contribution to the problem of intercultural communication should assume the form of a new opening of the academic discourse to its own historicity and cultural background, as well as a new acknowledgement that other cultural, but nonacademic, practices of "sense-formation" are equally important forms of human orientation and self-understanding (in their general function, in fact, not much different from the efforts of academic thought itself).

Such a reinscription of the universal claims of modern academic discourses into a variety of cultural contexts, with the object of providing new starting points for intercultural communication, is an enterprise that cannot be accomplished or even outlined in a series of a few books. Consequently, "Making Sense of History" should be regarded as something like a first attempt to map out one possible field of research—the field of "historical cultures"—that might help us to achieve this aim.

The idea of the book series was born in the wake of the successful completion of a research project on "Making Sense of History: Interdisciplinary Studies in the Structure, Logic and Function of Historical Consciousness—An Intercultural Comparison." This project took place at the Center for Interdisciplinary Studies (ZiF) of the University of Bielefeld, Germany, in 1994/95. It was partly supported by the Kulturwissenschaftliches Institut (KWI) Essen (Institute for Advanced Study in the Humanities at Essen). A selection of contributions to a series of conferences and workshops included in the project f orms the core of the different books of this series. The arranging, revising, and editing of the different texts occupied the next several years, with the first volume coming out in 2002. In the meantime the series has enlarged its perspectives by bringing in other projects of the Kulturwissenschaftliches Institut and of its partners all over the world.

I would like to express my gratitude to the staff of the Center for Interdisciplinary Study at the University of Bielefeld and of the Institute for Advanced Study in the Humanities at Essen. I also want to thank the editors and co-editors of each of the volumes in this series and, of course, all the contributors for the effort and patience they expended to make these books possible. Finally, my thanks go to Angelika Wulff for her efficient management of this series and to my wife Inge for her intensive support in editing my texts.

Introduction

Jörn Rüsen

For most academics all over the world the concept of history is deeply influenced by the feature of historical studies as an academic discipline. The world is full of very different manifestations of history: oral narratives, monuments, exhibitions, museums, films, street names, advertisements, not to mention the manifold presentations of the past in literature, music, and the Fine Arts. Nevertheless, at least in the minds of the professionals, history is the realm of the work of the historians. And here we get the impression of a great similarity and uniformity. The professionals all over the world follow similar concepts and rules. They share trends and changes, paradigms and shifts, controversies and school-formations. Yet their realm, the past as a matter of experience and interpretation, offers a totally different impression of diversity and multifariousness. Difference in space and time is overwhelming. We experience a permanent change of views on the essential nature of what history is about. Accordingly, the representation of the past in the cultural orientation of human life reflects this difference and variety to such a degree that it is difficult to identify one specific form as essentially historical.

This multifariousness of history in the representation of the past has already knocked at the doors of the professionals. It has challenged them to admit it and thus accord it influence in reshaping not only their work, but their ideas about their own work as well. There is no chance of excluding this irritating guest of difference and diversity in the house of the historians. Given that they reflect the roots of their own thinking and its functions in social and cultural practice they have to realize that difference, variety, and change have become the form and not only the content of their thinking.

This is evident in respect to the issue of historical identity. The question of who we and the others are, cannot be answered without telling stories, and since identity essentially is a matter of discriminating oneself from others, this differ-

Notes for this section begin on page 4.

ence has to be expressed by the narratives presenting and explaining identity. This is true for the professional narratives as well, since the logic of historical narration is determined by the issue of identity. Even its professional academic form does not follow an essentially different logic.[1]

Nevertheless, the essential relationship between historical thinking and cultural difference has not yet sufficiently become an issue in metahistory, let alone in the professional attitude of most of the historians all over the world. Their work is mainly committed to a national perspective. When it transgresses the limits of this perspective to venture into wider ones like European, Western, Asian, African, or even global history, the mode of realizing these wider perspectives does not normally reflect the related differences in doing history.

Against this attitude a growing awareness of a fundamental independence of historical thinking on the varieties and differences of historical thinking in practical life has started to move the discourse in historiography and metahistory.[2] This book wants to contribute to a discourse about and within history in which cultural diversity and change have become an essential element. It seems to follow the mainstream of current debates and discourses on difference in the field of culture and its related disciplines, the humanities. But it remains an open question whether it can only be read as a plea to end any idea of a unity of historical thinking and related claims for validity. Does it favor relativism? The only clear consequence that can be drawn from its findings is an effort to rethink the unity of historical thinking and its claims for validity by recognizing the cultural variety, difference, and change in historical thinking. The academic (to avoid the misleading word "scientific") mode of history within which difference and variety is presented should be understood as an element within this variety, difference, and change. The fact that this mode has been established in all modernized or modernizing countries all over the world indicates a unifying tendency of historical thinking into the future. This tendency will not dissolve the cultural multifariousness of historical thinking outside the realm of the professionals, but it should be understood and pursued as a comprehensive discourse about this multifariousness within which the recognition of difference is of main concern. To start such a motion means to inform the professional historians about different cultural foundations and manifestations of historical thinking. The contributions of this book may serve to deepen and enlarge this knowledge.

The first part of this book is related to time as the most important and fundamental category of historical thinking. History is a specific way in which humans deal with the experience of temporal change. The way they realize it essentially depends upon pre-given or underlying ideas and concepts of time. It is not the intention of this book to cover the whole realm of time concepts, but to make this realm visible, to open up its dimension in general and to give paradigmatic examples of difference in the cultural practice of making sense of the experience of time in human life. The first text of this part gives a general typological overview of the basic category that all sorts of history depend upon: time. The following two texts present time concepts of a human life form that we can call "archaic." Klaus E. Müller draws an outline of dealing with time in the widespread and long-lasting culture of early agrarian societies. Fritz Kramer's case study on time concepts of the Dimodonko people can be put into this outline, although it refers to field research in the late 1980s and early 1990s. The next essays refer to so-called world civilizations, providing an overview of the idea of time in China, India, and the Islamic world. Only the last chapter of this part refers to the West. Here it is literature that informs us about the modern understanding of time in the cultural orientation of human life.

The constellation of these contributions seems to be arbitrary. But it at least characterizes the essential elements of diversity in different ways. It includes a general theory and typology of time concepts that argues for anthropological universality and for a typological representation of diversity at the same time. The texts by Müller and Kramer indicate difference by presenting ideas of time, that can be characterized as contrasting with those of our time. In the essays treating China (Achim Mittag), India (Georg Berkemer), the Islamic world (Otfried Weintritt), and the West (Harro Müller), four examples of difference on the level of so-called world civilizations are presented.

The second part of this volume proceeds from time to history. It shows how sense generation related to the experience of temporal change ends in the interpretation of changes with a very specific character. Now time is incorporated into a chain of events to which people living in the present refer in order to understand their present life form and to know about its future perspective. Again the emphasis lies on variety and difference. Nevertheless, the first article of this part (by Aziz Al-Azmeh) denies the essential relevance of cultural difference in writing history, thus pointing to the danger of overlooking general trends and features of history across different cultures. The other texts in this part—case studies on the Islamic world, India, and Japan—go beyond simply presenting one comprehensive feature of historical thinking in these parts of the world, which would only confirm Spenglerian prejudices about the nature of cultures and their difference. Instead, they give insights into processes and functions, communicative forms, political strategies, and controversies in writing history. Abdullahi A. An-Na'im dissolves any stereotype that posits one single mode of historical thinking as being typical of Islam. B. D. Chattopadhyaya gives an instructive example of how cultural plurality is an important issue in the intersection between history and education in modern India. D. L. Sheth shows how the historical work depends fundamentally upon political trends and constellations in India. Michael Gottlob presents postcolonial Indian historical culture in a different perspective: in the intersection and tension of general cultural trends in India's struggle for its place in the modern world. Masayuki Sato and Shingo Shimada address the issue of identity in modern Japan. Sato describes the conflicts between tradition and modernity brought about by the introduction of modern historical studies into the higher educational system, and Shimada analyzes the political strategies of using history to create a modern concept of national identity.

In respect to the purposes of the historiographical and metahistorical discourse of today, how to relate similarities and differences to each other in getting a proper idea of what history is about remains an open question. But for the purpose of arguing with empirical evidence, it is useful to have examples illustrating historical discourses and discourses about history in different contexts. All the examples in this book make it evident that one cannot speak about historical discourses without referring to politics and its struggle for power. This struggle includes the question of how relevant modern historical thinking in the form of an academic discipline is for the politics of history. What can it bring about in politics, with its specific methodical strategies and claims for validity? Is it only a tool or even a weapon in the politics of memory and identity building of groups and nations?

Contrasting different contexts of acting history, by relating cultures and times in a comparative perspective will enrich our awareness of the power of history in human life, of its role in shaping different life forms with sense and meaning. At the end it should be evident that the variety of examples stand for the unity of mankind, in so far as all of them present the human effort to understand the present and expect the future by understanding and interpreting the past. The variability in realizing this cultural procedure of sense generation in conceptualizing time and history should strengthen efforts to understand the work of the historians and pursue the metahistorical discourse with a higher awareness of the "dignity of difference"[3] as a quality of the unity of humankind.

It took a rather long time to compile the contributions to this book. Originally they were published in German, and for the English publication they had to be revised. I would like to thank the authors for their patience with the long procedure of bringing this book to print, and express my deepest gratitude to Angelika Wulff for her enormous engagement in guiding its realization. Finally, I would like to thank Marion Berghahn for her wisdom in treating authors and editors.

Notes

1. Jörn Rüsen, *History: Narration – Interpretation – Orientation* (New York 2005); Jürgen Straub, ed. *Narration, Identity, and Historical Consciousness* (Making Sense of History, vol. 3) (New York 2005).

2. Eckhardt Fuchs and Benedikt Stuchtey, eds. *Across Cultural Borders: Historiography in Global Perspective* (Lanham 2002).

3. Jonathan Sacks, *The Dignity of Difference: How to Avoid the Clash of Civilizations* (London 2003).

Part I

TIME

CHAPTER 1

Making Sense of Time

Toward a Universal Typology of Conceptual Foundations of Historical Consciousness

Jörn Rüsen

The following argumentation is developed in the context of research dedicated to historiography in a comparative perspective.[1] Such a comparison can be easily done within a cultural context that is grounded on the same or at least on similar principles of understanding the past as history. Substantial comparative research and interpretation of Western historical thinking has been done. It is much more difficult to compare treatments of the past that lead to historical thinking in an intercultural perspective. Not much work has been done in this field; and such work as there is tends to take the most advanced form of historical thinking, namely the Western one, as a parameter, and look at other cultures in terms of similarities and differences. This is where the problem lies: one mode of historical thinking is taken as a parameter of comparison regardless of what the other mode of thinking may be—which can lead to a distortion of perspectives.

What can we do to avoid this distortion? First of all, there is a need to develop the parameter of comparison in a theoretical way, so that we can ascertain how the perspective of comparison has been worked out, what it makes visible, and what it tends to obscure or hide.[2] To keep the conceptual frameworks of one culture from dominating those of another, it is necessary to ground this theory by asserting fundamental and universal elements of man's relationship to the past.

This sort of anthropology of time concepts is abstract and lets all the differences vanish. However generalized it may be, it can serve as a starting point. In order to bring the differences into view, it is necessary to develop this

Notes for this section begin on page 17.

anthropology of time concepts into a typology of different categories, thus making these time concepts more concrete.

Time is a basic dimension of human life. It is embedded in growth and decline, birth and death, change and continuity. It is structurally distracted by contingency and can only be pursued by following temporal perspectives in the cultural framework of human activity and suffering. There is no cultural orientation of human life without a complex interrelationship of memory and expectation. Husserl has characterized the two underlying main intentions of the human consciousness as *retention and protention*.[3] The human mind[4] always mediates both by working through the experience of change and giving it a meaning by interpretation, which can function as a source and impulse for future perspectives.

History, as we are used to understanding it, cannot be found in all cultures and all times. But in every human culture the human mind conceptualizes time in a special way such that it distinguishes between different time dimensions related to past, present, and future (in a very variable way), and in every human culture the experience of the past is brought into a pattern of significance that makes temporal changes in the present day world conceivable and understandable, enabling people to guide their activities (and suffering) along the line of an idea of change, however vague it may be. Then they can understand what change means and how it is related to the human mind, its threats and hopes, its expectations and memories. All cultures share one mental procedure and cultural practice that brings about this interpretation by presenting change itself as sense bearing and significant for the purposes of human activity. This is the universal and fundamental mental strategy of telling a story.[5]

By *narration,* the human mind masters contingency.[6] It gives a meaning to a temporal sequence of events occurring in the changes of the world. So, human agency can be culturally guided by an idea of the temporal extension of the human world and the human self. This idea mediates the experience of change with the intentions, purposes, and values of the human mind into one meaningful totality of time. The culturally dominant way of presenting this totality is telling a metanarrative (or master-narrative), which in ancient societies had a cosmological extension and a religious setting. In later periods, it was normally realized as a universal history, which combined secular, religious, and very often ritual elements in a dynamic way.[7] Modern historical thinking tries to present such a totality of time in purely secular terms by using the mental power of human agency as a comprehensive and constitutive source of meaning and significance.[8]

This idea of narration as a source of historical thinking should be brought together with a complex concept of time. Doing so, I will enlarge the focus within which the work of historians appears in such a way that its acquaintance with other cultural practices and orientation strategies will be stressed and structurally interrelated.

I would like to start with an artificial distinction between experience and interpretation of time, concentrating the focus on *experience* (1). While there is no experience without interpretation and no interpretation without experience, it makes sense to distinguish between them: every interpretation is an interpretation of something, and every experience is a challenge to a certain kind of mental activity called interpretation. Interpreted experiences can function as cultural means of human activity and suffering.

The first step in interpreting the experience of time is giving it a *dimension,* a space in the realm of human consciousness where it gets a specific treatment—aesthetic, emotional, political, technological, cognitive, etc. (2). Within these dimensions, the experience of time eventually gets its *sense* by becoming related to our intentions, which guide our memories and future expectations (3). The way of generating this sense can be furthermore differentiated in respect to its *modes* (4). A special issue is the question whether there is a logic of *development* in time concepts over the long term across different cultures (5). Finally, I would like to stress the *limits* of making sense in dealing with time and the principles of change within the human treatment of time as a basic dimension of practical life (6).

My distinctions and specifications have a *typological* character. This means that they are artificially constructed in order to bring to light and make understandable complex phenomena and changes in reality. Such ideal types make specific characteristics visible by isolating them from others; the distinctive nature of the findings can be identified, described, interpreted and understood as configurations of these single specific attributes. Logically, types are clearly distinct and separated from each other; empirically, they rarely appear in a pure form, but mainly in mixtures, which have to be reconstructed.[9]

Experience

Starting with the *experience of time,* I would like to point out an anthropological universal that underlies every human concept of time, namely, the simple fact that humans must make sense of time in order to be able to live through the temporal changes of their world. They have to think about it, to give it meaning and significance in order to come to terms with the changes of their world and themselves. They have to bring about a viable interrelationship of the intentions of their consciousness and the real changes in their world, and they have to mediate their experiences of the past with their expectations of the future. Time as a matter of experience has the character of a challenge to the human mind. Here, it causes mental movement and activity by which it becomes appropriated (in the sense of interpreted) and gets the feature of meaning and significance. Rise and fall, birth and death, growth and decay, youth and old age—all these occurrences have to be put into a pattern of significance that lets those who experience it come to terms with them. I call this challenging character of time *contingency.* There is a structural difference between the pregiven change of the human world (the external one as well as the internal one) and the human self with its specific temporal dimension of remembering and expecting, which is loaded with values and norms. This difference has to be mediated, and history is the cultural practice of doing it.[10]

Every time experience has its specific character in this realm of mediation between contingency and its appropriation and shaping by human interpretation. And every mode of time experience has these two sides: the challenge of contingency and the response of human sense generation.

In this tense dimension of experience, time may occur in very different modes. I will enumerate just some of these modes: duration, standstill, fleetingness, improvement or worsening, return to origins, a rupture of time by an unusual event, acceleration, emptiness, contemporaneity of the noncontemporaneous (*Gleichzeitigkeit des Ungleichzeitigen*), the melting of horizons of understanding into a diachronic relationship. I would also like to mention here the *unio mystica* of present and past, in which both lose their difference and distance so that people of the present feel immediately contemporaneous to the people of the past. A specific time experience is related to the value of time: it may be short, fleeting, precious; time can be looked at as something to be used in a meaningful way; otherwise, one will miss a chance. The specific natural experience of time as emergence and passing, of birth and death, has already been mentioned.

The general time quality of contingency can be differentiated according to its frame of reference, in which the experience of temporal change gets a threatening power. So, in respect to a moralistic cosmology, contingency means misdemeanor; in respect to historical development, contingency has the character of an event; in respect to a concept of temporal development, contingency can have the character of an anachronism; in a concept of evolution (in the modern understanding), contingency has the character of accident; in respect to a theoretical order, it is a mistake, etc.

Dimensions

Interpreting these different time experiences, the human mind shapes time in the form of a dimension, as an extension of existence in respect to the different categories that are used to understand the human world and the human self. A prime example is, one can start with the fundamental difference between nonhuman and human time. In ancient societies, one would say that there is a difference between divine time and human time. We can discriminate between natural time and human time, and we can differentiate this discrimination further: natural time has physical, geological, and biological dimensions, and human time can be subdivided into a psychological, a biographical, a social, and a historical time.

Another dimension of time refers to the fundamental temporal structure of human agency. Here one may distinguish between presence as the presently living world, past as the realm of experiences, and future as the realm of projects, projections, and prediction. These three time dimensions of the human life-world can each be distinguished by a completely different time, where the constitutive principles of meaning and significance are at work. This reflects the traditional difference between secular and divine time. This other dimension contains those essential elements of sense that were used by the human actors in pursuing the time of their life-world. In archaic societies, this time is the time of origin (*arché*), which is the essence of everything. In other cultures this time is conceptualized as a final goal of temporal change (*eschaton*) with elements of an unpredictable future. Origin and future as temporal sources of sense can become identified or distinguished, and they can be mediated into exceptional moments of the present.

Sense

Dimensioning time in the described way is the first step in giving it sense, i.e., integrating it into the internal time dimension of the human mind where memory and expectation are loaded with values, purposes, threats, and hopes. In the shape of sense, time becomes a pattern of significance according to which people live their lives, organize their activities, stand their sufferings, place their life spans into intergenerational sequences, etc.

The shape of time in this pattern of significance depends upon the principles of significance, which are related to the temporal extension of human life between yesterday and tomorrow. These principles decide the way the human mind appropriates contingency by narration. By narration, time becomes, so to speak, a body of the human world and self. This body has integrated subjectivity and objectivity, values and facts, past, present and future, togetherness and otherness, identity and difference.

In this realm of sense, time has the mode of a moving force of the human mind, deciding upon the way that the people understand themselves and organize their activities in the course of time and come to terms with their suffering through temporal change. Here time is manifested as culture; it has the features of symbols and language, of images and concepts. Time gets a specific cultural manifestation, a mental power, with which the people pursue the temporality of their lives. In this realm of culture, where history gets its specific feature of sense generation and cultural orientation of human life, it is very useful to elaborate a typology of the different modes time can have, as such a mental power in human life orientation. In doing so, it becomes apparent that history is more than just one single type among others. In fact, it is a combination of different types of shaping time into a cultural pattern of human life orientation.

In the line of such a typology, one can distinguish between a mythical and a historical sense of time. It is *mythical* when it is grounded in the divine time of origin and future, strictly separated from the human time of practical everyday life. We speak of the *historical* sense of time when the order of temporal change in the human world is essentially brought about by an innerworldly chain of events. This historical sense can further be differentiated into different types of historical narration: traditional, exemplary, critical, genetic.[11] The exemplary type has an *allegorical* character that is widely known from the slogan "historia vitae magistra." Here the temporal chain of events stands for a metatemporal rule of human conduct.

The distinction between past, present, and future, which plays an important role in the mythical, historical, and allegorical sense of time, is given up in favor of another mode of this sense: the *mystical* one, where the three time dimensions fall together in a metatemporal moment.[12] *Kairos* is similar to this, since here the three-sided dimensions are also synthesized. Yet, it does not happen in a metatemporal moment, but in an eminent situation of historical time.[13] In the West, a well-known example of the concept of Kairos in religion is the idea of God's incarnation in Jesus[14] or in the secular world, say, in the enthusiasm over the American or French Revolution. In Chinese history, the "Grand Beginning" in 104 BCE could be characterized as Kairos, when a fundamental reform on the mental level took place and a new calendar was introduced.[15]

Across these differences, one can elaborate different types of the *religious sense* of time, including a general one where changes in the human world are significant for religious salvation, which is conceptualized as an occurrence on this temporal scale itself. A sacred sense of time can be manifested in different ways, such as a significatory relationship between different epochs, as it is conceptualized in the "*typological*" thinking where the Old Testament signifies the New, and vice versa. In a way, the various attempts in traditional China to conceptualize the past as a cosmologically ordered succession of two, three, or five cyclical periods[16] may be seen as examples conforming to the same mode of thinking. These "typological" time concepts include nature as well as morality, (and in the West) the end of time and eternity as well as the events of secular history. The religious essence of time covers the whole realm of temporal change of the human world. Its power is intensified through single events in which the forces of salvation accumulate. If this is the case, at the end of the human world, one speaks of an *eschatological* sense of time. This eschatological sense can be distinguished from the *apocalyptic* one, since it is not opposed to the time before, as is the case with apocalypses. Here, the time of salvation is completely separated from the time before, and there is no mediation between these two times. There is a gap between secular time and divine time at its end.

Such an apocalyptic element can be put into the significance of events within the course of secular time. In this case, one can speak of a *momentary* sense of time, where change and development come to a stop, and the chain between before and after is broken so that such a moment drops out of its historical connection with other moments and gets a significance of its own, which can be preserved and renewed by historical commemoration. Walter Benjamin has elaborated on this momentary sense of time.[17]

I would like to add three other types of temporal sense, all of which play a role in culture in general and historical thinking in particular. The sense of time is shaped by the *sequence of epochs.* The time of one's own life gets meaning from its relationship to this periodization. Another type is of *numerical character;* here, the numbers of years play the decisive role. The physical chronology changes into a magic of numbers, which shapes the course of time into a very specific texture of significance related to the arithmetic of chronology. It transforms the historical order of time into a natural one. Finally, I would like to mention another mode of giving the temporal status of events a specific significance: namely *synchronism* as an indication of an internal relationship of events. Synchronic occurrence of events means that they have something in common that is important to understanding what has happened.

Modes

In human life the sense of time works on different levels and in different modes. It can have the power of the unconscious or gain the structure of a highly elaborated philosophy or theology. In order to understand the variety of modes, I would like to propose a typology that distinguishes between three different ways in which the sense of time works in human life. One could call these ways the *functioning,* the *reflective,* and the *operative* modes. I found the best interpretation of these three modes in Paul Ricoeur's analysis of three kinds of mimesis in making sense of time by narration.[18]

The sense of time is an element in everyday life activities and sufferings. Human agency takes place under pregiven circumstances, which have their own temporal extension and their built-in symbolic interpretation. We are used to thinking of culture as "invention" or "construction." The level I have in mind (Ricoeur calls it "Mimesis 1") is the level of human life where this invention and construction are forces of living human life. Here, the people are invented and constructed by the way their activity is determined by culturally powerful orders of time.

On the second level (Mimesis 2), the people explicate and interpret their own activities and sufferings by reflecting on their inbuilt sensebearing elements. By doing so (Ricoeur stresses only the cultural practice of narration, since it is essentially related to time), they "invent" or "construct" the cultural orientation of their lives. Here, time is a matter of active practical sense generation, done in the specific realm of cultural practices. This is the place where priests, soothsayers, historians, chief-praisers, and other specialists in the sense of time have the say—in their communications with those who need an elaborated form of presenting the cultural framework of their practical life, apart from its simple pursuit.

The third level (Mimesis 3) combines the reflective activities of sense generation with the pursuit of activities to which the generated sense is dedicated. Now, the "objective" courses of human activities are deliberately pursued by subjective orientations. People deliberately intend what they do and they understand what they do in a cultural framework of meaningful human life. On this level, the constructions and inventions become real, so they can construct and invent the people and the symbolic world to which they belong.

I refer to this threefold time dimension of human life, precisely described by Paul Ricoeur, because one can understand the *historicity of human life* by looking at their interrelationship. Each level is necessarily interrelated with the other levels. There is no simple pursuit of human activity and suffering without reflection on their cultural motives and orientations, and such an orientation, of course, can only be given if those who follow it relate their cultural practices to this pursuit. In addition, this relation itself brings about an element of reflection within the pursuit of activity and suffering itself.

Reflecting on this interrelationship, we discover a specific temporality within it: this is the permanent distraction of time experience, which does not completely fit into the unreflected orientation of human life, making the reflection of the second level permanently necessary. Further, the product of this reflection merges as an input into the life practice, where it meets new challenges. This temporal dynamics transgresses the time dimension of level 2 where people are concerned to make sense of time and speak of history.

The temporal interrelation of the three levels of pursuing time in human life generates a historicity that even transgresses the realm of historical consciousness. This historicity is clearly different from historicity as it is reflected on level 2. Compared with the cultural practice of doing history (in the different types of sense of time), this transgressing historicity is "real," whereas the people who are making sense of the time of their lives give this sense an "ideal" form. This ideal form becomes imbedded in the practice of life, moving or at least influencing its temporal change. But one cannot say that this ideal construction of interpretation already includes the reality it produces. So, there is always a surplus of historicity in the pursuit of human life that transgresses the sense of time produced by the people who have to come to terms with their time experience.

History, thus, can be understood as a dynamic temporal extension of human life that has to be brought into a concept of sense, which is a part of this dimension itself. In this respect, history is always beyond the horizon of historians and those who work on concepts of time. It is this transcendent character of history that will keep the historians busy.

Developments

Conceptualization of time has changed substantially over time. On the level of theory of history, this change has been addressed with the question of a general direction. Is there a general development of time concepts that covers or goes across different cultures? Today the answer to this question seems to be a clear "no." The traditional philosophy of history of the Kantian, Hegelian, or Marxian kind seems to have completely lost credibility. The critique of its Western ethnocentricity has reached a crucial point (at least for cross-cultural comparison).

I think, nonetheless, that valid arguments remain for a different view: there is strong empirical evidence that at least in human ontogeny there are general genetic processes that have a clear direction, i.e., that are irreversible and have to be passed through by every human being. Without falling into the trap of onto- and phylogenetic parallelism, one can nevertheless argue that these early steps of ontogeny indicate a developmental logic of change in time concepts, the structure of which can be applied to the history of the human species.[19] The result of this application can be called a new philosophy of history. It differs essentially from the old one in its logic. It is *reconstructive and not teleological.*[20]

This irreversibility of change can be observed as well: in the early stages of cultural development, time is seen as an objective quality of pregiven things and events in the world (everything has its own time); in later stages and to a very high degree in modern culture, time is seen as a construction of the human mind. Human time has become distinguished from natural time, and even the latter is (epistemologically) understood as constructive (with an element of realism). The "objective" time of nature has become "rationalized" in the form of time reckoning by sheer numbers, losing its own pregiven meaning and becoming a matter of getting meaning by subjective construction. In addition, the "subjective" time of human life has become a matter of subjective competence: time in its projective function in human activity follows subjective rules (save time, use time for valuable purposes, don't waste time, etc.). In both cases, one can speak of a process of subjectivation (along the line of "rationalization" in the Weberian meaning) in so far as they have been exposed to the commands of the human mind (measurement in abstract numbers and objectives of the human will).

An example of the increasing separation of time concepts from their "objective" grounding is the well-known five hundred-year cycle, which played a significant role in Chinese historical thinking. Being originally tied to astronomical calculations and cosmological speculations, it was then transformed into a concept of continuity of the cultural tradition (by Han Yu, 768–824) and later into the highly abstract concept of the "correct succession of the Way" (*daotong*).[21] Today, one can discuss this issue of "rationalization" in close relationship to historical chronology: vis-à-vis the globalization process and clash of different cultural traditions of time reckoning in history, the best way to proceed is to use a scheme of extensive meaninglessness (in respect to dynasties, traditions of sacred time, etc.). This can be understood as a plea for simply counting years in a chronologically abstract manner.

Clearly, it makes sense to speak of a universal tendency of subjectivization (without a loss of realism). This tendency deeply influences the daily life of ordinary people. It has brought about the human attitude of being responsible for the time of one's own life and has irretrievably left behind the conviction that time is beyond the competence of human subjectivity.

Limits

Typologies of sense generation should pay attention to limits. I would not like to discuss the question of whether the idea of "sense" even allows a contradicting idea of senselessness. Ideed, Niklas Luhmann has made clear that it is not possible: speaking of limits of sense and senselessness presupposes the idea of sense and does not negate it. But this is not my point; rather I would like to elicit challenges to sense generation that cast pregiven time-related concepts of sense into doubt and urge (on level 2 of Ricoeur's threefold mimesis concept) the work of explication and in certain cases encourage alteration and change in the concepts of time itself.

Here I distinguish between three different modes of provoking pregiven time concepts on the level of their reflective treatment. A "normal" challenge provokes the potentials of interpretation in order to apply the already developed patterns to the new experience. In this case, the work of reflection and application confirms the time concepts. But some time experiences make it necessary to change the concepts, since they do not fit into the pregiven framework of interpretation. In this case, I have in mind "critical" time experiences. A prominent Western example is in the work of Herodotus (the "father of Western historiography"); it can be understood as a response to an orientation crisis, that came about because of the new power of Athens that arose after the Persian War.[22] A Chinese example of similar import is the breakdown of the traditional political order of the Western Zhou, conceptualized by religion, the response of which is Confucius' new way of thinking. In other words, religion became rationalized into morality. The *Spring and Autumn Annals* stand for this transformation (even more the *Zuo commentary*).[23] Another Chinese example is Sima Qian's work, understood as a response to the "critical" experience of the unification of the empire and the rapid fall of the Qin dynasty.

But time experience can have a radicalized critical status that does not change the pregiven framework of interpretation but destroys it without replacing it with a new one. In this case I would speak of a "traumatic" time experience. The best-known example of this sort of experience is the Holocaust, a theme that shows in itself how such an experience can be dealt with in the realm of historical thinking. With respect to China, one could think of a number of historical examples of chaos, turmoil, invasion by "barbarians," etc., that might have had a traumatic character; in fact, this has been argued for with regard to the collapses of the Jin and Song dynasties and the Mongol subjugation of all China, the Taiping Rebellion in the mid nineteenth century, and in more recent times, the Great Leap Forward in Communist China.[24]

Every sense concept has a limit that produces its permanent challenge. This can be easily explained by the interrelationship of the three levels I have described according to Ricoeur. The input of sense concepts of time into the process of human activity and suffering is already a limitation of their ability to interpret the human world, since they too become an element of temporal change and cannot be sufficiently interpreted by themselves. Time is always more than the concepts by which it is interpreted; it always transgresses them in the dynamics of the historicity of human life. But this limit is not an impermeable border; on the contrary: the permanent challenge of time experience that goes beyond the concepts of its understanding keeps this understanding in permanent movement to renew, apply, and change them according to this challenge. So, the limit of time concepts is part of their internal historicity. They always transgress their own limits in the course of the time we call history, and we all are a part of it and we try to make sense of knowing that. By doing so, we acknowledge historical time as always already ahead of us.

Notes

1. I would like to thank Achim Mittag for his extremely valuable comments, critique, and allusions to Chinese examples. The use of his proposals is, of course, exclusively my responsibility.

2. Jörn Rüsen, "Some Theoretical Approaches to Intercultural Comparison of Historiography," *History and Theory*, Theme Issue: *Chinese Historiography in Comparative Perspective* 35 (1996): 5–22.

3. Edmund Husserl, *Vorlesungen zur Phänomenologie des inneren Zeitbewußtseins,* 2nd ed., ed. Martin Heidegger (Tübingen 1980).

4. Speaking of "mind," I want to point out that its scope of meaning embraces what in the Chinese tradition was called *xin* (the heart-and-mind).

5. Cf. Paul Ricoeur's fundamental philosophical analysis: Paul Ricoeur, *Time and Narrative*, 3 vols. (Chicago 1984–1988). I have discussed narration as a fundamental operation of historical thinking in a closer relationship to historical studies: *Studies in Metahistory* (Pretoria 1993), 3–14; *Historische Vernunft: Grundzüge einer Historik I: Die Grundlagen der Geschichtswissenschaft* (Göttingen 1983); *Rekonstruktion der Vergangenheit: Grundzüge einer Historik II: Die Prinzipien der historischen Forschung* (Göttingen 1986); *Lebendige Geschichte: Grundzüge einer Historik III: Formen und Funktionen des historischen Wissens* (Göttingen 1989); *Zerbrechende Zeit: Studien über den Sinn der Geschichte* (Cologne 2001).

6. "Contingency" has a broad meaning. For my purposes I concentrate on "by chance," "occurrent by accident," "being out of order," "beyond a pattern of meaning." In Chinese, it fills the range of meaning between *cha* (inaccuracy), *bian* (deviation), and *luan* (disorder, chaos).

7. The best Chinese example of such a universal history is, of course, Sima Qian's *Shiji.*

8. Nevertheless religion remains an (often hidden) element of historical sense generation. See Jörn Rüsen, "Historische Methode und religiöser Sinn-Vorüberlegungen zu einer Dialektik der Rationalisierung des historischen Denkens in der Moderne," in Wolfgang Küttler, Jörn Rüsen, and Ernst Schulin, eds., *Geschichtsdiskurs 2: Anfänge modernen historischen Denkens* (Frankfurt/ M. 1994), 344–379.

9. The classical description of the form and use of ideal-types was given by Max Weber in "Die 'Objektivität' sozialwissenschaftlicher und sozialpolitischer Erkenntnis," in idem, *Gesammelte Aufsätze zur Wissenschaftslehre,* 3rd ed. (Tübingen 1968), 146–214; English in Max Weber, *The Methodology of the Social Sciences,* transl. and ed. Edward A. Shils and Henry A. Finch (New York 1949).

10. A remarkable confirmation in Chinese historical thinking can be found in the *Gongyang commentary* to the *Spring and Autumn Annals:* "To set to right things which have been thrown into chaos and to restore the world to order, there is none better than the *Spring and Autumn Annals*" (*Gongyang zhuan,* Aigong, 14th year).

11. Cf. Jörn Rüsen, "Historical Narration: Foundations, Types, Reason," *History and Theory* 26 (1987): 87–97; idem, *Studies in Metahistory,* 3–14, 63–84; idem, *Zeit und Sinn, Strategien historischen Denkens* (Frankfurt/ M. 1990), 153–230.

12. This also seems to be the case in Daoist ritual, the time aspect of which has been given elaborate treatment in Kristofer Schipper and Hsiu-huei Wang, "Progressive and Regressive Time Cycles in Taoist Ritual," in Thomas J. Fraser, N. Lawrence, and F. C. Haber, eds., *Time, Science, and Society in China and the West* (Amherst 1986), 185–205.

13. Here I refer to Paul Tillich, *Writings in the Philosophy of Religion,* ed. John Clayton (Berlin 1987), 53–72, 327–342; idem, ed., *Kairos* (Darmstadt 1926).

14. Mark 1: 15: "The time is fulfilled, and the kingdom of God is at hand."

15. See Michael Löwe, in idem, *Crisis and Conflict in Han China (104 BC to AD 9)* (London 1974), 17–36 "The Grand Beginning 104 BC".

16. For an overview, see Fung Yu-lan, *A History of Chinese Philosophy,* 2 vols., transl. Derk Bodde (Princeton 1952/53), vol. 2, 58–71; Joseph Needham, *Science and Civilisation in China,* vol. 2: *History of Scientific Thought* (Cambridge 1956), 232–261.

17. Walter Benjamin, "Theses on History," in Walter Benjamin, *Illuminations,* ed. Hannah Arendt (New York 1985).

18. Paul Ricoeur, *Zeit und Erzählung,* vol. 1: *Zeit und historische Erzählung* (Munich 1988), 90ff.

19. This is the strong thesis of Günter Dux, *Die Zeit in der Geschichte: Ihre Entwicklungslogik von Mythos zur Weltzeit* (Frankfurt/ M. 1989).

20. Günter Dux, *Historisch-genetische Theorie der Kultur: Instabile Welten: Zur prozessualen Logik im kulturellen Wandel* (Weilerswist 2000), passim.

21. For further remarks on this issue, see Achim Mittag, "Zeitkonzepte in China," in Klaus E. Müller and Jörn Rüsen, eds., *Historische Sinnbildung: Struktur, Funktion und Repräsentation des Geschichtsbewußtseins* (Reinbek 1997), 251–276, here 266ff and this volume chap. 4.

22. Christian Meier, "Die Entstehung der Historie," in Reinhart Koselleck and Wolf-Dieter Stempel, eds., *Geschichte - Ereignis und Erzählung (Poetik und Hermeneutik V)* (Munich 1973), 251–305.

23. Cf. Heiner Roetz, *Chinesische Ethik der Achsenzeit: Eine Rekonstruktion unter dem Aspekt des Durchbruchs zu konventionellem Denken* (Frankfurt/ M. 1992); idem, *Konfuzius* (Munich 1995).

24. See the articles by Hoyt Tillman, Achim Mittag, and Susanne Weigelin-Schwiedrzik, in Jörn Rüsen, Achim Mittag, and Helwig Schmidt-Glintzer, eds., *Collective Identity—Experiences of Crisis—Traumata: New Approaches to Chinese Historiography and Historical Culture in a Comparative Perspective,* (forthcoming).

CHAPTER 2

Concepts of Time in Traditional Cultures

Klaus E. Müller

Time and life measure one another, by way of passage, "caesuras," activities, and experiences. All this takes place within spaces, both concrete and imagined. Time, life, and space are indissolubly bound together, forming a complementary whole—an *experienced* space-time.

This is what determines the dimensions of measurement—for units for which constant conditions obtain. Among the Yakuts in Siberia, for example, a "clay pot" indicated the time needed to cook a meal (in, of course, a clay pot). This took a good two hours and at the same time corresponded to the distance of seven to twelve kilometers that a man could normally cover on foot in two hours. Other forms of reckoning, for example among the Mongols, were based on the saddling of a horse (2–4 minutes), the smoking of a pipe (3–5 minutes), or the time it took to milk a sheep (10 minutes).[1]

Activities that were repeated in periodic succession, for which constant conditions thus also obtained, served to divide the day. The Nkole in Uganda, for whom cattle provided the basis of subsistence, began the day with the milking time (6 AM); other important points of orientation are constituted by the resting time for people and animals (12 noon), fetching water (1 PM), the time when the animals leave the drinking trough and begin grazing again (3 PM), etc.[2] The Mongols, predominantly nomadic horse-herdsmen, divided their day according to the milking of their mares, usually performed six times.[3] To achieve greater precision, the economic was correlated with the *solar* "clock": the times at which the first rays of sun hit certain stretches of landscape in the west in the morning, and in the east in the evening, and when at midday, at its zenith, the sun fell upon the settlement itself or, more precisely, shone through

Notes for this section begin on page 31.

openings (or cracks) in house walls and in the course of the day, rising and falling, illuminated clefts and objects on the opposite wall. Correspondingly, peoples of the interior of Asia distinguished times of day in terms of "when the sun shines on the shelf with the holy pictures," "when the sun stands at the head of the bed," or "when the sun illuminates the floor [of the yurt]," that is, when at its highest point at midday the sun falls vertically through the opening for smoke in the roof.[4]

Of course, reckoning changed according to the position of a house and with the seasons so that supplementary correlations had to be made with *comprehensive* periods of time. The most common of these were constituted by *months* and, connecting several of these, *phases of the year* or (more rarely) a whole year. In the case of these larger units a greater measure of reliability was also sought by combining different—cyclical—systems. Months were generally reckoned in terms of the movement of the moon and internally divided up according to lunar phases (dark phase, new, waxing, full, waning),[5] but at the same time also according to the activities and natural occurrences characteristic of individual months—the appearance of certain types of bird, the blossoming of particular bushes and trees, the ripening of their fruits, the shedding of horns by domestic animals (hence the old German word *Hornung* for February), planting, the time "when the rice stands knee-high,"[6] harvest, and so on.[7]

Seasons were divided according to the movement and the "turning points" of the sun, but more concretely according to phases of warmth and cold, wet and dry periods, and further differentiated according to the heaviness of rains and to temperatures. They were also differentiated in terms of the longer-term cycles of activity involved in keeping livestock and agriculture, which were marked by corresponding festive rituals celebrating commencement, peak, and conclusion, and were often additionally linked to particular, seasonally specific natural occurrences such as the fishing forays and migrations of animals, the regrowth and the growth and ripening phases of wild vegetation, the turning of autumn foliage, etc.[8] Some peoples, such as the Mongols, employed bear, wolf, fox, and even marmot calendars as means of correlation, the latter subdivided, for example, according to phases of significant change in the animal—when it awakes from hibernation, loses its fur, puts on fat, and begins to gather dry grass for its den.[9] The solstices were determined according to the proven procedure already described of synchronizing the movement of light and spatial points. Eskimo groups in Labrador oriented themselves to the length and direction of shadows that were thrown by certain rocks at a given moment, while other peoples, such as Dayak groups on Borneo or Nuristani (or "Kafirs") in the Hindukush, used points on the horizon at which the sun appeared to rise or set "at a standstill" for several days.[10]

In this way there emerged calendars that were in some cases highly differentiated, the precision of which grew with the wealth of the combined systems of reckoning and increased where this was particularly required by the conditions of production. It was precisely here, however, that certain differences stemming from the changing vantage point of the viewers—that is, due to the different horizontal positions and altitude of their houses—and perhaps also from informally diverging traditions of time-reckoning, could create problems. Where dates were concerned that had relevance for the community as a whole, such as those marking the beginning of collective enterprises, arrangements had to be made that ensured coordination. These were formulated, for example, through consultation among the elders.[11]

To an extent, the fundamental inner axis of all the temporal systems of a society was constituted for the individual by his own *lifetime*, for the society possibly by that of its leader (king), since this was what determined the activities, tasks, and duties of all. Although possessed of its own caesuras, phases, and periods—such as gestation, birth, and "the first weeks," the first permanent teeth coming through, the first stages of walking, puberty (initiation), menstruation, marriage, the birth of the first child, and death—the individual lifetime nevertheless remained part of the whole since its caesuras and phases were *socially defined,* that is, scheduled for particular times within the framework of the common calendar, as in the case of initiations and weddings, and celebrated according to their respective significance. However, there was also another aspect. Questions regarding particular points of time within a lifetime, such as the day, month, or year of a birth, a wedding, or the death of a relative, were usually answered with reference to familiar occurrences or natural phenomena that were clearly remembered on account of their conspicuous peculiarity and that had been more or less simultaneous with the life event in question—an abundant harvest, an attack by an enemy, an earthquake, a lunar eclipse, etc.[12] That is to say, the biography bore, at least in basic form, traces of *event-history;* time was in these cases emptied of both its linear and its cyclical character and brought "to the point," so to speak, with the result that, potentially, the possibility of an aperiodic sequence arose. Augustine (354–430) was apparently the first[13] to later attach importance to this aspect. He saw here an essentially *subjective* moment, the "power of memory" (*vis memoriae*), which in this way created an individual "space of memory" (*spatium memoriae*) (*Confessiones* X.8 ff.).

Thus, in traditional societies, but also predominantly in advanced archaic civilizations such as China,[14] time was experienced concretely[15] and as multidimensional rather than unitary. The times of specific activities were distinguished from time reckoned with reference to climate, periods of office, and "object times" (the duration of the use of an implement or a house), and combined with one another as required. As a rule therefore, a range of differentiating concepts was employed.[16] Time was therefore always *systemically bound,* was reckoned as "systemic time" or, more correctly, consisted of a variety of system-specific times.[17] For the purpose of optimizing accuracy, certain of these were combined to form groups of periodically equivalent processes. However, a comprehensive *unitary* system of reckoning, a "world-time," so to speak, did not yet exist but emerged only later with the abstract construction of physical time.[18]

For traditional societies "world" meant the group to which one belonged, one's own village community, which was integrated within and delimited from what lay outside it by a unifying *consciousness of identity,* which also provided the framework for "total time," the body of all partial systemic times—that is, it constituted the systemic time, or better, *life-time* of the group in toto.[19] This would seem to point to the idea that such a society's conception of time was *fundamentally* structured by its form of *social organization*: activities and duties were always assigned according to the rules governing the division of labor to persons from certain age and status groups, in one case women, in another men; they changed or recurred periodically with the generations such that the intervals separating, for example, birth and marriage, the hunting expeditions of the men and the gathering expeditions of the women, puberty and admission to the council of elders, the first menstruation and marriage, or the initiations of sons, fathers, and grandfathers emerged as constants.[20] Time, in this view, thus had the *function* of regulating coexistence and above all the totality of its sequential processes in order to lend the social relationships of the whole group continuity by ensuring *foresight,* dependability, and stability.[21]

Such a conception can be found in the work of Émile Durkheim (1858–1917). According to Durkheim, the coordination of activities and the passage of action within a society do not reflect time (in the Kantian sense of an *a priori* transcendental category of consciousness), but *create* it—"it is the rhythm of social life which is at the basis of the category of time."[22] This position was adopted—initially—by Edward E. Evans-Pritchard (1902–1973), a prominent representative of functionalism who substantially took up Durkheim's position: perceptions of time are "functions of systems of time-reckoning and hence culturally determined."[23] In essence, for Evans-Pritchard, they expressed the distances within the structures of relationships between persons and groups or age-groups and generations.[24] On the other hand, it could hardly be overlooked that particular criteria according to which time was reckoned, such as coincidences of events or the phases of natural cycles, were purely environmental givens and thus superordinate or "external" to social organization—connections, for example, between the appearance of a constellation and the ripening period of a wild fruit, the first rains of the year and the regrowth of vegetation. These had to be respected in the division of working time, with planting coming directly after the commencement of rains or the bringing in of cattle when the leaves began to fall in autumn.[25] This also became clear to Evans-Pritchard in the context of a more detailed reworking of his field notes on the Nuer of the upper Nile. He now revised his position, making a distinction between a long-term, socially structured time and a shorter-term "oecological time," that is, the rhythms of cultivation and livestock-farming in conformity with natural conditions. In Evans-Pritchard's view, these two kinds of time were indissoluably entwined with one another, or, more precisely, ecological cycles were always integrated into more fundamental socially determined "structural time."[26] This thesis was adopted by, among others, Meyer Fortes (1906–1983), another prominent representative of functionalism.[27]

Conceptions of time thus undoubtedly have important social functions. They contribute to the stabilization of normative social relationships, ensuring the continued existence and strengthening the identity of the group. However, in this regard, these conceptions themselves require continuous support through confirmatory, institutionalized action. In traditional societies, this need was served above all by periodic rituals marking important caesuras—the beginning or resumption of a new phase of life, a phase of cultivation, a season. At the time of such caesuras, the time between "sequential segments" was, as it were, annulled, or, put another way, stood still;[28] for a limited time "initial conditions" prevailed—as in macrocosmic terms at the time of creation—that offered the ideal prerequisites for the exertion of a steering influence on the following period. Birth celebrations, initiations, weddings, the driving out to pasture of livestock, the beginning of planting and harvest, the laying of foundation stones, and other such initiating acts were constantly accompanied by rituals that confirmed bonding, strengthening, fertility, and tradition, but that over and above this had the goal of keeping the "circling" processes in society and nature stabile and strictly *reversible* over the "breaks" between temporal continua by means of a magical "re-enactment of cyclical character" in order to avoid the development of threatening aberrations.[29] Rhythmical figures, for example in dance, music, and architecture, as already realized by the Greeks, order and limit movement, and "maintain the ties which bind people together."[30]

Order, however, requires not only confirmation but also *reason*, legitimation, sanctioning, necessarily building on that which precedes, which is linked to that which follows—the *ordo explicandus*—by a continuous causal relationship. The basis for this was provided in traditional societies by the creation myth, in which a motionless, structureless (chaotic) mass of mud and water, the primordial sea, was set in motion by the creator, divided into fluid and solid, light and dark, above and below, and then formally further differentiated; the cosmos took on form, space came into being. Through the subsequent creation of sun and moon, winds, flowing waters, plants, animals and people, the divine propelling impulse, captured in the periodic recurrence of processes, was maintained within the world; henceforth events were determined by time, by *space-time*. Toward the end of the creation period "cultural heroes" instructed the people (i.e., according to the usual ethnocentric perspective, the ancestors of one's own group) in all things constitutive of their future existence, equipping them with culture and domiciling them in their life-time, in the specific constellation of elements of their "systemic time."[31] Herewith all was explained, legitimated, and in consequence linked to obligations, and inviolably sanctioned by the authority of God.

Total cosmic time, or "world-time," was thereby composed in the traditional view of three clearly distinct phases: the creative primeval time, in which everything accomplished by gods and cultural heroes obtained irreversible validity; the immediately experienced present, in which mistakes, deviations from rules, could occur, but could also always be reversed, "adjusted" through rites of atonement, penalty measures, or sacrifices to ancestors and gods; and the time between these two, the past or "ancestral time."[32] The present, the "peak time" according to the ethnocentric view, was of the greatest interest. It was determined by one's own experience, thus corresponding to biographical time, which lent it synchronically a frame of orientation and diachronically, or in "axial" terms, a certain linear quality—the life of each individual formed a part of the genealogical tree of the group, the roots of which reached back to the primordial time. Its borders became blurred both "below" and "above" as experience meshed, via one's own memory and that of others, in particular that of elders, with the "upper layer" of the past; the near future was "observable" via children and grandchildren and, given loyalty to tradition, also foreseeable.[33] Only the past was of little interest. The ancestors had lived the life of the people of the present; since one still existed and experienced well-being, then either nothing of any far-reaching consequence could have occurred or possible aberrations had been successfully reversed by the ancestors.[34] Memory therefore remained shallow. Normally it hardly extended any further back than three generations;[35] indeed, it was not seldom the case that this timespan was equated with the entire past, the period of time stretching from the end of the primordial time to the dawn of the present.[36]

The past nevertheless remained "alive" for people—by way of continual contact with the dead within the framework of ancestor worship, and the constant, cyclical return of these ancestors by way of reincarnation, which as a rule took place after three to four generations. Moreover, the example set by the ancestors and their ever-present surveillance weighed oppressively on their descendants "above" in a quite perceptible manner. Cases of carelessness or indiscretion were promptly punished with illnesses, crop failures, or other disasters. "The past," according to the philosopher Otto Friedrich Bollnow (1903–1991), "is the pressure which weighs on the present and which besieges it with demands."[37] Only in differentiated, hierarchically structured, traditional societies did the phase of the past markedly extend, stretched, as it were, by the growing genealogies of leaders and rulers, whose greater need of legitimation necessitated longer ancestral lines of nine, twelve, even fifteen generations—as, for example, in the case of the Kuba of the Congo—periods of time that spanned many centuries and, according to the ideal, were supposed to be directly conjoined with primordial time by genealogies at whose beginning stood a god or a hero.[38] The concrete meaning of genealogization is illustrated by the physicist(!) Ernst Mach (1838–1916), using the following—in his words—"trivial example." The statement that the time of the pharaohs was some 3,600 years ago, Mach argues, merely provides us with an abstract, purely numerically fixed period of time that does not enable us to imagine anything in particular; it is empty. Only if one adds temporal contours, he argues, only if one links beginning and end with a continuous ancestral line, does such a period take on a comprehensible concreteness and directly "affect" us: "Let us take an ancient Egyptian of 60 who fathers a son; the latter, reaching the same age, does the same as his father etc.; the sixtieth descendant of this line, which we can easily imagine drawn up on the wall of a medium-sized room, already belongs to the present."[39] Were *we* the descendants of the ancient Egyptians, then the sequence of generations shown leading back into the past to the point of "origin" would form the axis of *our* "group-time."

It was not only the past that appeared "shallow"—as was at least generally the case in less differentiated traditional societies—but also the *future.*[40] It extended hardly more than a year beyond the planning of customary activities and only in a biographical respect possessed a greater albeit also limited reach, with its "horizon" constituted by the death of all the individuals of a generation, its "radii" by the birth, initiation, wedding, fatherhood or motherhood, etc., of each individual, bound to an imaginary line. Beyond this there was some notion of the life horizon of the following generation, but barely any at all of the generation of the grandchildren. All this, as discussed above, was, given loyalty to tradition, foreseeable and therefore not of especial interest. Furthermore, unlike the—near—past, the events of which possessed a concretely experienced objective reality,[41] the future had only a virtual reality; its temporality became indistinct, or "diluted" the further away it was. It was, according to the African ethnologist of religion John S. Mbiti, "practically eliminated, because the events lying within it have not taken place, are unrealized and therefore cannot represent time."[42]

In regard to both looking back and looking forward, one's own experienced *present,* the time central to the ideology of identity, remained decisive. Augustine, in his justifiably famous reflections on the problem of time in the *Confessions,* also caught this issue precisely. For him neither past nor future existed in themselves; strictly speaking, one should rather speak of a "present of things past, a present of things present," and "a present of things future"; "for these times form a kind of trinity *in the mind,* and I do not see them anywhere else" (XI.20, my emphasis).

Societies that lived strictly according to tradition perpetuated the divine order of creation and endeavored to do so in precise detail. Tradition must have thus appeared to them as customs unbroken over the ages, present reality "as an organic continuation of the mythic past,"[43] "free of breaks and abrupt changes—the grandchildren replace the grandfather and continue the traditional order in the same manner."[44] Nothing, in their view, had changed; or if it had it had been successfully corrected by those who had gone before. Where they existed at all, irreversible processes took place only in the chaotic exospherical foreign world. Although de facto change took place in one's own world, and was also perceived to do so, the elders in particular, who were chiefly responsible for the maintenance of tradition, either integrated it into myth, that is, declared it to be part of an ancient heritage, or, if they did not completely ignore it, played it down. Even the experience of decades of colonialism often proved unable to shake this view. For the Lugbara in Uganda it still remained a certainty at the beginning of the 1950s that there had been no fundamental changes to their traditional way of life since the primordial period. Even schooling did not shake their conviction "that Lugbara society is an unchanging one and independent of outside change."[45] According to this view, typical of traditional societies with a largely stabile identity, time—at least in the modern linear-dynamic sense—seemed to have been annulled, to virtually "stand still."[46]

Here the extensive practice of reading events in terms of cycles "topologically" played a major role. All existentially relevant processes appeared to lead back to the "point" where they had begun—or, more accurately, there was, as is the case for systemic time in general,[47] neither a "beginning" nor an "end." Such points could only be determined with any exactness, that is, in the sense of an unambiguous causal relationship, in regard to individual segmential stretches, that is, short-term phases. The changes that could be observed here—regular (e.g., planting or harvest) and irregular (e.g., taboo infringement or illness)—which could be fixed in exact temporal terms by sequential relationships such as "before and after" or "earlier than" and "later than,"[48] were repeatedly annulled and reversed; thus, seen in terms of the cyclical totality, they lost their "historical" significance.

Where such changes did acquire relevance was in the context of the observation of extra-systemic processes. According to the premises of the—always ethnocentrically oriented—ideology of identity, the cultural divergences of other, foreign ethnic groups could only be comprehended as deviations from the correct path, as "aberrations." In this regard, one's own mythology often cited a particular occurrence, an instance of abnormal behavior or a serious taboo infringement in primordial time that subsequent generations had obviously not been able to undo, with the result that the "barbarians out there" had found themselves on the path of irreversible processes of change that would one day necessarily lead to their downfall. Their erroneous development possessed a beginning and an end, linked by a process not open to investment with cyclical character, advancing from change to change, that is, their progress was event-determined and linear. From one's own point of view, "history"—which necessarily meant a history of disaster—was something only others could have.

This way of conceiving of time and "history" forms the basis of the persistent misunderstanding that the people of traditional societies thought "cyclically" or possessed a cyclical understanding of time and the world. What is correct is that a cyclical principle is given preference in both respects but primarily for reasons of the stabilization of the social order and tradition. Moreover, cycles offer "economic" advantages: they can be more easily grasped, controlled, and predicted and can be synchronized with one another (as a result, biologists attribute to them a "high" so-called selective advantage).[49] Of course, in such societies, *linear* processes are also recognized and distinguished from other processes. In macrotemporal terms, they correspond to the time from creation, especially the creation of the first human being—that is, the forefather of one's own society—until the group's own present. In microtemporal terms, all individual segments of the mesotemporal-cycles are understood in this way: the fashioning of an implement, putting animals out to pasture, the time from planting until harvesting, individual lifetimes and phases of maturity (childhood, youth, etc.). In quasi-geometrical terms, it would be more exact to define them as segments of a curve that exhibit minimal curvature since they always remain part of a cycle. Only within the limited span of these "quasi-straight lines" can causal relationships and processes of change and development clearly and usefully be defined—the inverse of the modern Eurocentric conception of time and history.

Traditional conceptions of time thus represent combinations of a linear, irreversible "axial time" from the primeval beginning to the present, short-term quasi-linear segmental sections, and a changing plethora of mesotemporal cyclical periods.[50] Everything that occurs in this context is socially structured and spatially bound. Growth, for example, is defined according to criteria of height or the passing of particular points of maturity (or status positions), work processes are predominantly linked to movements specific to location and surface, the paths of the sun and moon can be described in terms of particular points in the landscape marking rising and setting positions or with which the sun and moon come into close proximity at given times of day and night; even the "cycle" of the year, with its succession of months and seasons, was represented in terms of its relationship to space, with the "disc" of the earth being pictorially divided, for example, into segments corresponding to seasonal changes.[51]

Toward the end of the creation period, according to the traditions of many peoples (in particular those of the Australian Aborigines), monster-like primeval beings appeared in particular locations—or "world points", in the language of the mathematician Hermann Minkowski (1864–1909)—passed through the land, giving it form and fashioning "world lines," and then departed again at other, later sacred world points, for example by descending into the earth.[52] Later the spatial distribution of the individual groups of one *ethnos* was analogously traced back in legends to phases of migration and the founding of localities by their first ancestors—thus linking "world points" and "world lines" to one another.[53] This connection finds expression in many languages in the use of the same terms in both temporal and spatial adverbial phrases, as for example in the case of the German *vor* (prior to/in front of), *nach* (to, as in. travel *to* another location/after), *weit zurückliegend* (a long way behind/a long time ago), etc.[54]

Strictly speaking, one should refer to a five-dimensional "social space-time." Embedded in the respective frameworks of concrete groups, this social space-time is also systemic time, which, however, is de facto constituted, according to the culture in question, by a variable range of more or less correlated partial systemic times. It follows that it is not only across different cultures, as has often and justifiably been postulated,[55] but also within one and the same culture, according to activity, time of day and time of year, condition, age, etc., that time is perceived and experienced divergently[56] or "multitemporally" (cf. election campaign periods and periods of office, parliamentary terms, economic cycles, etc. in modern industrial societies[57]). That is to say, time possesses variable qualities.

Apart from broader differences between times of day and times of night, times of war and times of catastrophe, "filled" time that is characterized by intensive activity and "empty" time in which little change occurs[58] pregnancies, for example, were attributed a different meaning from that given to periods of residence in an initiation camp or a festival period over several days. The first half of the day with the ascending, "strengthening" sun and the period of the waxing moon were generally regarded as propitious; phases of descent were ominous. But above all a particular quality was attributed to the transitional periods of dawn and dusk, periods of the assumption of an office or of a change of status, periods between segments of life and periods between seasons.

Time and space came into being with the creation of the cosmos and are thus specific qualities of this world.[59] Therefore they do not exist outside the material world. The sphere outside this world is inhabited by incorporeal, purely spiritual beings that can neither fall ill nor age, and is ruled by acausality[60] and timeless eternity—the other world represents an inversion of this world. However, the two are not rigidly separated but merge in various ways, giving rise to particular phenomena in border and transitional zones. Moreover, matter (that is, space-time) is not equally distributed over the world. In a number of places it becomes "thinner" and even "tears open" with the result that passage points are created through which the spirit world streams, spiritual forces move back and forth freely, and the souls of those in dream-states and of shamans can effortlessly reach the "other world." Such "space-time points" are formed, for example, by caves and ravines at midnight or at daybreak, or by towering mountain areas and shrines in the period "between the years," that is, on the eve of the new year.[61] If one crosses such points, then it is only under particularly favorable conditions—at points where matter is at its "thinnest" concentration—that one immediately reaches the spirit world; for example, when the soul of the shaman rises directly into the upper world through the "hole in the sky" at the zenith, that is, at the Pole Star.

Usually, however, it is necessary to traverse "border landscapes," which, at least to begin with, are ruled by relations that are "close to the earth," as is the case, for example, in the realm of the dead. There the dead still participate in time because, as ancestors, they remain tied to their descendants on earth; they oversee, bless, and if need be punish, and will one day be reborn among those on earth. Spirits in the bush and the forest can approach people in visible form and communicate with them at ponds, in caves, on mountaintops, etc., especially at midnight or at any other "transitional time." While dreaming, people experience a phase during which their soul returns from the spirit world and becomes increasingly "immersed" in materiality, that is, shortly before they wake, the images they perceive become increasingly confused, distorted, and broken until they are unrecognizable—and at the same time appear to follow each other more and more rapidly, almost blurring into one another.[62] And when one is completely transported to the spirit world or, as in the case of shamans, consciously chooses to go there, one's experience on returning is analogous. The "traveler" realizes that during what seemed a short absence—a few hours or days—months, often years, sometimes centuries have passed on earth.[63] The perception of time "becomes distorted" corresponding to the changing density of space-time in the transitional sphere between this world and the next; while moving to the "other side," the experience of *dilatation* increases; while returning, time, due to the relationship of inversion referred to between the worlds, seems to *accelerate.* Although traditional societies did not recognize a consistently standardized systemic time, they were acquainted with a—largely—unified system of partial systemic time qualities.

"History" did not leave such societies untouched. Following the rise of the advanced archaic civilizations, they were—in some cases peacefully but in border regions predominantly by violent means—drawn into *other* systemic times incompatible with their own. Breaks, changes, and innovations occurred within the culture and the *Weltanschauung,* which could in part be integrated into a cyclical schema and in part not. In the latter cases, "historical" events—that is, irreversible, unique occurrences—"lit up," as it were, the darkness of the past. If these led to lasting social or even ethnic differentiation, they took on foundational significance and were retained in the traditions subsequently handed down. If several such instances were involved, the coordination and continuity necessary to the ideology of identity required that they be linked together, that is, in a chronologically linear sequence—a process intensified in proportion to the degree of differentiation. Legends pertaining to a particular territory expanded into epics of conquest, and the genealogies of the leading lineages were filled out with histories of events: the great feats of warriors, victorious battles, the founding of fortresses and towns, instances of wondrous deliverance from desperate situations, and so on, all of them significant "caesuras" or "turning points" that brought with them momentous changes and that fitted together to form segments of a goal-oriented process leading straight to the triumph of the victorious power. Accordingly there was an intensification of the sensation of "movement" as well as the "length," the duration of time. On the one hand, with the accumulation of events,[64] time grew back into the past;

on the other, with the introduction of a teleological perspective, it also increasingly acquired a future perspective.

For the older and literally "backward" communities, the "dissident" groups had of course joined the barbarians outside the cyclical world and had thereby actually acquired "historicity," that is, had acquired, together with an origin that could be fixed in post-primordial time and that was based on change, a temporal end that could be predicted with certainty.

The development of such a differentiated, historicized consciousness of time is correctly attributed to the advanced archaic civilizations: "Without the aid of a mathematical system which was already to some extent developed and an increasingly exact consideration of points and periods of time beyond the basic calendrical schema, this task could not be mastered or if so then only imperfectly."[65] However, this should not be interpreted as an evolutionary achievement—in traditional societies there existed neither the occasion nor the necessity to precisely reckon units of time in terms of hours or, for that matter, minutes.[66]

The consistent, quasi-"vectorial," linear extension of time via the interconnection and sequencing of events within the advanced archaic civilizations can ultimately be traced back to the ancient Israelites, from whom the new form of historical consciousness, mixed with relevant concepts drawn from the philosophy of the Greeks (e.g., Aristotle), spread via Christianity through Occidental thought.[67] For the rising industrial societies of modernity, the "heirs" of this development, it thus appeared as a hallmark of *their* history. Combined with a Eurocentric viewpoint, its subsequent development gave rise to the doctrine of evolution and the euphoria of progress.

It was from this developmental schema that Claude Lévi-Strauss (born 1908) derived the distinction between "cold," traditional societies, which are static and unreceptive to change, and "hot," dynamic modern societies, which because of the increasing compression of events are "heated up"[68]—a somewhat crude dichotomy that he later attempted to qualify,[69] which, moreover, due to its rendering ontological what are purely typological categories, is clearly problematic from an analytical point of view.[70]

In "dynamic" societies—or more correctly, societies that are becoming increasingly differentiated—there emerges another effect that strengthens this impression of dynamism: the experience of a growing mass of irreversible events, of constantly increasing situational changes, gives rise to the fiction of temporal acceleration.[71] In order to avoid being, as it were, swept up in the maelstrom of time, this experienced sense of time moving faster is countered by both private and public attempts to invest time with a cyclical character. Periodic meetings, commemorative and jubilee celebrations, annual municipal festivals, periods of office, congressional terms, and so on aim to create "cold" islands of *return*—to halt time, at least in places, amidst growing confusion.[72]

For some this is not enough. They seek refuge in fundamentalist positions, whose claim to universal validity seemingly brings the flow of time to a standstill and compels that which is flowing away to "return"—thus laying the restless, "harassed" individual once again in the arms of those (indigenous, Hindu, Jewish, Christian, Islamic, etc.) societies that hark back to a "primordial state" in which everything that is thought and done constantly revolves "around the central point" and in that time is not divided up and accelerated, but is periodically created and exhausted again. However, archaic societies no longer exist. Their approach to time can *not* be regained through reversion, since "their" time and "our" time involve highly different systemic times between which smoothly alternating conversions are not (or no longer) possible.

Translated by Joseph O'Donnell

Notes

1. Natalija L. Shukowskaja, *Kategorien und Symbolik in der traditionellen Kultur der Mongolen,* Studia Eurasia 4 (Berlin 1996), 35f.

2. John S. Mbiti, *Afrikanische Religion und Weltanschauung* (Berlin 1974), 25; cf. J. Ki-Zerbo, "General Introduction," in J. Ki-Zerbo, ed., *General history of Africa,* vol. 1 (Berkeley 1981), 1–23, here 18; Janet Hoskins, *The Play of Time: Kodi Perspectives on Calendars, History and Exchange* (Berkeley 1993), 67.

3. Shukowskaja, *Kategorien,* 35.

4. Shukowskaja, *Kategorien,* 34f; T. O Beidelman, "Kaguru Time Reckoning: An Aspect of the Cosmology of an East African People," *Southwestern Journal of Anthropology* 19 (1963): 9–20, here 13f. Ki-Zerbo, "General introduction," 18; Hoskins, *The play of Time,* 65; Schuyler Jones, "Zur Religion des Waigaltales," in Karl Jettmar, ed., *Die Religionen des Hindukusch* (Stuttgart 1975), 150–162, here 159.

5. Jan Vansina, *Oral Tradition: A Study in Historical Methodology* (London 1965), 100; Beidelman, "Kaguru Time," 14; Mbiti, *Afrikanische Religion,* 26.

6. Cf. Nevill Edward Parry, *The Lakhers* (London 1932), 193.

7. Edward E. Evans-Pritchard, *The Nuer: A Description of the Modes of Livelihood and Political Institutions of a Nilotic People* (Oxford 1940), 100; E. Fettweis, "Orientierung und Messung in Raum und Zeit bei Naturvölkern," *Studium Generale* 11, no. 1 (1958): 1–12, here 8.

8. Leslie Spier, "Inventions and Human Society" in *Man, Culture, and Society,* ed. Harry L. Shapiro (London 1971), 296–318, here 315; Parry, *The Lakhers,* 191; Evans-Pritchard, *The Nuer,* 95f.; Vansina, *Oral Tradition,* 100; Beidelman, "Kaguru Time," 15; Jones, "Zur Religion," 159.

9. Shukowskaja, *Kategorien,* 34.

10. Fettweis, "Orientierung," 6; Jones, "Zur Religion," 158.

11. Cf. Jones, "Zur Religion," 159.

12. Cf. for example John Henry Hutton, *The Sema Nagas* (London 1921), 260; note 1; Christoph von Fürer-Haimendorf, *The Chenchus* (London 1943), 127.

13. Rudolf Wendorff, *Zeit und Kultur: Geschichte des Zeitbewußtseins in Europa* (Opladen 1980), 98.

14. James H. Zimmermann, "Die Zeit in der chinesischen Geschichtsschreibung," *Saeculum* 23, no. 4 (1972): 332–350, here 333.

15. Godfrey Lienhardt, *Social Anthropology* (London 1966), 48; Mbiti, *Afrikanische Religion,* 21, 24; Beidelman, "Kaguru Time," 19; Ki-Zerbo, "General Introduction," 18; cf. André Leroi-

Gourhan, *Hand und Wort: Die Evolution von Technik, Sprache und Kunst* (Frankfurt/ M. 1980), 391, 393.

16. Cf. for example Edmund R. Leach, "Zwei Aufsätze über die symbolische Darstellung der Zeit" in Wilhelm E. Mühlmann and Ernst W. Müller, ed., *Kulturanthropologie* (Cologne 1966), 392–408, here 392; Beidelman, "Kaguru Time," 11; Hoskins, *The Play,* 58f., 78.

17. Wolfgang Deppert, *Zeit: Die Begründung des Zeitbegriffs, seine notwendige Spaltung und der ganzheitliche Charakter seiner Teile* (Stuttgart 1989), 65; Werner Bergmann, *Die Zeitstrukturen sozialer Systeme: Eine systemtheoretische Analyse* (Berlin 1981), 35, 98f.

18. Deppert, *Zeit,* 220.

19. Cf. Donald J. Wilcox, *The Measure of Times Past: Pre-Newtonian Chronologies and the Rhetoric of Relative Time* (Chicago 1987), 252: "Identity and Time have a close and indissoluble relationship."

20. Cf. Günther Schlee, *Das Glaubens- und Sozialsystem der Rendille, Kamelnomaden Nord-Kenias* (Berlin 1979), 86; W. E. H. Stanner, "The Dreaming, an Australian World View," in Peter B. Hammond, ed., *Cultural and Social Anthropology: Selected Readings* (New York 1966), 288–298, here 295, 297.

21. Cf. Andreas Kuntz, "Erinnerungsgegenstände: Ein Diskussionsbeitrag zur volkskundlichen Erforschung rezenter Sachkultur," *Ethnologia Europaea* 20 (1990): 61–80, here 62.

22. Émile Durkheim, *The Elementary Forms of the Religious Life* (New York 1968), 488, 22f. Cf. Alfred Gell, *The Anthropology of Time: Cultural Constructions of Temporal Maps and Images* (Oxford 1992), 4f.; Nancy D. Munn, "The Cultural Anthropology of Time: A Critical Essay," *Annual Review of Anthropology* 21 (1992): 93–123, here p 94f.

23. Edward E. Evans-Pritchard, "Nuer Time-reckoning," *Africa* 12 (1939): 189–216, here 209.

24. Evans-Pritchard, *The Nuer,* 105ff.

25. Cf. Jones, "Zur Religion," 161.

26. Evans-Pritchard, *The Nuer,* 94ff.; cf. Gell, *Anthropology of Time,* 15 ff.

27. Meyer Fortes, "Time and Social Structure: An Ashanti Case Study" (1949), in Eugene A. Hammel and William S. Simmons, eds., *Man Makes Sense: A Reader in Modern Cultural Anthropology* (Boston 1970), 104–125, here 109.

28. T. A. Bernštam, "Budni i prazdniki: povedenie vzroslych v russkoj krest'janskoj srede (XIX-nacalo XX v.)," in Albert K. Bajburin, ed. *Etniceskie stereotipy povedenija* (Leningrad 1985), 120–153, here 139.

29. Cf. Gell, *Anthropology of Time,* 27, 51ff.; Klaus E. Müller, "'Prähistorisches' Geschichtsbewußtsein: Versuch einer ethnologischen Strukturbestimmung," *ZiF-Mitteilungen* 3 (1995): 3–17, here 6.

30. Werner Jäger, *Paideia: Die Formung des griechischen Menschen,* vol. 1 (Berlin 1936), 174.

31. Müller "'Prähistorisches' Geschichtsbewußtsein," 3f.; cf. Mary Dillon and Thomas Abercrombie, "The Destroying Christ: An Aymara Myth of Conquest," in Jonathan D. Hill, ed., *Rethinking History and Myth: Indigenous South American Perspectives on the Past* (Urbana 1988), 50–77, here 61.

32. Müller, "'Prähistorisches' Geschichtsbewußtsein," 6f.; Charles Hudson, "Folk History and Ethnohistory," *Ethnohistory* 13 (1966): 52–70, here 56, 58; Mbiti, *Afrikanische Religion,* 21.

33. Cf. Mbiti, *Afrikanische Religion,* 21, 28f.; Hudson, "Folk History," 56.

34. See Müller, "'Prähistorisches' Geschichtsbewußtsein," 7.

35. See Müller, "'Prähistorisches' Geschichtsbewußtsein," 7; Robert Redfield, "Die 'Folk'-Gesellschaft," in Mühlmann and Müller, ed., *Kulturanthropologie,* 327–355, here 332.

36. Cf. for example Lothar Stein, *Die Sammar-Gerba: Beduinen im Übergang vom Nomadismus zur Seßhaftigkeit* (Berlin 1967), 128; E. A. Alekseenko, "Predstavlenija ketov o mire," in I. S. Vdovin, ed., *Priroda i celovek v religioznych predstavlenijach narodov Sibiri i severa* (Leningrad 1976), 67–105, here 70.

37. Otto Friedrich Bollnow, *Existenzphilosophie* (Stuttgart 1969), 107.

38. Cf. Richard Thurnwald, *Psychologie des primitiven Menschen* (Munich 1922), 277; Ian Cunnison, *History on the Luapula: An Essay on the Historical Notions of a Central African Tribe* (Cape

Town 1951), vi, 8; Diedrich Westermann, *Geschichte Afrikas: Staatenbildungen südlich der Sahara* (Cologne 1952), 16f., 406; Hermann Baumann, "Mythos in ethnologischer Sicht," Part 1, *Studium Generale* 12, no. 1 (1959): 1–17, here 16; Jack Goody and I. Watt, "The Consequences of Literacy," *Comparative Studies in Society and History* 5, no. 3 (1963): 304–345, here 308; Vansina, *Oral Tradition*, 153; William C. Sturtevant, "Anthropology, History, and Ethnohistory," *Ethnohistory* 13 (1966): 1–51, here 29.

39. Ernst Mach, *Erkenntnis und Irrtum: Skizzen zur Psychologie der Forschung* (Darmstadt 1968), 162.

40. Robert Redfield, *The Primitive World and Its Transformations* (London 1968), 125.

41. Cf. Alfred North Whitehead, *Abenteuer der Ideen* (Frankfurt/ M. 1971), 348.

42. Mbiti, *Afrikanische Religion,* 21.

43. Alekseenko, "Predstavlenija ketov," 70.

44. Meyer Fortes, "Bewusstsein," in Raymond William Firth, ed., *Institutionen in primitiven Gesellschaften* (Frankfurt/ M. 1967), 93–106, here 97; cf. Redfield, *Primitive World*, 125.

45. John Middleton, *Lugbara Religion: Ritual and Authority among an East African People* (London 1960), 26; See Müller, "'Prähistorisches' Geschichtsbewußtsein," 6ff.

46. Cf. Evans-Pritchard, *The Nuer,* 108; Fürer-Haimendorf, *Chenchus*, 247; Stanner, "Dreaming," 297; Mary-Elizabeth Reeve, "Cauchu Uras: Lowland Quichua Histories of the Amazon Rubber Boom" in Hill, *Rethinking History and Myth,* 19–34, here 25; Gell, *Anthropology of Time,* 22.

47. Deppert, *Zeit*, 245: "It is an analytic truth that within a systemic time a beginning and an end of this systemic time cannot be named." Cf. Fortes, "Time and Social Structure," 106f.

48. Cf. Josef Meran, *Theorien in der Geschichtswissenschaft: Die Diskussion über die Wissenschaftlichkeit der Geschichte* (Göttingen 1985), 30.

49. Bergmann, *Zeitstrukturen*, 151.

50. Evans-Pritchard, *The Nuer*, 95, 108; Zimmermann, "Die Zeit, 336ff.; Munn, "Cultural Anthropology," 102; Müller, "'Prähistorisches' Geschichtsbewußtsein," 7; cf. Bergmann, *Zeitstrukturen,* 173.

51. Werner Müller, "Raum und Zeit in Sprachen und Kalendern Nordamerikas und Alteuropas," *Anthropos* 68, nos. 1–2 (1973): 157–180, passim.

52. Cf. Munn, "Cultural anthropology," 101, 113f.; Ronald W. Clark, *Albert Einstein: Leben und Werk. Eine Biographie* (Esslingen 1974), 91.

53. Müller, "'Prähistorisches' Geschichtsbewußtsein," 8; cf. also Berthold Riese "Zeitstrukturen in Mesoamerika," in Klaus E. Müller and Jörn Rüsen, eds., *Historische Sinnbildung: Problemstellungen, Zeitkonzepte, Wahrnehmungshorizonte, Darstellungsstrategien* (Reinbek bei Hamburg 1997), 240–250.

54. Helmut Gipper, "Die Kluft zwischen muttersprachlichem und physikalischem Weltbild," *Physikalische Blätter* 12, no. 3 (1956): 97–105, here 99; Hoskins, *The Play,* 64; Riese, "Zeitstrukturen in Mesoamerika," 240–250.

55. Cf. for example Wilhelm E. Mühlmann, *Geschichte der Anthropologie* (Frankfurt/ M. 1968), 214; Lienhardt, *Social Anthropology,* 48; Reeve, "Cauchu Uras," 25.

56. Klaus Mainzer, *Zeit: Von der Urzeit zur Computerzeit* (Munich 1995), 120.

57. Cf. Hoskins, *The Play,* 78f.; Allen R. Maxwell, "Kadayan Ideas of Time," *Sarawak Museum Journal* 27, no. 48 (1979): 1–14, here 6ff.

58. Cf. Leach, "Symbolische Darstellung der Zeit," 398.

59. Klaus E. Müller, "Epistemologische Grenzfälle: 'Höhere' Erkenntnis in traditionellen Gesellschaften," *Zeitschrift für Parapsychologie und Grenzgebiete der Psychologie* 32, nos. 3–4 (1990): 137–152, here 139; cf. Erhard Scheibe, "Die Einheit der Zeit," in Erich Fries, ed., *Festschrift für Joseph Klein zum 70. Geburtstag* (Göttingen 1967), 53–72, here 58; Mainzer, *Zeit*, 26.

60. Cf. Carl Gustav Jung, "Synchronizität als ein Prinzip akausaler Zusammenhänge," in Carl Gustav Jung and Wolfgang Pauli, eds., *Naturerklärung und Psyche* (Zurich 1952), 1–107, here 31, 106f.

61. Klaus E. Müller, "Reguläre Anomalien im Schnittbereich zweier Welten," *Zeitschrift für Parapsychologie und Grenzgebiete der Psychologie* 34, nos. 1–2 (1992): 33–50, here 39.

62. Müller, "Reguläre Anomalien," 37.

63. Cf. Waldemar Bogoras, "Ideas of Space and Time in the Conception of Primitive Religion," *American Anthropologist* 27, no. 2 (1925): 205–266, here 231f.; Müller, "Reguläre Anomalien," 36.

64. Cf. Wilcox, *Measure of Times,* 252.

65. Wendorff, *Zeit und Kultur,* 19f.

66. Fortes, "Bewußtsein," 98; Mbiti, *Afrikanische Religion,* 24; Jones, "Zur Religion," 159.

67. Wendorff, *Zeit und Kultur,* 26; Mainzer, *Zeit,* 27.

68. Claude Lévi-Strauss, *Race et histoire* (Paris 1948); cf. idem, *"Primitive" und "Zivilisierte"* (Zurich 1972); idem, *Das wilde Denken* (Frankfurt/ M. 1973), 251ff.; Gell, *Anthropology of Time,* 23.

69. Cf. for example Claude Lévi-Strauss, *Strukturale Anthropologie* (Frankfurt/ M. 1967), 116, 133f; Lévi-Strauss, *Das wilde Denken,* 9th ed. (Frankfurt/ M. 1994), 271f.

70. Cf. Gell, *Anthropology of Time,* 52f.

71. Müller, "'Prähistorisches' Geschichtsbewußtsein," 16; Bergmann, *Zeitstrukturen,* 231f.; cf. Fortes, "Time and Social Structure," 106f.

72. Cf. Kuntz, "Erinnerungsgegenstände," 62.

CHAPTER 3

Time, Ritual, and Rhythm in Dimodonko

FRITZ W. KRAMER

Dimodonko was a small, de facto autonomous Nuba land west of the White Nile in the far south of the Nuba Mountains of the Sudan until 1992, when it became the target of a Muslim Jihad militia, as did the entire territory of the Nuba. The last years before this "ethnic cleansing" constitute the situation I describe here, using the "ethnographical present." My information stems from a field study conducted in 1975 and 1987/88.[1] The structures I intend to describe existed in similar fashion in the wider area of the southern Nuba; I shall refer to them occasionally, but not to possibly comparable situations of the northern Nuba people and definitely not to those of African or peasant societies in general.

The Kodonko, the more than two thousand inhabitants of Dimodonko that could be called a "village," or rather a "land," represented until 1992 what used to be called in traditional anthropology a "people without history" or a "culture without a written language." In some respects one could consider their culture a good example of what A. D. Nock characterized in the following way: "a small social unit with elementary needs and interests and only insignificant contact with other cultures which are superior either materially or intellectually or have a cult or a religion which could excite curiosity and draw attention—a social unit where, in the words of a playwright, nothing happens except that you hear the clock strike three and wait for it to strike four."[2] I would like to show that time in Dimodonko actually was marked, as if with the stroke of a bell, by ritual and rhythm; and that because such "strokes" did not follow each other with the regularity of a clock, waiting for them was full of drama. When looking at this situation more closely we notice a web of time,

Notes for this section begin on page 43.

ritual, and rhythm where the past is present and history as awareness of origin is constitutive.

The Kodonko are typical millet farmers of the African savanna; long ago they abandoned the growing of cotton, which had connected them marginally to the global economy during colonial times. The few cattle, goats, and pigs they own are for their private consumption, as are the sesame, peanuts, vegetables, and tobacco they grow. Iron spearheads and ax blades are acquired from Arabs or other Nuba in exchange for millet: young boys are in charge of small herds, some of the women are expert potters, some of the men excel in building artistic roof trusses over the mud huts. The important task for all adult Kodonko, however, is the cultivation of millet: the sowing, the caring for the young plants in the fields, the harvest, and the threshing. An early type of millet is sown on small fields in the mountains, a late type on big fields in the foothills. While the latter type yields the most important stocks, its growth is dependent on the seasonal rainfalls. All elementary needs and interests turn around the early and late millet harvests. In conversations with the Kodonko the discussions always turn to millet. This happens with such regularity that outsiders consider it monotonous; yet for the Kodonko the subject is full of drama. Millet is the measure of time and at the core of all symbols.

Millet moves from the field to the storehouse. In early autumn it reaches its maturity as standing grain. It is visible until it disappears in the storehouses after the harvest and the threshing. In the spring what remains of it returns to the fields as new seed material. Millet needs its own time, but it renews itself only through labor. The time of the millet determines the time of labor.

The Kodonko speak of time in terms of a precious belonging, just as we do. In Krongo, the language of Dimodonko, *uuso,* which I translate as "time," is something like an object one has or does not have. One may say, for example: "I have time (uuso)" to volunteer for something, and "I have no time" when one is busy with something else. Time is quantified by speaking of "much" and "little" time, although it is not measured or counted, but only estimated while looking at the position of the sun and the shadows, but especially at the growth of the millet. The word uuso has still another meaning; it indicates the period between dawn and the highest position of the sun at noon, the time when there is daylight and the heat not extreme. In the morning one can labor and exert oneself; morning is the useful time of the day and thereby related in meaning to uuso, time that one has or does not have.

In addition to time that can be determined quantitatively, uuso, there is time which can be determined only qualitatively, *nabu,* time indicating duration. While among the times of the day morning is called uuso, all others are called nabu, like *nabu tirifiriifi,* dusk. *Nabu alimi,* the "cold time," or *nabu miiri,* the time of hunger, is nothing that one has or does not have, but something that is or is not. In other expressions the word nabu can mean "space," in the concrete sense of "space of a definite quality" or "location." From this word the agent noun *kanabu* is derived, which describes the locality as an anthropomorphic subject and must be translated as "spirit" or "local spirit," *genius loci.* Contrary to nabu "locality," nabu "time" is not conceived anthropomorphically, even though the start of a time, nabu, as a time with a definite characteristic is seen in connection with kanabu, "spirit." This holds true especially with respect to the most important turning points in the yearly cycle of millet, the start of the harvest of the early millet on the fields in the mountains and the start of the sowing of the late millet on the fields in the foothills; both activities are determined by nature, but are nevertheless marked by special publicly validated rituals.

The word *tanyaara,* which I translate as "ritual," describes a celebration where people gather to drink beer and honor spirits. The presence of the spirits and the slightly intoxicating drink of millet beer are the basic elements of this feast, but in addition the Kodonko almost always engage in other activities, such as beating of drums, dancing, singing of songs, excesses and animal sacrifices, wrestling matches and spear throwing—anything that serves to prepare the community for the coming time, nabu. Tanyaara is a sign making people aware that a natural threshold has been reached, the happening of something that cannot be brought about by man, like the start of the rainy season or the ripening of the millet; it reaffirms what has occurred even though a hint of magic may play a role in it.

Every year the ritual of the first harvest is brought from village to village as in a relay race to celebrate, everywhere, the start of the harvest of the early millet from the fields in the mountains. Songs and wrestling matches turn it into a spectacular feast, but at the core, marking the time, it consists of a ceremonial funeral of a millet stalk. A millet priest—or priestess—breaks the first stalk, buries it, and hangs the millet ears under the roof of his house, just as a human corpse is buried in the ground after death while its immortal spirit is kept in a calabash tied it under the roof of a special house, the wailing house. The ritual symbolism thus emphasizes the clear separation between the mortal plant in which the grain renews itself every year, and the grains of the seemingly everlasting, immortal millet.

Every year in May, in a similar, even bigger relay race, the sowing ritual is brought to the wider region to celebrate the sowing everywhere, in particular the sowing of the late millet on the fields in the foothills. The relay race is carried by the rhythms of drums, rattles, songs, and dances until it arrives at one of the four sacred groves that are cult sites shared by Dimodonko and the neighboring Tabanya. If a good harvest is expected, a sacrifice of an antelope or duiker that has been caught during a battue is offered there to the "millet maids," spirits of the late millet who reside inside the drums and under the trees of the grove. (It is said that catching an antelope is a good omen; however, in reality it is nothing but an expression of mostly well-founded hopes, considering that in periods of drought lasting several years or in times of war they abandon the battue.) Other rites, such as the lighting of a new flame with a fire gimlet, are also part of the ceremonies, but the core of the ritual that marks the time consists of the presiding priest or priestess uncovering especially large and beautiful grains of the late millet under the trees; he or she then distributes most of the grains among the farmers present, who mix these grains with their seed.

The symbolism of both rituals reveals an understanding of time that a purely cognitive analysis would not bring out. It is easy to decipher the newly drilled fire as a symbol of purification and renewal, and antelopes can be related to renewed vegetation because these animals appear in the mountains or nearby at the start of the rainy season. It is no coincidence that the Kodonko compare their velvety, reddish-brown fur with the flawless skin of young girls and the matte glow of full millet ears. In the ritual of the first harvest this symbolism of fertility and growth turns into a ritual of transitoriness and death. Becoming and dying conflicts with perfecting what is seen as permanent. The stalk that the millet priest breaks and buries has been grown from the same ears that he harvested first the previous year and tied under the roof of his house; and the grains that the seed priest uncovers under the trees in the sacred grove have been cultivated from a share reserved from those he found there the previous year. Ordinary millet is grown to be consumed or sown again; ritual millet is kept separate and always in a special protected place so that, at least symbolically, it possesses an immutable identity and permanence. Every millet and seed priest has taken it over from his or her predecessor, together with the office. The Kodonko, therefore, imagine a continuous chain of tradition, not of words but of millet, which links the beginnings of their culture to the present. The relationship between the transitoriness of the plant and the permanence of the grain is repeated in the relationship between the mortality of man and the immortality of the social group, the public body.

In Dimodonko the social structure is determined, as in other lands of the southern Nuba, by matrilinear descent. There is first the clan, which gives its name to all its subdivisions, which I call "sections." Only the exogamous sections, however, can own land; ritual offices with their insignia can be inherited only within the sections; and only within the sections can certain kinds of witchcraft be performed effectively, as is characteristic of matrilinear societies in Africa.

Residency rules confer a special character to matrilinearity in Dimodonko and in several of its neighbors. Children grow up with their parents, but before they reach puberty, boys move in with those uncles on their mother's side who have early on declared them their heirs or successors. Girls move to their husband's house after marriage, with the result that in general only the young and adult men, not the women and small children, live in the area and the territory of their section. However, men may also on occasion leave their land because of a quarrel or a feud; they then ask for asylum and arable land elsewhere. If a section is dispersed in this way or if it becomes simply too large, it can divide into sections having equal status; they are then allowed to intermarry. Since they have the same clan name, they can be distinguished by the name of the place where they have their wailing house. Each section has its own wailing house; for a death brings together all members of a section, even the women and those who have resettled elsewhere. The burial takes place in the only cemetery of the section, even in effigy in case someone has died and was buried far away, in which case funeral rites can be performed only on a piece of cloth or material worn by the dead person h at the time of death; the spirits of the dead are kept in calabashes and hung for a while under the roof of the wailing house.

The death of a person is accompanied by a series of rituals. The last and most important of them, the death ritual, is not granted to all dead, but only to those who were initiated into manhood or womanhood through puberty rites, those who did not ally themselves with the spirits of the wilderness or commit an unforgivable sin such as incest or murder of a family member. The death ritual, during which the calabash containing the spirit of the dead person is taken from the wailing house to be smashed on the backs of cattle, which then are speared and distributed to the assembled section members, elevates these honorable dead persons to ancestors who, together with all ancestors of the Kodonko and their neighbors, reside in the house of the rain priest and ancestor medium. The living can come before this house and call on the ancestors, who seem to answer from inside. The voices of the past are not lost; they can be heard at a place where they intervene in the fate of the living with counsel and advice.

In the wailing house there is a clay pot filled with ashes. For the most part the ashes stem from ordinary hearth fires and from branches that the young men of the section have won as a prize in wrestling matches and have burned for the wailing house. The ashes in the wailing house are replenished from time to time and used to mark the cattle that are speared during the festival of the dead; but they represent above all the unity and continuity of the section, for like the ritual grain of the seed and millet priests, some of the ashes that lend significance to the rest have come down from the beginning of history. When a section is divided, one of the new sections remains in possession of the ash pot while the others move away, each with a handful of ashes in order to start new ash pots in their new wailing houses. Thus the ash pots of all sections of a clan seem substantially to be offshoots of a single, original ash pot of the clan. Retracing the history of the ashes therefore means retracing the history of the clan. The ashes, i.e., what remains after the fire has consumed the wood, symbolize the clan in its branches and connect the clan's mythical beginnings to its present in an uninterrupted tradition.

Like the rituals of which they are a part, the rhythms of the drums, songs, and dances mark thresholds in time. Each period of a qualitatively definite time, nabu, is marked by the rhythms of the ritual that inaugurates it. For the

Kodonko the song, *ntuwi*, is the essence of rhythm. When at the start of the rainy season the time for the seed ritual approaches, the drums can be heard in the house of the seed priest, and it is believed that the "millet maids," the invisible spirits of the late millet, are beating these drums. Wooden and clay drums can also be found in the wailing house of the section; they are heard during the festival of the dead. During the seed rituals—and those performed at funerals—the dancers form two concentric circles, men the inner, women the outer one; these circles enclose the lead singers, choir, and drums. However, during those rituals where women dominate, such as initiation for girls, the rhythms are only sung and clapped, and the female dancers line up in straight rows that move back and forth. The songs sung in the concentric circles are serious and solemn; those sung in the rows are exuberant and full of mockery. The circle, linked to drums and animal sacrifices, has the shape of the permanent, of the ritual grain of the seed priest—when the seed priest buries the stalk he is surrounded by a circle of wrestlers—and the shape of the ash pot in the wailing house. The moving row, the dance of the girls and women, has the shape of renewal, of regeneration.

In Dimodonko and other lands of the southern Nuba songs are judged first by applying esthetic criteria. Their composition and presentation are the responsibility of the *kotoosu,* professional male and female poets who have been publicly tested and formally initiated. Most often the kotoosu specialize early on in one of the twelve types of songs; from time to time they organize a contest in order to discover the best of each type. Lay persons who have come to listen as well as the kotoosu themselves are the judges, assessing the linguistic and musical perfection, as well as the newness and surprising elements of the song. A choir that is invited to sing at a festival or a ritual is expected to present as many new songs as possible. But new songs are not always as catchy as some old ones, so even famous poets either take recourse to their own older compositions, or pay homage to competitors by accepting their songs into their repertoire. As a result, the totality of songs performed during a ritual has depth from the point of view of time; it links the past and the present.

In the acephalous society of Dimodonko public criticism is decisive for the preservation of the moral and political order. The most important media of criticism are the kotoosu, who observe carefully and disclose obvious deviations from the distinct form of their songs. Songs of the circular pattern most often reveal mysterious events, dangerous alliances between the spirits and healer-witches, grain, seed, and rain priests; songs of the row pattern target typical human frailties, often the follies of men. Kotoosu always prefer discreet, ambiguous allusions, of which the listeners themselves have to make sense. This openness furthers the relevance of the past; old songs may often be related not at all to past events but, just like the new ones, to what is currently relevant to the listeners. All songs, old and new, follow traditional patterns of interpretation, even if in the new ones mention is made of cars and airplanes.

Recurring events and configurations are not only the seasons and blunders, but also wars, epidemics and famines, treason, collaboration, and meanness when dealing with those in need. As a result the songs appear from an outside perspective as archives of historical knowledge; from the inside, however, they are more like an inventory of something that is possible at all times.

The oral tradition in which the past is valid as something past is less rich than the songs and of surprisingly little depth from a temporal perspective, even for African conditions. It begins with an event, probably a famine, which we can date to approximately 1830, although the Kodonko, like other southern Nuba, tell about it as if it were the mythical origin of death.[3]

"In olden times," so goes one version, "people lived in Tanbilli. People were born and death was unknown. And so there were more and more people and they had cattle in abundance. Then people wondered what good all those cattle were. Someone said: "Let's do something; we will make something that looks dead, and then we will spear cattle." They went around a house looking for a lizard in the roof. They killed it. They put the corpse on a sleeping mat and covered it with a cloth. Four men took the bier, one at each corner, and carried it in a circle around the cattle. They began wailing, speared the cattle and buried the dead lizard; they ran around the wailing house calling the dead. Thereupon death went among the people and many died. When the others saw this they fled from death. Some went this way, others that way."

African myths about the origin of death deal implicitly with the origin of culture; they narrate how humanity—their own ancestors, their own clan—lost its state of innocence by committing a folly and was thus forced to lead the life it is leading at present. Even though the state of innocence may seem attractive, the limitations and obligations imposed by culture are part and parcel of being a human being. The myth characterizes what we call culture by an object or an act, for example, a hoe or the theft of cattle.[4] In the case of the Kodonko myth and its variants, which are often linked to certain clans, this act is the death ritual or one of its partial rites, for example the burial on the sleeping mat, the spearing of cattle, wailing chants, sexual excesses, or spear throwing contests—ritual acts that serve to wipe out the guilt of the living with respect to the dead.

It seems that each death ritual repeats the original mythical event with which time begins. We cannot speak, however, of a cyclical renewal of time through ritual because the myth of the origin of death is not a myth of creation, and the purpose of death rituals is to elevate of the dead to the status of ancestor—the striving for perfection to reach permanence—while the seed ritual coincides in fact with the yearly renewal of the vegetation and the ritual of the first harvest coincides with New Year.

While the history of the origin of death is a myth for us, for the Kodonko it is the beginning of authenticated history. The Krongo word *barakuunyong,* which must be translated "myth" according to anthropological linguistic usage, is however not applied to the events in Tanbilli. These are considered to be *nyaama ngafiidi,* literally "something that happened." The geographical name Tanbilli contains the word *billi,* "long time ago," but refers to an uninhabited mountainous region east of Dimodonko from which many of the southern Nuba emigrated around 1830 in order to settle gradually in their present-day territories; and the loss of cattle could be the remembrance of a famine. The explanations, which in the complete versions follow the last sentence of the quoted text, are, however, entirely historical. After the actual mythological part each section tells its own history, the history of migrations following the flight from Tanbilli, how the clan X first moved to A, how thereupon one section moved to B, another to C, etc.—until the current distribution of sections has been achieved. The narrations of the clans are not synchronized, and the Kodonko would not be able to date them, but in the background there is definitely a concept of the linear progress of time that reminds us of the genealogical time of other societies.[5] This linear time is not projected into the future; rather, it links the present with the beginning. It seems that the oral tradition of migrations and divisions of sections corresponds to the genealogy of the ashes in the wailing houses, and among the sections, where the offices of the seed and millet priests are inherited, to the genealogy of the ritual millet in the sacred groves and the houses of the millet priests.

Nyaama ngafiidi means history, if one understands by "history" a continuous linking of the present with the past and the beginnings of time. It is remembered in the oral tradition of the sections and represented and authenticated by the continuity of ashes and millet. The oral tradition tells of wars, famines, epidemics, and narrates in moving terms the slave raids that happened at the time of Mahdi and his khalif. Those, however, were contingent catastrophes disturbing the course of events; they forced the Kodonko to interrupt their rituals and with them the progress of time. The Kodonko failed to see in them history as a meaningful drama with a beginning and an end, as a project planned by god, the spirits, or men. In Dimodonko the good—or better—order of the world is not the purpose of history. The order of the world is realized in the regular progress of the rituals; too-powerful events only disturb it.

The ordering of time through ritual and rhythm in historical reality is perhaps more flexible than it seems. The events that made the Kodonko flee Tanbilli are in their consciousness not a "historical turning point" but the beginning of time fusing with the origin of death, and the older tradition that must have existed in Tanbilli seems wiped out. Prehistory is also for us unknown at first, but we know that after great catastrophes such as slave raids, other peoples of the Sudan patched together their cultures from fragments of tradition to create a new order.[6] The culture of Dimodonko does not give the impression of being an original, homogeneous entity, but rather a patchwork whose parts fit together only with difficulty. There are many gaps and repetitions. It is reasonable to speculate that the exodus from Tanbilli was not the first time that the southern Nuba rearranged their traditions in order to adjust them to changed conditions. Historical civilizations, the Egyptians, the Christian Nuba tribes, and the Muslim Arabs have at all times made slave raids in the central Nile Valley; the name Nuba came from Old Egyptian and entered Arabic through Greek and Latin. Ancestors of the southern Nuba are believed to be recognizable in the physiologically detailed representations of prisoners during the XVIII dynasty, substantiating the claim that the ancestors of the Kodonko suffered a long history of persecution and flight. The rituals involving ashes and millet grains would then appear in a different light: they would make it possible to erect a new order and at the same time form a link to what was lost after each destruction, each exodus. It could very well be that in the current flight from the Jihad against the Nuba, some have taken along ashes and millet grains.

Translated by Anne D. Cordero

Notes

1. See Fritz W. Kramer and Gertraud Marx, *Zeitmarken: Die Feste von Dimodonko,* Sudanesische Marginalien 7 (Munich 1993).

2. Arthur D. Nock, *Conversion* (Oxford 1933 [1961]), 4ff.; quoted in Carsten Colpe, "Die Zeit in drei asiatischen Hochkulturen," in A. Peisl and A. Mohler, eds., *Die Zeit,* Schriften der Carl-Friedrich-von Siemens Stiftung 6 (Munich and Vienna 1983), 225–256, in original p. 228.

3. Cf. Ronald C. Stevenson, *The Nuba of Kordofan Province* (Khartoum 1984), 39.

4. Cf. Fritz Kramer, *Der rote Fez* (Frankfurt/ M. 1987), 29–40.

5. Cf. Fritz Kramer, "Über Zeit, Genealogie und solidarische Beziehung," in Fritz W. Kramer and Christian Sigrist, eds., *Genealogie und Solidarität,* Gesellschaften ohne Staat 2, (Frankfurt/ M. 1978), 9–27.

6. Cf. Wendy James, *'Kwanim Pa: The Making of the Uduk People* (Oxford 1979); Joachim Theis, *Nach der Razzia: Ethnographie und Geschichte der Koma,* Sudanesische Marginalien 3 (Munich 1995).

CHAPTER 4

Time Concepts in China

ACHIM MITTAG

In a magnificent manner, [the Sage] comprehends the beginning and end [of all things], and how the six basic constellations [of the 64 hexagrams] are accomplished in their time. Therefore he is able to mount the carriage drawn by the six dragons at the proper time and to travel through the skies.

Confucius (551–479 BCE)[1]

Time—but what idea has a Chinaman of time? Time does not enter into the essence of his ordinary conceptions of a day, or, at all events, the idea is so very vague that the conception of it seems but an inchoate one.

James Dyer Ball, 1892[2]

Introduction

Whether amused or irritated, nineteenth-century travelers to China and China experts frequently remarked upon the indifference of the Chinese toward time: they often turned up for appointments half an hour, three or four hours, sometimes even days late without the slightest trace of embarrassment; the majority of the population set their time according to the position of the sun, some even according to the pupils in cats' eyes that change according to the time of the day; in some cities time was kept by means of very unreliable water clocks or by burning joss sticks. Mechanical clocks were only to be found in front of public buildings in big cities, and when they went wrong or broke, they were not repaired or removed but another clock was set up next to them, so that

Notes for this section begin on page 59.

sometimes there were half a dozen clocks next to each other, each showing a different time.[3] No wonder, then, that a report on the new clock tower in the French concession in Shanghai in an 1880 issue of Wu Youru's (?–ca. 1893) widely read *Shanghai Magazine* (*Danshizhai huabao*) is full of admiration for European inventiveness. The clock could be heard within a radius of a mile, consequently enabling hundreds of thousands of people to have *one and the same* time of day.[4]

This report reflects the almost naïve Chinese enthusiasm for any kind of automatic apparatus, and in mechanical clocks in particular, an enthusiasm from which the Jesuit mission of the seventeenth and eighteenth centuries clearly profited. Before opium, European clocks—initially brought to China by the Jesuits, subsequently produced by Jesuit clockmakers and then also by converted and trained Chinese—were one of the few foreign products actually in high demand. Such clocks were generally bought and sold as curiosities and prestige objects for the wealthy upper classes; the imperial collection alone comprised over 4,000 masterpieces of European precision engineering, one of them an automatic tiger made especially to ensure safe hunting for His Majesty.[5]

The introduction of the mechanical clock and the spread of its use in China did not appear to have any fundamental effect on the normal way of dealing with time until the first half of the twentieth century. As late as the 1930s, in his internationally famed book *The Importance of Living,* Lin Yutang (1895–1976) praised the ideal of the Chinese flâneur in contrast to the Westerner and his chronic lack of time. Lin knew full well even then that China would not be able to escape for long from the Weberian "iron cage" of the modern and thoroughly rationalized world of machines, and yet he dreamed of a breath of the old Chinese art of flâneurship floating across Broadway.[6]

Within the space of one lifetime, or so it seems, Lin Yutang's prophecy (for China) and dream (for the West) have become reality. In the China of today, the saying "Time is money" (*shijian shi jinqian*) has replaced Maoist revolutionary slogans to become the magic formula for breathtaking economic growth. Writing on Macao, the former Portuguese colony that returned to China in 1999, a newspaper report has it that "the European conqueror of once is pulling out, skeptical and pensive, and is being overtaken by its most diligent students who, free of all scruples, retain its activistic economic mentality alone." The author of the article adds the worried question: "Will the new Prometheus China swallow us up with our lethargic hesitancy? Or will we infect her with our scepticism, our contemplative distance to ourselves …?"[7]

These questions indicate a completely new take on China, one that makes Hegel and Ranke's talk of China in terms of "eternal standstill," "static China," "the empire of timelessness," and the "timeless Orient" seems like a reminiscence of the pristine age of our own modern world.

Often repeated and often criticized, this idea of a slumbering Middle Kingdom generally served as a starting point for discussing China's sense of time.

However, these discussions tended to overlook the fact that, to some extent, the image of a "static China" corresponded to China's self-perception during the late imperial period, actively promoted by the educated elite in the seventeenth and eighteenth centuries to the best of their ability. This underlines the need for inquiring into the *longue durée* of Chinese time perception throughout the ages. Over the past ten years there has been a good deal of research done in this field,[8] yet a comprehensive history of the Chinese sense of time is still being missing.

After some initial remarks on the semantics of the term "time" (*shi*) in premodern China and on the three most frequently discussed models of the Chinese sense of time—cyclical, linear, and spiral forms—as well as on the distinction between "outer" and "inner" time cycles in Daoist ritual, I shall make some brief observations on time concepts in Chinese historical thinking, the focus being on an "inner" time pattern in Chinese historiography.

From a "Plane of Time" to the Philosophy of Timeliness: On the Semantics of Time in Premodern China

The Chinese language does not inflect; i.e., the dimension of time within a sentence is not indicated by the verb. On occasion, this can lead to misunderstandings in everyday conversation (for instance in the two questions, which are formulated identically, as to how long a person has already been staying in a certain place or how long he or she is going to stay in that place). This "openness" of the Chinese language in respect of the past, present, or future has been termed a "plane of time," a notion that is rightly said to be at the root of the Chinese perception of time.[9]

This notion somehow evokes the image of the sheet of white paper to which Mao compared China during the Cultural Revolution. Yet such an association would be wrong: in the course of over three thousand years of documented history, the Chinese language has developed a host of linguistic and communicative devices that make up for any lack of verbal time indicators and give the spoken and written language clear time dimensions. They may, of course, not always be put to use. In poetry this is often and quite intentionally the case. Yet even a cursory glance at Chinese poetry through the centuries reveals the immense variety of linguistic means the Chinese language has at its disposal for portraying moments in time or subtle temporal differentiations.[10]

The use of adverbs and particles denoting time—today, tomorrow, next year, still, already, temporarily, immediately, etc.—is perhaps the most important and most effective of these linguistic means. To a large extent, they can be found in the oldest Chinese written testimonies known to us, the oracle bones and tortoise shells inscribed for the purpose of divination from the later Shang and early Zhou periods (ca. thirteenth to eleventh century BCE).

In oracle bones inscriptions we encounter yet another Chinese technique of indicating time: dates. Apart from a few rare specifications of the month, consultations of the oracle are dated by the day, the day in a cycle of sixty (six weeks of ten days each). This is done by combining two symbols from two sets of ten and twelve symbols respectively, the so-called "Heavenly Stems" and "Earthly Branches" (*tiangan dizhi*).

After the Shang period, no sacrificial or ritual activities were associated with this cycle of sixty, and yet it continued to play an important role in temporal orientation well into the twentieth century. It is still evident today in the importance accorded to celebrations of sixtieth birthdays as well as in the fact that prominent historical events such as, e.g., the Hundred Day Reforms of 1898 or the Revolution of 1911, continue to be referred to according to the sixty-year cycle. In Chinese historiography, the sixty-year cycle was not only used for counting days; from the year 4 BCE on, it was also used as a means of counting years.[11] A more common designation for years, however, was in accordance with a ruler's chosen reign period. Introduced in the year 113 BCE, these reign periods originally lasted for only a few years; from the late Yuan dynasty onward (1271–1368), however, emperors preferred to rule under one and the same reign period for their whole reigns.

Tendencies to specify exact dates and to place events within an exact time frame, along with a preference for annalistic records, are salient characteristics of Chinese historiography.[12] As markers of time, however, dates in China have attained a much broader significance beyond historiography. In fact, the Chinese have justly been said to have an "obsession with dates" (Wolfgang Bauer). Thus, Chinese fairy tales and ghost stories, for instance, do not begin with the words "once upon a time" but with "in such and such a year of the reign period so-and-so." Mythical tales are also placed within a historical temporal framework. There are abundant other examples of the significance of providing dates and exact time references in China. Suffice it to say that the Chinese attach great importance to dates as the necessary markers on the language-induced open "plane of time." Indicating either individual points in time or periods of time, these markers provide coordinates of temporal orientation.

In early Chinese, the word *shi*, the first component of the currently used term *shijian* that is the equivalent of our word "time," denotes both a point in and a period of time. This is clearly expressed in the Mohist canon (ca. fourth century BCE), where we read: "Of the times [*shi*] in a movement some have duration, others do not have duration."[13] The various connotations of the term *shi* over the centuries, then, can be traced back to these two basic aspects: point in time and duration. It always had a concrete meaning in premodern writings. *Shi* in the sense of time as an abstract concept in itself only gained currency in everyday language toward the end of the nineteenth century, under the influence of the West and its modern notions of time.[14] The following is a brief sketch of the semantic field of the term *shi* in the premodern written language.[15]

In the *Shuowen jiezi* (*Explaining Pictographs and Analyzing Compound Characters;* around 100 CE), the ultimate basis of Sinolinguistics, *shi* is glossed as "season" in the sense of the four seasons.[16] According to an early legend, each season was under the protection of one of the sons of Xihe, who had been entrusted by the legendary emperor Yao with the duties of watching the heavens, maintaining the calendar, and "instructing the people during the seasons" (*shou shi*).[17] This legend indicates that in the Confucian system of knowledge, *shi* had its place within the category of "heaven," one of the three basic categories of heaven, earth, and man.

Taking "season" as the original meaning of *shi,* we can roughly distinguish five different semantic fields: 1) seasonal rhythms, sequences, frequency;[18] 2) the course of time, times (in the sense of good and bad times), epochs;[19] 3) auspicious or inauspicious times, fate's preordained hour;[20] 4) time as a precious commodity (in the sense of *tempus fugit*);[21] 5) the right moment, a (favorable) opportunity or happy chance, time as a decisive element in strategic planning.[22]

The latter set of meanings is based on the idea of favorable and unfavorable constellations with which human beings must come to terms. This idea lies at the heart of the *Book of Changes* (*Yijing*), the Confucian classic that not only ranks first as to the amount of the exegetical literature amassed throughout the ages, but also has exercised the greatest influence on the Chinese sense of time and space.[23] In the *Yijing* or, more precisely, in its early commentarial layer dating from about the fourth or third century BCE, there occurs a repeated call for action in accordance with the times and at the right moment: "Success lies in attaining [this or that goal] at the [correct] moment," or: "Where actions fit the time, their success is great."[24] A similar note is struck in the *Doctrine of the Mean* (*Zhongyong*) (ca. third century BCE), one of *Four Books* (*Sishu*) canonized in late imperial China, where we read: "The superior man acts in harmony with time and maintains the Mean."[25]

If we may generalize, we can conclude that the notion that one "ought to go with the times" (*qu shi*) lies at the heart of Confucian ethics. Interestingly, this notion of going with the times also found its way into poetry. Liu Xie (ca. 465–522), China's first great literary critic, says of the writing of good literature: "Success will be his who follows the changing times, and he will have no need of fear if he can take advantage of his opportunities."[26]

To keep in step with time and to act in accordance with the times requires a foregoing "examination of the times" (*shen shi*),[27] in order to know when things should happen—sowing, harvest, slaughter, hunting, marriage, etc. This, in turn, demands a precise knowledge of the course of natural events and of the calendar. Thus we can read in the explanation of the Hexagram no. 49 "Ge" attributed to Confucius, "The marshes are burning: the image of chaos. The superior man comprehends the course of time through the organization of the calendar." With an eye to the Chinese commentarial tradition, Richard Wilhelm added the following remarks: "The fire below and the water [my translation: marshes] below fight and destroy each other. In the same way, and through the course of the year, there is a struggle between the powers of light and darkness, the effects of which are revealed in the change of seasons. Human beings are able to control changes in nature if they can comprehend the regularity of these changes and organize the course of time accordingly."[28]

Controlling nature by acting in accordance with the times is also the ruling principle in the ancient "monthly ordinances" (*yueling*), which are divided according to the twelve lunations of the year and include astronomical observations with notes about plant and animal life, agricultural necessities, and sacrificial and ceremonial duties to be performed by the king. These "monthly ordinances" are commonly seen as the prototypes of farmers' calendars.[29] In addition there were almanacs (rishu) in which hidden dangers and auspicious actions were specified for individual days.[30] Even in today's China, "predicting the time"—determining an auspicious date for important events—plays an important role in people's private lives.

Time Models

Cyclical, Linear, Spiral Time

The Chinese understanding of time has been shaped by two closely interrelated areas of life: agricultural activities, and religious practice in the form of ancestor worship and supernatural beliefs. The agrarian character of Chinese society brought with it a distinctive sense of the celestial constellations, the cyclical course of nature, and the changes of the seasons. Moreover, the Shang divination culture encouraged the keen observation of such extraordinary and irregular phenomena as solar and lunar eclipses, comets, and rainbows. Yet it was only after the rise of Heaven (*tian*) as the supreme deity of the Zhou ruling house in the wake of the Zhou conquest of the Shang (eleventh century BCE) that astronomy and calendrical science came to full fruition, first at the royal court and later, during the Spring and Autumn period (722–481 BCE), at the courts of the various feudal states.

Being in use from the Shang period onward, the lunisolar calendar needed constant adjustments to harmonize the solar year with the lunar monthly cycles, which kept the Chinese calendar makers preoccupied. The specificity of the lunisolar calendar gave also rise to calculations of cosmological cycles such as the "great year," a cycle of 23,639,040 years computed by Han court astronomers. In the post-Han time, inspired by or under the influence of the Buddhist idea of *kalpas,* i.e., periods during which a universe is formed and destroyed, even more extended cycles were computed.[31] Suiting the educated elite's taste for numerical speculations, these cycles had no practical use whatsoever. More important for everyday life was the fourth-century division of the day into twelve double hours. Yet another development of great significance for agriculture as well as for festivals and religious and ritual practices was the use of the traditional calendar that divides the solar year into twenty-four phases (*qi*) of fifteen or sixteen days each.[32]

One of the basic charcteristics of Qin/Han thought (221 BCE–220 CE), strongly influenced as it was by the concepts of yin and yang and the doctrine of the Five Elements, was the idea of the interconnectedness of all phenomena. This idea gave rise to a mode of thought that placed time within a relation with space and things of the animate and inanimate world. A foremost example of this thinking is the theory that correlated the sequence of dynasties to that of the Five Elements (earth, wood, metal, fire, water). This theory gained prestige as it was believed that after the founding of the empire in 221 BCE, the First Emperor, Qin Shihuangdi, had made it into a state doctrine by adopting the element water, along with the color black and the number 6, for his new dynasty. In various modifications, this theory played a significant role in Han political thought.[33] Thereafter its influence diminished rapidly, yet it was revived for legitimation purposes time and again, for instance under the non-Chinese Jin dynasty (1115–1234).[34]

In his discussion of Chinese concepts of space and time in his great work *La pensée chinoise* (1934)—one of the seminal treatments of the topic—Marcel Granet emphasized the influence of correlative thinking on the Chinese sense of time. He was of the opinion that the Chinese saw time as a rhythmic sequence of discrete time units, each with its own symbolic value and each standing in relation to other symbolic values—space, color, notes in music, etc.—in short, as a "liturgical order."[35]

Joseph Needham, the great pioneer of the study of the history of science and technology in China, came to quite opposite conclusions in his mammoth project on the research of the history of the Chinese sciences and technology. In a long essay under the title "Time and Eastern Man" (1965), he marshaled evidence that spanned early philosophical writings, historiography, methods of measuring time, concepts of utopia, evolutionary models, the deification of inventors, and the core-idea of "sciences" in China, in order to demonstrate that in numerous areas the Chinese were able to free themselves from backward-looking thinking and that a linear concept of time had been present, even dominant, in China since ancient times.[36]

At the center of Needham's broad argumentation lies China's contribution to horological engineering, in particular to the development of the mechanical clockwork. Needham argues that the Chinese technique in making clocks with driving-wheels anticipated the invention of the verge-and-foliot escapement in early fourteenth-century Europe, which was essential for the triumph of European clockmaking.[37] This thesis of a "stimulus diffusion" from the East to the West is based upon a meticulous study of the water-clock in the huge clock tower built under the supervision of the prominent scholar-official Su Song

(1020–1101) in the Northern Song capital of Kaifeng. Needham was especially fascinated by this clock's escapement mechanism, described as "celestial balance" (*tianheng*).[38]

Granet's and Needham's views have both met with criticism.[39] Derk Bodde took on the role of umpire in the controversial issue of whether a cyclical or a linear perception of time had been dominant in China. At the close of his study on the Chinese sense of time, he offered the following résumé: "Naturally, the evidence pro and con cannot be quantitatively weighed. Nonetheless, on the cyclical side, the evidence appears to me quite sufficient in quantity and clarity to justify the conclusion that, until quite recently, Chinese cyclical thinking was considerably more widespread and influential than was Chinese linear thinking. Conversely, evidence for the linear side seems to me harder to find, more scattered, and less convincing."[40]

Helwig Schmidt-Glintzer came to a similar conclusion: "On the whole it [time] was not linear, but neither was it imagined to be definitely cyclical; the two concepts were intertwined."[41] The late Wolfgang Bauer went one step further than the mere statement of the coexistence of cyclical and linear time concepts. Noting that ancient China, regardless of the many cyclic concepts, had indeed brought forward various teleological and linear time concepts such as the doctrine of the Three Ages (see below), Bauer proposed the idea that the Chinese sense of time was characterized by a concept of time in spiral form. Ultimately, however, this idea of time as a spiral is not satisfying for the simple reason that there is little evidence for it in textual sources, images, or other forms of representation.

The Beginning of All Things, Eternity, and the Distinction between "Outer" and "Inner" Time

If we pause at this point and ask ourselves how Chinese scholars from all ages would answer if we were to question them about their perceptions of time and their opinions on how the Chinese perceptions of time were discussed above—as cyclical, linear, or spiral time—we would soon realize the hopelessness of entering into a discussion with them. The reason is that there was no philosophical discourse about the issue of time in China comparable to that in the Western tradition—a discourse that began with the ontological question about the existence of time (Aristotle), which then, in the Middle Ages, was transformed into one about the theological question about the creator of time (St. Augustine), before turning into a discourse about the constitutive premises of time within the consciousness of man (Kant, Husserl).

To be sure, similar questions were addressed in Chinese texts on rare occasion. One need only think of the passage in *Zhuangzi* (ca. third century BCE), in which the author reflects on levels of existence preceding non-being, described by Laozi as the "mother of the entire world" (*Daode jing* XXV):[42]

> If there is a beginning, then there was a time when the beginning was not there and beyond that a time that preceded the time when the beginning was not there. If there is an existence, then non-existence preceded it, and non-existence was preceded by a time when non-existence had not started and, furthermore, a time in which the non-beginning of non-existence had not yet begun. Non-existence came into being immediately, without one being able to say whether this existence of non-existence was part of existence or of non-existence. Now I have an expression for it—non-existence—but one cannot say whether what I am expressing by it actually makes sense or not.

This passage reminds us of a characteristic concept in Chinese thought, namely the concept of the universe having been formed by evolutionary creation without a personalized creator.[43] Being incompatible with this concept, the Judeo-Christian idea of creation by God *ex nihilo* became a major bone of contention during the seventeenth century between Jesuit missionaries and Chinese converts on the one hand and Confucian scholars on the other.[44]

The concept of evolutionary creation may also go some way toward explaining the fact that problems of how time came into being and whether it exists—problems that so haunted Western philosophers—were just not feasible to the Chinese. "The sage cannot make time," as the *Spring and Autumn of Lü Buwei* (*Lüshi chunqiu;* third century BCE) succinctly states; he can merely act in accordance with it.[45] It was only this practical aspect of moving in accordance with the times and waiting for the right moment that occupied Chinese thinkers throughout the ages.

The idea of eternity was no less alien to the Chinese than the personalized God at the beginning of all things. True, in early Zhou bronze inscriptions the "eternal" is already inchoately conceived as indicated, for instance, by the term "ten thousand years" (*wansui*) that is used to express good wishes and exhortations to obey one's feudal duties and be loyal. However, the term implies only a relative or limited eternity. A telling example of the concept of a finite eternity from later times is the Treaty of Nanjing of 1842, which put an end to the First Opium War. According to the English version, China had to cede Hong Kong to Britain "forever," while the Chinese version speaks only of "a very long time" (*changyuan*).[46]

Admittedly, with the rise of Heaven (*tian*) as the Zhou supreme spiritual power, the concept of an immutable heavenly order came into being. It found its "classic" expression in the much-cited words of Dong Zhongshu (ca. 179–104 BCE): "Heaven does not change; neither does the Way" (*tian bu yi, dao yi bu yi*). However, this concept did not entail a philosophical debate, at least not prior to the influx of Buddhist thought.

Even if one hesitates to agree that the lack of a concept of eternity in Chinese thought was the crucial difference between Chinese and Western perceptions of time,[47] there can be no doubt that the tension between eternity and finiteness as it existed within the Western tradition, the tension between "outer time"—time with regard to God's creation—and "inner time"—time within our consciousness—did not exist in China.

It may, then, be even more surprising when we do encounter the distinction between "outer" and "inner time" in the Chinese tradition. In this context, "outer time" refers to the cosmic or calendrical time, time governed by the natural processes and governed by the duality of yin and yang, time that moves on relentlessly, whereas "inner" time refers to the time of prenatal or preworldly existence, the time of the immortals and of the fairy-like regions beyond our everyday world, a time span not measurable by calendrical time units.

The two different orders of time are clearly borne out in Daoist ritual. Whereas the various time cycles symbolically laid out in the construction of the altar with its three precincts represent a model of the universe and hence the various rituals performed during the altar's installation belong to the temporal order of "outer time," the subsequent ninefold ritual of circulation known as "practicing the *Dao*" (*xingdao*), which is repeated nine times, is thought to enact the ninefold transformation of one's innerself, and hence is governed by "inner time." Moving from "outer" to "inner time" requires thorough knowledge of secret arts such as the "hidden period" (*dunjia*) and the "irrational opening" (*qimen*). All the marvelous "tricks" in Chinese popular literature, such as covering huge distances in seconds or becoming invisible before one's enemy, are grounded in the esoteric ability to step out of the normal time and reenter it at any desired moment.[48]

Yet the idea of the two orders of time was not limited to Daoist ritual and folk belief; it also found its way into China's "great tradition" of Confucianism, albeit not in a discursive form but in the form of two different arrangements of the eight trigrams and sixty-four hexagrams that form the basic framework of the *Book of Changes.* According to the conventional arrangement known as "Postdating Heaven" (*houtian*) the hexagrams are arranged in inverse pairs, as in the received text of the *Book of Changes*, whereas according to the arrangement of "Antedating Heaven" (*xiantian*) in a gradually shifting way, starting with Hexagram no. 1 "Qian," which consists of six unbroken lines, and ending with Hexagram no. 2 "Kun," comprising six broken lines. This latter arrangement of "Antedating Heaven," elaborated in the writings of Zhou Dunyi (1017–1073) and Shao Yong (1011–1077), lies at the heart of Song cosmological thinking. Some six hundred years later, Leibniz recognized it as congenial to his binary system.[49]

If we ask about the implied time concepts, then the arrangement of "Postdating Heaven" may be said to represent "outer" or calendrical time, or in contrast to the arrangement of "Antedating Heaven," "inner time," i.e., the process of gestation and mutation not conforming to calendrical or clock time. In the Daoist tradition, the emblematic number for this internal process of transformation is nine. Thus, for example, we are told that Laozi, literally "the old child," was carried in his mother's womb for 81 (9 x 9) years.[50]

A future task in studying time concepts in China, then, could be to take this distinction of "outer" and "inner time" as a guideline to look into other realms of Chinese culture beyond Daoist ritual.[51] In the following section I will try to do this with regard to Chinese historical thinking. After discussing cyclical time concepts, which can be categorized as models of "outer" time and which have heretofore been at the center of scholarly attention, I will offer some remarks on a pattern of "inner" time in Chinese historiography.

Time Concepts in Chinese Historical Thinking

It is generally thought that Chinese historical thinking was dominated by cyclical time concepts and that dynamic and linear time concepts were the exception that proved the rule.[52] Indeed, much evidence can be marshaled in support of this general view, from Mencius (ca. 372–289 BCE), who stated that "It has been a long time since the creation of the world; [since then] there has been order and then chaos" (*Mengzi* IIIB.9), to Wang Fuzhi (1619–1692), who characterized the course of history as an unending "cycle" (*xunhuan*) of order and chaos.[53] It appears to me that Chinese historical thinking has brought forth three different types of cyclical concepts, to which we shall turn presently.

The "Dynastic Cycle"

According to the "dynastic cycle," the founding of a dynasty was achieved by a strong, charismatic ruler, followed by a gradual decline and a restoration or "mid-term resurgence" (*zhongxing*) before the dynasty was doomed under a weak or disdainful ruler. This pattern of the "dynastic cycle" is *in nuce* already included in the *Book of Documents* (*Shangshu*) and the *Book of Odes* (*Shijing*). Exerting a great influence on the composition of the Twenty-Five Dynastic Histories, the "dynastic cycle" facilitated generalizations, schematic depictions, and stereotypes such as the stereotype of the "bad last minister." Opinions among Western scholars differ as to the validity of the "dynastic cycle" as well as its usefulness as a heuristic model.[54] Yet, as a matter of fact, there still has been no comprehensive study that investigates the changing historiographical significance of the "dynastic cycle" throughout the Dynastic Histories.

The Cyclic Sequences of Dynasties

As mentioned above, models describing the sequence of dynasties in terms of the Five Elements and similar categories were widespread during the Qin and Han periods. Granet had these models in mind when he wrote: "[Chinese historians] are convinced that they have found the correct temporal sequence when reconstructing the past as soon as they can fit events into the rhythmic framework of a liturgy. … They assume that dynasties with different virtues take turns in power in a cyclic sequence."[55]

Here Granet overlooks the ideological character of these historical models, which often were used to underscore the legitimacy of the ruling dynasty or to justify claims to the throne. Also, they were used by historians as a means of criticizing a ruler. An example is provided by Sima Qian's (ca. 145–90/85 BCE) discussion of the Han dynasty in his *Shi ji* (*Records of the Grand Historian*). Referring to the "Three Teachings" (*sanjiao*) propounded by Dong Zhongshu, Sima explains therein the sequence of the Three Dynasties—Xia, Shang, and Zhou—as a succession of the virtues of loyalty and honesty, respect and refinement. He goes on, pointing out that the Han had recognized the requirements of the times, rectified the major failings of the previous Qin dynasty, especially its strict laws and harsh punishments, and thus had rightly obtained the Mandate of Heaven. As has been convincingly argued, Sima Qian is giving vent here to his anger about the emperor Wudi (r. 141–87 BCE), whom he constantly criticized for his large-scale expansionist policies.[56] Thus, this example teaches us to be wary of generalizing about the Chinese sense of time on the basis of such constructions.

World Cycles

There are also historical models that may have been influenced by Buddhist concepts of *kalpas.* An example is provided by the above mentioned Song scholar Shao Yong, who conceived of the entire history of mankind as a part of extended cosmological cycles, his world-cycle being calculated at 129,600 years. These models were of little significance for contemporary history writing, yet it should be noted that Shao Yong's historical thinking provided an impetus for the writing of world-history in the Hunan School around Hu Hong (1105–1161) in the early Southern Song dynasty.

"Compartmentalized Time"

Besides the various cyclic time concepts, we must briefly discuss the notion of "compartmentalized time," which has been seen as characteristic of Chinese historiography with its strong penchant for annalistic records. According to the proponents of this notion, neither the Chinese in general nor the Chinese historians in particular imagined time as a "flow," but instead saw it as "compartmentalized" by small time units. Hence, Chinese historiography presumably remained at the level of "chronography" and did not rise to the level of proper, i.e., reflective, "historiography."[57]

In refutation of this thesis one could argue that annals in general and the later canonized *Spring and Autumn Annals* (*Chunqiu;* henceforth *Annals*) in particular are only *one* origin of Chinese historiography. Another, and for this matter, an older one, are the Odes, especially the *Ya* Odes (note that the character *ya* were been interpreted as a synonym for *ji,* "to record").

It can be safely assumed that a number of these *Ya* Odes were been handed down in the *Book of Odes* from the Zhou founding era (eleventh century BCE), notably those which report the early history of the Zhou people—their origin, migration, occupation of territory, and settlement—as well as the recent past, especially the victory over the Shang dynasty. With regard to these Odes, the early twentieth-century scholar and poet Wen Yiduo (1899–1946) advanced the thesis that all historical records in China had originally been in rhyme form—a thesis that gains some credibility in the light of new textual archeological findings such as the bronze basin of the Scribe Qiang (*Shi Qiang pan*) with a rhymed inscription 282 characters long, that has earned the epithet of "the first conscious attempt in China to write history."[58]

Be it as it may, annalistic records are only one side of early Chinese historiography; another pertains to ballad-like songs as well as narratives built around them, like those exemplified in the *Zuo Commentary* (*Zuo zhuan,* ca. fourth century BCE), which is more than a commentary to the *Annals*, since it alone breathes life into the skeleton of the *Annals.* In short, the notion of "compartmentalized time" only refers to one particular form of Chinese historiography and overinterprets the aforementioned phenomenon of the ubiquity of dates in Chinese writings.

The Five Hundred-Year Cycle as an "Inner Time" Pattern

We finally come to an "inner time" pattern in Chinese historiography, which can be seen in the cycle of five hundred years after which a "Sage" (*sheng*) is said to appear to renew the mythical sage-emperors' rule.

This cycle is first mentioned by Mencius: "From the time of Yao and Shun to Tang more than five hundred years have passed. … From the time of King Wen to Confucius more than five hundred years have passed."[59]

Clearly, the number of five hundred years is being used emblematically here, but its use is by no means as arbitrary as has generally been assumed. As a recent study has shown, it refers to a cycle of 516.33 years as the time that it takes for a conjunction of the planets Jupiter, Saturn, and Mars. This very conjunction was obviously observed in the years 1576 BCE and 1059 BCE, and in both cases, the demise of the ruling house and a Sage's reappearance, i.e., Tang and Wenwang viz. Wuwang, respectively, followed suit.[60] Again, a Sage had to appear after the conjunction of the planets in the year 543 BCE, to receive the Mandate of Heaven and replace the meanwhile politically bankrupt Zhou ruling house.

Mencius' view that Confucius was that sage and that he had received the Mandate of Heaven to at least safeguard the cultural tradition finds support in one of Confucius' own sayings in the *Analects* (*Lunyu*). There it is recorded that Confucius, in a precarious situation for him and his disciplines, exclaimed: "After the death of King Wen, were not these cultural refinements [*siwen*; or: this writing] entrusted to me? If Heaven had wished them [it] to perish, they [it] would not have been entrusted to me!" (*Lunyu* IX.5).

As indicated in the translation, the term *siwen* may refer to either the cultural tradition in the broadest sense, or—as *siwen* was alternatively understood—to the *Annals.* The five hundred-year cycle, the stylizing of Confucius as a "Sage," the interpretation of the *Annals* as his legacy—all this formed a tightly knit web of concepts that had a decisive influence on the historical thinking during the Han period. We shall presently take a brief look at this web from two angles by focusing on, firstly, the exegetical tradition of the *Annals* known as the *Gongyang* School, and, secondly, Sima Qian and his *Shi ji.*

Firstly, venerating Confucius as the "uncrowned king" (*suwang*) and as the herald of the coming Han empire, the *Gongyang* School interpreted the *Annals* as Confucius's political will aimed at "setting to rights the world which has been thrown into chaos and restoring the proper order" (*bo luan shi, fan zhu zheng*; *Gongyang zhuan,* Ai 14th Year). The most significant time concept of the *Gongyang* School was that of the "Three Eras" (*sanshi*), which gained prominence with Dong Zhongshu. Dong understood the "Three Eras" as structuring the period covered by the *Annals* (722–481 BCE): the period Confucius had himself lived through (541–480 BCE); the period Confucius had known from oral reports (626–542 BCE); and the period for which Confucius depended on written records (722–627 BCE).

It was He Xiu (129–182 CE), author of the standard commentary to the *Gongyang Commentary,* who gave an eschatological touch to the notion of the "Three Eras" in that he interpreted the Spring and Autumn period as a three-phase ascendancy from "chaos" to "incipient tranquility" to "great peace" (*taiping*).[61] From the decline of the Han dynasty in the late second century CE onward, the concept of "great peace" was time and again remodeled into religious and political slogans in often Daoist-inspired millenarian movements and popular uprisings.[62] It is worth noting that the notion of the "Three Eras" resurfaced during the resurgence of the New Text School in the late eighteenth century and that a hundred years later it formed the basis for Kang Youwei's (1858–1927) utopian vision of a "Grand Unity" (*daotong*).[63]

Secondly, after Mencius it was Sima Qian who made explicit reference to the five hundred-year cycle, albeit through his father Sima Tan (d. 108 BCE). In the autobiographical section of his afterword to *Shi ji,* Sima Qian quotes his father exclaiming that five hundred years had passed since Confucius and that someone ought to continue his work.[64] Although Sima Qian assures readers that—using Confucius' own words—he merely intended "to hand down and not create" (*shu er bu zuo*),[65] it is quite evident that he understood the project of writing *Shi ji* as continuing Confucius' devotion to "these cultural refinements." Hence, it is no surprise that in later ages the *Shi ji* was ranked next to the Confucian classics.[66]

After the Han, the idea of the cyclic return of "Sages" once again regained prominence through Han Yu (768–824). In his pivotal treatise *Yuan dao* ("On the Origin of the *Dao*"), Han Yu conceives a genealogy of Sages that have transmitted the *Dao* since the early beginnings of the Chinese civilization.[67] What is new about Han Yu's adoption of the "inner time" pattern of old are three points: first, the formerly astronomically determined period of five hundred years becomes insignificant; second, following Confucius, Mencius is included in the genealogy of Sages; third, there is a great hiatus of over one thousand years from the appearance of the last Sage to the present.

This third point is important: It introduced the notion of the *Dao* having become interrupted or dispersed for a long time. By the same token, antiquity, i.e., the period from the legendary emperors Yao and Shun down to Mencius, attained the somewhat timeless normative quality of a "classical period" clearly separated from the postclassical, historical period from about the founding of the empire onward. In this sense, Han Yu's genealogy of Sages—Yao, Shun, Yu, Tang, Wenwang, Wuwang, Zhougong, Confucius, Mencius—is a true "inner time" cycle. Is it to be wondered at that it is also structured by the number nine?

All scholarly speculations revolving around the "correct succession of the *Dao*" (*daotong*) in Song Confucianism took Han Yu as their starting point. It was through Zhu Xi (1130-1200) that the *daotong* doctrine finally gained its authoritative form. This doctrine was a key tenet of the "Learning of the Way," (*daoxue*) which rose to become what has rightly been termed a "state orthopraxy," not only in late imperial China but also in the other East Asian societies.[68]

According to Zhu Xi's conception of the *daotong* doctrine, the hiatus between the last Sage and the reappearance of Sages in recent times, namely Zhou Dunyi and the Cheng brothers (Cheng Hao, 1032–1085, and Cheng Yi, 1033–1107), even grew to more than one thousand years. This had a tremendous impact on Zhu Xi's view of history. For Zhu Xi there were brighter and less bright periods during the twelve hundred years of imperial history, yet compared to the "classical period" it was just a "dark age."[69] Zhu Xi's influential view of history found its way into a work inspired and initiated by him, the *Tongjian gangmu* (The String and Mesh of the Comprehensive Mirror for the Aid of Government). Quasi-canonized under the Manchu emperors Kangxi and Qianlong (r. 1662–1722 and 1736–1796, respectively), it certainly played its part in the tendency toward "dehistoricization" that is observable for the seventeenth and eighteenth centuries. Through Joseph-Anne Marie de Moyriac de Mailla's translation of the *Tongjian gangmu* (*Histoire générale de la Chine,* 13 vols., Paris 1777–1785), this Chinese self-view of their own history finally settled into Hegel's notion of a "static China."

Concluding Remarks

From our glimpses into the five hundred-year cycle as an "inner time" pattern of Chinese historical thinking there are two points that are worth noting. First, it is interesting to see that this "inner time" pattern itself changed over time. This is most obvious in that that the astronomical aspect regarding the five-year cycle becomes insignificant. Secondly, as is elucidated by the remarks in the last section, the "inner time" pattern influenced and shaped views of history. An analogous relationship between "inner" and "outer" time has been observed with regard to Daoist ritual: "[It] passes from the cycles of outer time through a regressive movement into inner time.... However..., inner time in principle not only precedes outer time but also determines outer time."[70]

If a conclusion is to be drawn from our inquiry, I am inclined to point out that time concepts in China cannot be neatly classified under categories such as cyclical, linear, spiral, or "compartmentalized" time. Rather, time concepts in China seem to be more akin to the situation, sketched at the outset that surprised nineteenth-century European observers: one clock next to the other along the street. In other words, one time concept, once formulated, did not replace an earlier one; it instead complemented earlier ones and was, in its turn, then complemented by later ones or taken up again at a later date. This process of continuous accretion resulting in a syncretism quite alien to Westerners does indeed appear to be a characteristic feature of Chinese intellectual history.[71]

Notes

(Note that the works from the Twenty-Five Dynastic Histories, in particular *Shi ji, Han shu,* and *Wei shu,* are cited according to the standard Zhonghua shuju editions.)

1. From the "*Tuan* Commentary" attributed to Confucius on Hexagram no. 1 "Qian" of the *Yijing.* Adopted from Richard Wilhelm, *I Ging: Das Buch der Wandlungen,* 1st ed. 1924 (Munich 1989), 342 (with a slightly different translation). The image of the heavenly journey with a team of dragons refers to the myth of the sun goddess Xihe, who mounts her chariot drawn by six dragons each day with one of her ten sons, i.e., the ten suns, to cross the sky from one end to the other; see Robert Shanmu Chen, *A Comparative Study of Chinese and Western Cyclic Myths* (New York 1992), 20. The image expresses an idea, frequently repeated in commentaries on the *Yijing* (ca. fourth or third century BCE), that by acting in keeping with the times and with the course of nature, the Sage would be able to order the ways of the world and effect perfect harmony.

2. James Dyer Ball, *Things Chinese: Being Notes on Various Subjects Connected with China* (London 1892), 709.

3. See Silvio A. Bedini, *The Trail of Time: Time Measurement with Incense in East Asia. Shih-chien ti tsu-chi* (Cambridge 1994), 5ff., 18.

4. See Wu Youru, *Wu Youru huabao,* Shanghai Magazine, 3 vols. (Shanghai 1983), vol. 2: *Haiguo congtan tu,* 12a.

5. See Catherine Jami, "Western Devices for Measuring Time and Space: Clocks and Euclidian Geometry in Late Ming and Ch'ing China," in Chun-chieh Huang and Erik Zürcher, eds., *Time and Space in Chinese Culture* (Leiden, New York, Cologne 1995), 169–200, here especially 188.

6. Yutang Lin, *Weisheit des lächelnden Lebens: Das Geheimnis erfüllten Daseins,* American orig: *The Importance of Living* (1936), German translation 1960 (Reinbek bei Hamburg 1988), 178ff. and 198f.

7. Mark Siemons, "Die Welt gehört demjenigen, der nicht fühlt," *Frankfurter Allgemeine Zeitung,* 17 November 1994, R2.

8. See Julius Thomas Fraser, Nathaniel Lawrence, and Francis C. Haber, eds., *Time, Science, and Society in China and the West,* The Study of Time V (Amherst 1986); Günter Appoldt, *Zeit- und Lebenszeitkonzepte in China: Eine Untersuchung lebenszeitbezogener Vorstellungen und Handlungsstrategien im vormodernen China anhand einiger ausgewählter autobiographischer Texte* (Frankfurt/

M. 1991), with a detailed discussion of previous studies on the sense of time in China (44–75); Derk Bodde, "Cyclical and Linear Time," in idem, *Chinese Thought, Society, and Science: The Intellectual and Social Background of Science and Technology in Pre-modern China* (Honolulu 1991), 122–133; Bedini, *Trail of Time;* Chen, *Comparative Study;* Huang and Zürcher, eds., *Time and Space.*

9. See Wolfgang Bauer, *Chinas Vergangenheit als Trauma und Vorbild* (Stuttgart 1968), 59–60; Helwig Schmidt-Glintzer, "Zeitbewußtsein im älteren China," in M. Horvat, ed., *Das Phänomen Zeit* (Vienna 1984), 27–41, here 32–33.

10. See James J. Y. Liu, "Time, Space, and Self in Chinese Poetry," *Chinese Literature, Essays, Articles, Reviews* 1, no. 2 (1979): 137–156.

11. A good example of the structuring function of the sixty-year cycle in Chinese historical writing is the drawer system developed by Li Dao (1115–1184) for his continuation of the great chronicle, the *Comprehensive Mirror for Aid in Government* (*Zizhi tongjian*). This drawer system consisted of ten wooden boxes of twenty drawers each. They were marked with the running symbol combination of the sixty-year cycle and contained, in strict chronological order, the notes and materials for one particular year; see Yang Lien-sheng, "The Organization of Chinese Official Historiography," in W. G. Beasley and E. G. Pulleyblank, eds., *Historians of China and Japan* (London, New York, Toronto 1961), 44–59, here 58–59.

12. There are, of course, unintentional and intentional wrong datings (symbolic dates). One should also be reminded that the biographies in the Twenty-Five Dynastic Histories are mostly wanting in dates; see James H. Zimmermann in "Die Zeit in der chinesischen Geschichtsschreibung," *Saeculum* 23, no. 4 (1972): 332–350, here 341ff.

13. Quoted after Christoph Harbsmeier, "Some Notions of Time and of History in China and in the West: With a Digression on the Anthropology of Writing," in Huang and Zürcher, eds., *Time and Space,* 49–71, here 51.

14. The modern term *shijian* (literally: "time-space") is a neologism that, like many other notions in the humanities, was imported from Japan (*jikan*); see Shingo Shimada, "Überlegungen zur gesellschaftlichen Zeitlichkeitsregelung in Japan," in Joachim Matthes, ed., *Zwischen den Kulturen? Die Sozialwissenschaften vor dem Problem des Kulturvergleichs* (Göttingen 1992), 375–382, here 381.

15. I am mainly following Harbsmeier, "Notions of Time," who has prepared the ground for a comprehensive study of the semantics and history of the notion of *shi.*

16. In classical writings, the expression for three seasons (*san shi*) occasionally occurs as the designation of the three growing periods of the year—spring, summer, and autumn. In Western research literature one constantly encounters the notion of "sowing time" as the purportedly original meaning of *shi*, which seems to have originated with Hellmut Wilhelm, "Der Zeitbegriff im Buch der Wandlungen," in idem, *Sinn des I Ging* (Düsseldorf, Cologne 1972), 7–33, here 21. Wilhelm interpreted the pictograph *shi,* i.e., the right-hand part of the character *shi,* as a "sole of the foot," which has some semantic connection with planting, and a unit of measure. The explanation is questionable since the signifying component of the character is the one on the left—the character for "sun," "day," "noon"—the right side of the character probably having merely phonetic function.

17. From a commentarial gloss by Ma Rong (79–166) in *Shangshu,* "Yaodian"; trans. by James Legge, *The Chinese Classics,* 5 vols., 2nd ed. (Oxford 1893–1895 [Hong Kong 1960]), vol. 3, 18. In some versions of the legend, Xihe appears as the mother of the sun or the ten suns, while in the writings of the pre-Han period Xi and He are taken as two brothers. According to the *Bamboo Annals* (*Zhushu jinian*) and the Old Text-*Shangshu* chapter "Yin zheng," the brothers fell victims to alcohol, the calendar became disordered, and a solar eclipse occurred. Subsequently the two brothers had to be made to see reason during a campaign.

18. Besides "season," *shi* also denotes other calendar units, e.g. *shichen,* "the double hour." In the sense of "seasonal rhythms," *shi* is also defined as "the phases [*qi*] of flow and rest in the Yin and Yang powers"; see *Baihu tong* XXXVI ("Si shi"); translated in Tjoe Som Tjan, *Po hu t'ung: The Comprehensive Discussions in the White Tiger Hall,* 2 vols. (Leiden 1949–52), vol. 2, 597.

From this comes the meaning "constant, repeated" (*shichang, shishi*): "The Master said: Is it not pleasant to learn with a constant perseverance and application?" (*Lunyu* I.1).

19. See, e.g., *Mengzi* IIB.13: "That was one time, and this *now* is *another.*" Moreover, since the Chinese medieval period (third to seventh century), the notion that "the times have changed" is commonly expressed by a four-character phrase that occurs in many variations, for instance, *suishi biangai* (as in the preface to the *Wenxuan*). For other variations, see Harbsmeier, "Notions of Time", 59 and n. 9. In the sense of "era," *shi* has been used as a synonym for *shi* ("lineage, generation, world"): "time of peace" (*qingshi,* literally: "pure time"); "a contemporary" (*shiren*).

20. Auspicious or inauspicious fate gives time a special quality for man: "What I call time is not the flow of (calendrical) time (*shiri*), there are truly happy and disastrous times for man (*youli buli shi*)"; *Shi ji, juan* 45; cited from Harbsmeier, "Notions of Time," 52. Everyone is accorded a particular life span: "Whether my death comes sooner of later—that is a predetermined moment." (*Zuo zhuan,* Wen 13th Year). "The life accorded to the individual lasts longer or shorter—how can one know the time of one's fate!" ("Funiao fu," in *Wenxuan, juan* 13).

21. "For the Sage, it is not one foot of jade that constitutes a treasure, it is an inch of shadow [on the sundial]; for time is hard to get but easy to lose"; *Huainanzi* I ("Yuan dao xun"). "The appearance of time flies forth like an arrow, the days and months rush by like weavers' shuttles"; quoted from Harbsmeier, "Notions of Time," 58. On the basis of this and other examples, Harbsmeier concludes, 57–58, that in ancient China, there was a maxim of *carpe diem.* On the economy of time as apparent, e.g., in sayings such as "to be stingy with one's time" (*ai ri*), and on the regulation of work and leisure, see the seminal study by Yang Lien-sheng, "Schedules of Work and Rest in Imperial China," *Harvard Journal of Asiatic Studies* 18 (1955): 301–325 (repr. in Yang Lien-sheng, *Studies in Chinese Institutional History* [Cambridge, MA, 1961], 18–42).

22. In this meaning, *shi* frequently appears as a counterpart to *shì* ("favorable situation, position") particularly in the context of the military and strategy: "Prince Guang of Wu said: 'This is the right moment. We should not miss it!'" *Zuo zhuan,* Zhao 27th Year; Legge, *Chinese Classics,* vol. 5, 719, 722. Someone who misses the right moment in battle and does not make use of a favorable situation (*shi*) is acting wrongly, like, e.g., Duke Xiang of Song in the battle of Hong in 638 BCE: "What counts in the Way [of the ruler] is the favorable moment. Making use of it leads to advantageous positions [*shi*]" (*Guliang zhuan,* Xi 22nd Year). According to *Mengzi* IIB.1, during an attack on a city, the time factor, a time advantage or a favorable opportunity (*tianshi*), is of less import than topography, i.e, the advantage resulting from the terrain (*dili*). The concept of *tianshi* (literally "heavenly time"; here "heaven" is understood as referring to the natural order of *shi,* i.e, all phenomena of calendrical time) became an important category in military strategy in China. One good example is provided by the great statesman of the Tuoba-Wei dynasty Cui Hao (381–450); see his biography in *Wei shu* (Zhonghua ed.) 35/807–828 passim. Acting at the right time and in accordance with circumstances is a distinguishing mark of the sage (*shengren*): "The sage is like the wind blowing at the appointed time [*shifeng*]" (*Shangshu,* "Hongfan"; translated in Legge, *Chinese Classics,* vol. 3, 340. This corresponds with Mencius' characterization of Confucius as the "timely one amongst the sages" (*sheng zhi shizhe; Mengzi* VB.1).

23. See H. Wilhelm, "Der Zeitbegriff"; Li-chen Lin, "The Notions of Time and Position in the *Book of Change* and Their Development," in Huang and Zürcher, eds., *Time and Space,* 89–113.

24. "*Tuan* Commentary" to Hexagram no. 4 "Meng" and no. 14 "Da you"; translation adopted, with slight changes, from R. Wilhelm, *I Ging,* 373 and 415.

25. *Zhongyong* II.2; translated in Legge, *Chinese Classics,* vol. 1, 386.

26. *Wenxin diaolong* XXIX ("Tongbian"); translated in Vincent Yu-chung Shih, *The Literary Mind and the Carving of Dragons* (Hong Kong 1983), 325.

27. This is a title of a chapter in the *Spring and Autumn of Lü Buwei* (*Lüshi chunqiu;* ca. third century BCE).

28. Wilhelm, *I Ging,* 183.

29. See Wolfram Eberhard, "Beiträge zur kosmologischen Spekulation Chinas in der Hanzeit," in idem, *Sternkunde und Weltbild im alten China (gesammelte Aufsätze)* (Taipei 1970), 11–114, here 85ff.; Joseph Needham, *Science and Civilisation in China,* vol. 3: *Mathematics and the*

Sciences of the Heavens and the Earth (Cambridge 1959), 194–195; Francesca Bray, *Science and Civilisation in China*, vol. 6: *Biology and Biological Technology*, part II: *Agriculture*, directed by Joseph Needham (Cambridge 1984), 52–55.

30. Two such almanacs were among the spectacular finds of texts written on bamboo strips in the grave of Shuihudi in Yunmeng (Hubei province). The grave was closed in around 217 BCE and excavated in 1975; see Michael Loewe, "The Almanacs (*jih-shu*) from Shui-hu-ti: A Preliminary Survey," in idem, *Divination, Mythology and Monarchy in Han China* (Cambridge 1994), 214–235.

31. See Nathan Sivin, "Cosmos and Computation in Early Chinese Mathematical Astronomy," *T'oung Pao* 55 (1969): 1–73.

32. For the basics of calendar science in pre-modern China, see Needham, *Science and Civilisation in China*, vol. 3, 390–408.

33. See Michael Loewe, "Water, Earth and Fire: The Symbols of the Han Dynasty," in idem, *Divination, Mythology and Monarchy*, 55–60.

34. See Hok-lam Chan, *Legitimation in Imperial China: Discussions under the Jurchen-Chin Dynasty (1115–1234)* (Seattle and London 1984).

35. See Marcel Granet, *Das chinesische Denken: Inhalt, Form, Charakter* (Munich 1963 [originally published 1934]), 63–85, esp. 71ff.

36. Joseph Needham, "Der Zeitbegriff im Orient," in idem, *Wissenschaftlicher Universalismus: Über Bedeutung und Besonderheit der chinesischen Wissenschaft* (Frankfurt/ M. 1979), 176–259.

37. Joseph Needham, *Science and Civilisation in China*, vol. 4: *Physics and Physical Technology*, part II: *Mechanical Engineering*, with the collaboration of Wang Ling (Cambridge 1965), 440ff. and 532–546; Joseph Needham, "The Missing Link in Horological History," in idem, *Clerks and Craftsmen in China and the West* (Cambridge 1979), 203–238. On the discussion of Needham's controversial thesis among scholars of the history of science, see Antonino Forte, *Mingtang and Buddhist Utopias in the History of the Astronomical Clock: The Tower, Statue and Armillary Sphere Constructed by Empress Wu* (Rome and Paris 1988), 2ff.

38. Joseph Needham, Wang Ling, and Derek J. Price de Solla, *Heavenly Clockwork: The Great Astronomical Clocks of Medieval China* (Cambridge 1960). The authors refer (p. 74) to the fact that, according to Su Song's description, a similarly functioning water clock had been constructed at the Tang court in Chang'an in 725 under the direction of the tantric Buddhist monk Yixing (673–727). There may have been an even earlier model of a large clock that had been constructed under the rule of the Empress Wu (r. 684–705) in connection with the "Hall of Light" (*mingtang*); see Forte, *Mingtang and Buddhist Utopias*, 12ff.

39. The main point of criticism against Granet was that his concept of "liturgical time" corresponds to an ideal pattern that would only have existed for a limited period during the Qin/Han period, if at all. In contrast, Needham was criticized above all for the fact that, on the basis of his *idée fixe* that modern science could have developed in exactly the same way in China, he attempted to prove the existence of the requisite concept of dynamic progressive time in Chinese tradition and thereby ignored a number of issues that would have contradicted his theory of the dominance of linear time in China. See Zimmermann, "Die Zeit," 333ff.

40. See Bodde, "Cyclical and Linear Time," 133.

41. Schmidt-Glintzer, "Zeitbewusstsein," 31.

42. Quoted, with minor alterations, after Richard Wilhelm, *Dschuang Dsi: Das wahre Buch vom südlichen Blütenland* (Jena 1912), 16–17. This text passage is also taken up in the *Huainanzi* (second century BCE).

43. See Erik Zürcher, "'In the Beginning': 17th-Century Chinese Reactions to Christian Creationism," in Huang and Zürcher, eds., *Time and Space*, 132–166, here 134–140.

44. See Zürcher, "'In the Beginning,'" 142ff.

45. Translated according to Richard Wilhelm, *Frühling und Herbst des Lü Bu We*, 1st ed. 1928 (Düsseldorf 1971), 356.

46. See Anthony Dicks, "Treaty, Grant, Usage or Sufferance? Some Legal Aspects of the Status of Hongkong," *China Quarterly* 95 (1983): 429–454, here 444f.

47. As in Wolfgang Kubin, "Zeitbewußtsein und Subjektivität: Zum Problem der Epochenschwelle in China und dem Abendland," in Hans-Jürgen Gawoll and Christoph Jamme, eds., *Idealismus mit Folgen: Die Epochenschwelle um 1800 in Kunst und Geisteswissenschaften. Festschrift zum 65. Geburtstag von Otto Pöggeler* (Munich 1994), 325–336, here 332; Shu-hsien Liu, "Time and Temporality: The Chinese Perspective," *Philosophy East and West* 24 (1974): 145–153, here 147 (also referred to by Kubin).

48. See Kristofer Schipper and Hsiu-huei Wang, "Progressive and Regressive Time Cycles in Taoist Ritual," in Fraser, Lawrence and Haber, eds., *Time, Science, and Society,* 185–205. At this point I would like to thank Professor Schipper for his kindness during our meetings in Leiden, which had a decisive influence on the development of this essay.

49. See Fung Yu-lan, *A History of Chinese Philosophy,* 2 vols., transl. Derk Bodde (Princeton 1952/53), vol. 2, 434ff.

50. See Schipper and Wang, "Time Cycles," 197.

51. In this context, mention must be made of Wolfgang Kubin's interesting suggestion to read the famous elegy "On Encountering Trouble" ("Lisao"), attributed to Qu Yuan (340–278? BCE), as a "dropping out" of time; see Kubin's "Zeitbewußtsein und Subjektivität," 334.

52. On this point, see John Meskill, ed., *The Pattern of Chinese History: Cycles, Development or Stagnation?* (Boston 1965).

53. See *Siwen lu waipian,* 1st ed. 1956 (Beijing 1983), 33.

54. Approvingly, Edwin O. Reischauer wrote: "It must be admitted ... that there is considerable validity to the Chinese concept of the dynastic cycle, if one interprets it as a somewhat superficial political pattern that overlay the more fundamental technological, economic, social, and cultural developments." See "The Dynastic Cycle," in Meskill, ed., *The Pattern of Chinese History,* 31–33, here 31. For a critical view, see Hans Bielenstein, "Is There a Chinese Dynastic Cycle?" *Bulletin of the Museum of Far Eastern Antiquities* 50 (1978): 1–23.

55. Granet, *Das chinesische Denken,* 72 and 74.

56. See *Shi ji* (ed. Zhonghua) 8/175; translated and commented upon by Wai-yee Li, "The Idea of Authority in the Shih chi (Records of the Historian)," *Harvard Journal of Asiatic Studies* 54 (1994): 345–405, here 402ff. On the "Three Teachings," see *Hanshu* (ed. Zhonghua) 56/2518 as well as *Baihu tong;* translated in Tjan, *Po hu t'ung,* vol. 2, 555ff.

57. See Otto Berkelbach van der Sprenkel, "Chronographie et historiographie chinoises," *Mélanges de l'Institut des Hautes Études Chinoises* 2 (1960): 407–421.

58. Edward L. Shaughnessy, *Sources of Western Zhou History: Inscribed Bronze Vessels* (Berkeley, Los Angeles, Oxford 1991), 1. The Scribe Qiang bronze basin was unearthed in 1975 in a Western Zhou tomb dating from the ninth century BCE. Its inscription, which bears some linguistic similarity to the *Ya* Odes, is divided into two sections: in the section to the right the first seven Zhou kings are introduced, and in the section to the left the vessel's donor's ancestors over five generations are introduced. For a translation, see Shaughnessy, *Sources of Western Zhou History* 3–4 (translation) and 183–192 (translation with detailed explanations), and Ulrich Lau, *Quellenstudien zur Landvergabe und Bodenübertragung in der westlichen Zhou-Dynastie (1045?–771 v. Chr.)* (Nettetal 1999), 188–189 (translation) and 184–204 (transcription and glosses).

59. *Mengzi* VII, B.38; translated in Legge, *Chinese Classics,* 501–502. A further passage reads: "More than 700 years have gone by since the beginning of the Zhou dynasty. The five hundred years have passed and the time is ripe." (*Mengzi* II, B.13). Chun-chieh Huang also points to the significance of the five hundred-year cycle for Chinese historical thinking; see "Historical Thinking in Classical Confucianism—Historical Argumentation from the Three Dynasties," in Huang and Zürcher, eds., *Time and Space,* 72–85, here 82–83.

60. See Michael Friedrich, "Tradition und Intuition: Zur Vorgeschichte der Schule von Chu Hsi," in Helwig Schmidt-Glintzer, ed., *Lebenswelt und Weltanschauung im frühneuzeitlichen China* (Stuttgart 1990), 1–43, here 2–3 and the references there.

61. The most important passages are translated in Fung, *History of Chinese Philosophy,* vol. 2, 81–83; see also Anne Cheng, "La 'Maison des Han': avénement et fin de l'histoire," in *Extrême-Orient / Extrême-Occident,* IX: *La référence à l'histoire* (Paris 1986), 29–43.

62. See Christoph Zuncke, *Eschatologie und historische Wirklichkeit in der frühen chinesischen Tradition,* unpublished M.A. thesis (Munich 1993), 65ff.; Christine Mollier, *Une Apocalypse taoiste du Ve siècle: Le livre des incantations divines des grottes abyssales* (Paris 1990).

63. See On-cho Ng, "Mid-Ch'ing New Text (*chin-wen*) Classical Learning and Its Han Provenance: The Dynamics of a Tradition of Ideas," *East Asian History* 8 (1994): 1–32; K'ang Yu-wei, *Ta T'ung Shu: Das Buch von der Großen Gemeinschaft,* transl. Laurence G. Thompson, ed. Wolfgang Bauer (Düsseldorf and Cologne 1974).

64. *Shi ji* 130/1336. This passage has been translated and commented upon by Wai-yee Li, "The Idea of Authority," 360.

65. *Lunyu* VII.1; See also Helwig Schmidt-Glintzer, "Traditionalismus und Geschichtsschreibung in China: Zur Maxime 'shu erh pu-tso,'" *Saeculum* 28 (1977): 42–52.

66. As, e.g., by Zheng Qiao (1104–1162), who in turn placed his own monumental historical work, the *Universal Records* (*Tongzhi*), in a line of succession starting with Sima Qian; see my "Randnotizen zu Zheng Qiao (1104–1162) und seinem Tongzhi," *Chinablätter* 19 (1994): 62–88, here 72–73.

67. See Charles Hartman, *Han Yü and the T'ang Search for Unity* (Princeton 1986), 145–162.

68. See William Theodore de Bary, *The Message of the Mind in Neo-Confucianism* (New York 1988); Thomas W. Wilson, *Genealogy of the Way: The Construction and Uses of the Confucian Tradition in Late Imperial China* (Stanford 1995).

69. This view was expressed by Zhu Xi particularly in his great debate with Chen Liang (1143–1194). See Hoyt Cleveland Tillman, *Utilitarian Confucianism: Ch'en Liang's Challenge to Chu Hsi* (Cambridge, MA, and London 1982), 153ff. On Zhu Xi's view of history, see also Conrad Schirokauer, "Chu Hsis Einstellung zur Geschichte," in Schmidt-Glintzer, ed., *Lebenswelt und Weltanschauung,* 45–54; Schirokauer, "Chu Hsi's Sense of History," in Schirokauer and Robert P. Hymes, eds., *Ordering the World: Approaches to State and Society in Sung Dynasty China* (Berkeley 1993), 193–220.

70. See Schipper and Wang, "Time Cycles," 198 and 197.

71. This has been pointed out by Paul U. Unschuld in "Therapie der Welt und ihrer Krankheiten," in Schmidt-Glintzer, ed., *Lebenswelt und Weltanschauung,* 177–187, here 180ff.

CHAPTER 5

Aspects of *Zeitdenken* in the Inscriptions in Premodern India

GEORG BERKEMER

Introduction

The present chapter is about the change of the *Zeitdenken* in premodern India in a single group of sources.[1] This group contains the epigraphical material that forms the most important textual base for the historian of premodern South Asia.[2] It comprises a large number of historical documents that—due to the material they are written on—are termed "inscriptions." It is not exactly known how many of these texts have been found so far, or how many of them have been published or at least entered in one of the numerous catalogues and find-lists. Estimations are generally between 50,000 and 150,000.

The overwhelming majority of these are votive inscriptions, documents of pious donations by individuals to temples and other religious institutions. Apart from these "private" donations, there exist the official, politically important texts of kings and nobles. These inscriptions reveal not only the individual religious inclinations of the donors, but also allude to their propagandistic and legitimatory functions in the state ritual. They sometimes contain only material of the kind termed "tradition" in the sense of *historische Textkritik,* such as panegyric passages eulogizing the issuing king (*praśasti*) and genealogical information (*vaṃśāvali*) in varying combinations. In many cases the inscription was issued to document a pious donation and ends with a *"Überrest"* in the form of the deed to a piece of land or a village. A third form are those inscriptions that can be found in temples of regional or all-Indian importance where royal families and local elites have inscribed whole series of small epigraphs. These texts, which represent the patronage a family exercised over a shrine for

Notes for this section begin on page 80.

generations, are valuable to the historian as they contain long-term information. The whole corpus of texts inscribed onto the walls of such a temple—often in vernacular languages—thus assumes the character of an archive, is a valuable secondary source besides the "official" inscriptions that provide us, in flowery Sanskrit verse, with the court poet‘s version of the royal ancestry and king‘s achievements.[3]

A Short Typology and Chronology of South Asian Inscriptions

Inscriptions with religious and political content were issued by important dynasties all over South Asia for almost two thousand years, from the third century BCE to the sixteenth century CE. There were many regional varieties and also developments in style and content over this large span of time. A brief outline of these developments will be given here. Their location and their more or less public presentation reveal some information about the century when they were issued and the underlying world-view. Considering the place of installment, together with the material the inscription is made of, allows us to distinguish three types of epigraphs (excluding Islamic inscriptions).

First, there are stone inscriptions on pillars, steles, rocks, or stone tablets and slabs. This is the oldest type of South Asian epigraphical material. Stone inscriptions contain eulogies, commemorative texts, prayers, legal documents, and votive texts on statues, stupas, and buildings. Often combinations of such elements occur.

Temple inscriptions are texts containing information about donations to Hindu temples that are directly engraved into the walls or floors of the building itself or into stone slabs that are affixed to a wall. Direct engraving into building parts is less prevalent and mainly found in South and East India. Slabs are used mostly in North and West India, where temple inscriptions are the most widespread type in the early middle ages (500 BCE–1000 CE), while in the East and South a third type, copper plate inscriptions, predominate. In all regions it is the dominant type of epigraph that was used to preserve the deeds of royal donations.

In a wider sense, donations to religious institutions like Buddhist and Jaina monasteries, which are of importance predominantly before 400 CE, can be categorized together with this type of epigraphs. Such sources are short texts inscribed on the donated object itself or on small copper plaques, often attached to the object itself. Written deeds of land or money transfers to these institutions are rare, but numerous clay seals with imprinted cord marks have been found during excavations at Buddhist sites. Imprints of cords indicate that the seals were attached to perishable materials, just as metal seals were attached in later centuries to copper plate inscriptions.

The third type of epigraph, the copper-plate inscription, was mostly used to record religious endowments, especially land, to individuals or groups of persons. The plates were given together with the land to the donees, usually Brahmins, and a palm leaf copy was retained in the archive of the issuing noble or king. These copper plates are the typical product of medieval South Asian epigraphy, and thanks to their durable nature, contents, and large numbers they are of very high importance for the historian. In many cases they contain the only information enabling the historian to date a political event, even if the fact of it is known from other, e.g., literary sources. Copper plates are most often dated in a detailed and standardized manner that reveals the influence of Indian astronomical and astrological systems of time and shows how important the employment of specialists, i.e., Brahmins learned in these fields, was for the courts of the kings. The inscriptions are dated according to some or all of the following features: the day of the lunar month, the day of the week and solar month (named after the Greek zodiac), regnal year of the king, and dynastic or regional era. Often we also find included in the date an astronomical event such as the beginning of the solar year, an eclipse, or a festival that is dated according to certain celestial constellations.[4]

Stone Inscriptions	**Temple Inscriptions**	**Copper Plate Inscriptions**
• on rocks, pillars, stone slabs • contain edicts, eulogies, etc. • since ca. 260 BCE	• donations to Hindu temples • on the walls of the building • since c. 4th cent. CE	• land donations to mostly Brahmins • since 1st cent. CE, frequent since 5th cent. CE
	• Donations to Buddhist or Jaina sacred institutions • votive plaques on stone, brick, votive objects, seals, or copper plates • since 3rd cent. BCE	

This typology according to material and place of installation shows an implied chronology that can be made more precise by using thematic criteria and leads to a periodization in three phases of development. Since innovations were usually rather slowly transferred through a non-intentional "diffusion" via political marriages, wars, or migrations of religious specialists, chances of development of distinct regional and dynastic epigraphical styles were high. As a general rule, inscriptions on steles and rocks as well as inscriptions connected with Buddhist sites were already widespread in ancient India, while inscriptions in Hindu temples and on copper plates are typical medieval items.

Our first chronological phase comprises the whole time of ancient and classical India from the beginning of epigraphical records around 250 BCE[5] up to the fifth century CE. The social contexts of these early epigraphs are rather heterogeneous. Among them are: Buddhism, Brahmanism, Persian and Hellenistic influences, and Central Asian traditions. For a long time South Asian inscriptions displayed an insecurity of style and many experimental forms as they struggled for an inner structure and an external form.

Ancient and classical India are therefore characterized by a diversity of themes in dynastic inscriptions, out of which four general topics can be abstracted. Along with the most important theme of gift-giving and sacrifice, there are edicts for the right conduct in life (*dharma*) of monks and lay people. A third theme is the memory of the dead, and the rarest, the fourth, is the royal rendering account to himself and the people.

The documentation of rituals of gift-giving to monks, Brahmins, or temples is, besides the pure eulogy (*praśasti*)—which did not exist in ancient India—the only form of record that occurred frequently in medieval India. Preceding these inscriptional records in ancient times were mentions of royal sacrifices[6] and votive inscriptions to religious institutions, either in connection with or on the votive gifts themselves. But there were also documents about donations to monasteries that were written on perishable materials such as bark or palm leaf.

Dharma (virtue, morality, religion) is the oldest theme in Indian epigraphs. It dates from ca. 250 BCE, in the edicts of the Maurya emperor Aśoka (ca. 258–235 BCE).[7] This beginning of Indian epigraphical traditions can be seen as a successful synthesis of Persian and Hellenistic forms and indigenous Buddhist contents. The propagation of the *dharma* as an epigraphic category of its own does not survive Aśoka's reign; however, the ties of later texts to the norms of a then Hinduistic *dharma* endure throughout the history of this form of historical sources.

The earliest known inscriptions written on hero stones (*yạṣṭi*) in the context of the commemoration of the dead are from the first century CE. They contain the oldest written genealogical information. Given the existence of older, oral genealogical texts and also more recent forms of hero stones in illiterate communities of Indian society, we can assuredly assume that hero stones are part of an indigenous tradition of ancestor worship. Often the cause of the erection of such a stone can be determined by examining oral traditions about the setting of a megalith and analyzing narrative abbreviatures that may not always be written texts, but also depictions of the heroic event.[8]

Royalty rendering account to the gods and the people is the rarest of the mentioned themes. It occurs as a continuous account of works and achievements—in pure form, as it were—only once, in an inscription by the king Khāravela of Orissa (first century BCE).[9] Religious motives like ritual gift-giving are not part of its contents, even though the account of the good works (*rājadharma*) done by the king are the cause for issuing the inscription. Here, time is not constructed out of a sequence of rituals. The text simply enumerates the years of reign and renders the events therein. Recourse to the past is done by merely mentioning what had happened. This narrative theme was not continued after Khāravela; in the middle ages, narrations of royal heroism or good deeds were interwoven with the *praśasti*, the poetic and legendary eulogy.

Aśoka's early *dharma*-edicts and Khāravala's account were written in Middle Indian, or Prakrit languages, and in this regard as well there was no successor tradition, for later epigraphy was predominantly in Sanskrit, sometimes also in vernacular languages. Closely linked with the Sanskrit language was the Brahminic culture of knowledge that penetrated epigraphic traditions in the first century CE. At this time, the first cosmological and chronological details were mentioned, and thus we encounter information on time that exceeds the mere regnal years of the earliest stages of the development of epigraphy. From ca. 400 CE onward, inscriptions as a medium of royal propaganda became common throughout South Asia.

In the second phase, from the fifth to the tenth century, the complete set of characteristics of the typical medieval epigraphic tradition came into being not only in India, but also in Southeast Asia. In this period copper plate inscriptions occurred in increasing numbers, and stones were displaced as the most important material for writing inscriptions. Copper plate became the standard medium for displaying royal largesse and heroism and presenting dynastic information and propaganda to rivals and subjects. In this way, inscriptions were used as a repository of very specific knowledge of the past that was thought fit for public display. They are biased, in so far as they express a king or royal family's one-sided viewpoint of present and past political events, but in the absence of other material they are still the most important source for the historian of the period.

At the same time, the epigraphic tradition began to a large extent to express the world-view of Hinduism, integrating much of the religious, philosophical, literary, and moral value system of the Brahmin priests and functionaries. Texts, now mostly in Sanskrit, were from the end of this phase on the productions of learned poets who claimed a much higher social rank as scribes and administrators of the royal chanceries.

The third period, from the tenth to the sixteenth century CE, comprises both apex and decline of the epigraphical tradition. The eulogies of the grand royal charters were now epic poems that tended to become increasingly standardized. In the course of the ongoing penetration of Hindu religious ideas into more and more remote areas of the subcontinent, a growing number of lesser rulers adopted the custom of writing inscriptions using the same epigraphical style as the great kings of the political and cultural centers. Popular temple cults required the patronization of temples by great as well as little kings, and both groups issued inscriptions in Sanskrit and vernacular languages in order to document their piety and largesse to the masses of pilgrims. Rare is the city or a sacred site of that time from which no inscriptions are noted in the epigraphical catalogues. But as they increased in both size and number,

Indian inscriptions slowly lost their function as repositories of knowledge of the dynastic past. Genealogies were replaced by poetic descriptions of ancestors who inevitably were the descendants of gods and the heroes of the Golden Age. Historical information meanwhile found its place outside of the epigraphic tradition in other forms of literature, such as biographies and family chronicles. While inscriptions were transcribed in the 20th century, the end of the third phase also saw the end of the development of the epigraphic tradition. There was again a tendency toward a more matter-of-fact treatment of royal donations in smaller and less elaborate texts. For instance, the standardized eulogistic epithets on the present ruler and his ancestors were replaced by statements of military achievements and titles derived therefrom.

The Structure of Time

Genealogy and date are the most important temporal markers in South Asian inscriptions. The Indian epigraphical tradition began when Indian rulers were exposed to cultural traditions that came from the Persian Empire and later from the Hellenistic states of the Near East and India. The innovations adopted by the Maurya emperor Aśoka from the Persian system of representation of power were subjected to a filter-like process of cultural selection: the oldest inscriptions show that aspects of the royal and state ritual connected with the writing of epigraphs were incorporated into the indigenous system of rulership, while the art of stonemasonry and systems of time reckoning were not taken up by the writers of the inscriptions.

Kings in ancient India also did not use genealogical information for state propaganda.[10] This may relate to the differences in the roles of genealogic experts in the early states of South Asia as compared to elsewhere, as well as to the relatively low degree of complexity of early Indian states, with the possible exception of the largest, the Mauryan Empire. The first inscriptions in South Asia occurred approximately 300 years before the emergence of a dynastic era in the first century CE, and it took—despite intensive contacts with the Hellenistic world and the Roman Empire—a further 400 years for any dynasty to become widespread throughout the subcontinent.

On the other hand, is it obvious from literary and archaeological sources that the ongoing penetration of the subcontinent by institutions of state society and high religion (especially Buddhism) brought forth a highly subdivided social system and increasing professional specialization. As part of this development, dates and other moments or periods of time are given in increasingly complex ways in the forms of astronomical details, genealogies, and eras. From the first centuries CE onwards, royal courts not only dated their inscriptions in regnal years, as Aśoka did, but also mentioned months and days, religious festivals, solstices, equinoxes, the entrance of a celestial body into a sign of a lunar house or the zodiac, etc. This process was a slow one that more than once drew inspiration from non-Indian exemplars—especially, again, the Greek rulers of the Near East and India.

Together with the involvement of astronomers, poets furthered the simultaneous tendency of epigraphic writing toward literary styles and topics. The status of the genre was heightened by using Sanskrit to produce poetic texts (*kāvya*) instead of spoken or semi-literary Middle Indian languages (Prakrits), which were seen as more fit for administrative purposes. Part of this process of Sanskritization was a reduction in content, as the above mentioned themes of gift-giving and sacrifice became the only dominant ones in political epigraphy.[11]

These changes in form and theme accompanied a replacement of matters of administration and historical narrations by topics laden with religious content. Renderings of the past in medieval inscriptions often allude to mythical tales from Hindu sacred texts. As of the early middle ages the inscriptions feature the typical "Purāṇic" *Zeitdenken,* which is a constituent part of Hindu culture up to the present day. The main characteristics of this form of *Zeitdenken* are: (1) space and time of gods and human beings are a continuum in which things slowly deteriorate from an ideal toward a final disaster; (2) conditions worsen in each of the four subsequent eras of the aeon (*yuga*) until, after a final deluge, a new creation starts the cycle again; (3) within the cycle, connections between the beings of these worlds and times are encoded in genealogical or intellectual lines of descent, i.e., as lineages of ancestors or successive generations of teachers.

This world-view finds its place in the Purāṇas, still popular collections of texts whose oldest written version was compiled in Sanskrit in ca. 400 CE. Their earliest contents were common knowledge in North India in the form of "old tales" (*purāṇa* means "old"), which included oral myths, epics and bardic songs, court annals, and chronicles. However, the predominant material in the current Purāṇic editions consists of legendary tales and philosophical treatises. In the centuries between ca. 400 CE and today, such materials in the form of, e.g., didactic dialogues, *vitae* of holy men, and hymns to the gods were added. There are also legendary accounts and explanations of the emergence and spread of regional religious movements and the importance of pilgrimage sites.[12]

It is important to note that the Indian epigraphic tradition was part of a political propaganda apparatus that used more mythical and epic than historical literary forms. In this context of a predominantly Hindu discourse, it was molded itself by the structure of Purāṇic writing. Indian authors have characterized the ideal type of this structure by using the term *purāṇa-pañca-lakṣaṇa*. These "five characteristics of the Purāṇa" are explained by the ancient Indian Sanskrit dictionary Amarakośa: Purāṇa is supposed to contain the chapters "Creation" (*sarga*), "Re-creation" (*pratisarga*), "Lineage" of gods, vedic sages, and epic heroes (*vaṃśa*), epochs of the creation or "Epochs of the Manus" (*manvantara*), and finally the "Genealogy" (*vaṃśa-anucarita*) of the kings of the solar and lunar lineages. This definition of the ideal Purāṇa comes probably closer to some of the great medieval inscriptions than to the real contents of any of the eighteen Great Purāṇas (*Mahā-Purāṇas*) of today.

Exemplary of such an inscription are the copper plate inscriptions of the Eastern Ganga Dynasty of Orissa (498–1434 CE). After 1100 CE there appears a standardized genealogy in the form of a great poem (*mahākāvya*) in Sanskrit as the first part of the text, which records in all documents of this type a land donation to Brahmins.[13]

The Ganga *praśasti* is a poem that, like many other similar historical texts, is subdivided in a linear way so that each section—after a preliminary mythical and legendary account of the divine and remote human ancestors—tells the deeds of one king. Accordingly, a new section was added and later modified for each new ruler after his ascent to the throne. Its form imitates the epic and Purāṇic literature, but also its contents are molded according to such examples in so far as it is almost completely devoid of historical narratives. Information about historical facts is omitted in favor of narrative abbreviatures that allow the educated readers of the time to fill the gaps with their own knowledge about episodes from epic and Purāṇic texts. In contrast to other *praśasti,* there are few allusions to classical poetic works in the form of hidden quotations and stylistic imitations.

The first section alludes to a creation myth in which the god Viṣṇu brings forth Brahmā, Brahmā brings forth Atri, and Atri brings forth the moon (verses 1–4). Next it is stated, in a self-reference that alludes to a whole section of the history of Indian literature, that the heroic deeds of the kings of the moon-dynasty (*candra-vaàça*) are being sung in the Purāṇas and poems throughout the whole world, implying that it is not necessary to repeat all this here again (v. 5). The same abbreviature is again used in verse 6, where it is said that the deeds of the hero Arjuna alone have filled a whole *Mahābhārata,* the longest poem in the world. The poet then skips over twenty four generations of legendary kings by simply mentioning their names. They are said to already be rulers of the last, most cruel and shortest of the four aeons, the *kali-yuga.* From the last of these, the list of the "immediate" ancestors branches off, and the reader is told in the same Purāṇic style about the immigration of the early heroes of the Ganga family to Eastern India (verses 7–11). There follow more than 100 verses that relate the deeds of the twenty historical kings who ruled in Kalinganagara[14] and later in Cuttack, the old capital of Orissa. True to the style of the text, this section is again full of comparisons with gods and heroes of bygone ages that are intermingled with details of historical events, mostly wars and religious donations. No mention is made of any of the historical Ganga kings who ruled prior to 1019 and who are known through their own copper plate inscriptions.

This turn of epigrapical writing toward the world-view of the Purāṇas explains why there was never a long-lasting development beginning with the more informative themes in the royal inscriptions. The medieval *praśasti* do not present the information with an outside observer in mind or in order to explain anything. They are not written, for instance, from the point of view of an individual who wants to justify himself before a god; neither does a king or a royal official ever attempt to explain the decisions proclaimed as reactions to experiences of contingency (*Kontingenzerfahrung*), such as wars or natural disasters. Instead, they conform to the inevitable standardized picture of the king in his role determined by the Hindu *dharma,* which surpasses any individual characteristics of an individual ruler. Authors of medieval South Asian inscriptions did not intend to inform; few aspects of daily politics and public life, or archival materials on which administrative decisions were based, are mentioned.

History here is not presented as an "inventory of facts," nor is it a chain of causal relations between a past event and its present consequences. Instead, the historic past comes second to idealized tales that are the literary representation of a legendary past and an ethical world of royal *dharma.*[15]

With the growing length and complexity of medieval South Asian inscriptions, room was available for the treatment of phenomena of time in complex ways within the paradigm of Purāṇic cosmology. This means foremost that ideas about time were based on the Kalpa cycle. But time is being treated within the framework of poetry, not as a historical or philosophical problem.

Genealogy and Era

One of the most elementary ways of dealing with time in inscriptions is the insertion of a passage about the moment in time when something happened or about the duration of an event. This can be done either in the form of a comparison, in which something happened at or near the time when another well-known event was taking place, or by using countable units of time of various scale. Early South Asian inscriptions (third century BCE to first century CE) contain no statements about time or only very simple ones. Later on, however, dating systems and conventional formula were developed, culminating around 1000 CE in an all-Indian system of dating—though never without regional variants.

"History" has been present in South Asia since Vedic times in the form of oral texts such as bardic epics and dynastic traditions about the ancestors of one's own lineage and their marriages, and also about the *guru-paramparās*, the lines of teachers of sacred texts and their disciples. Epigraphical sources are not only the first written texts found so far, but also the oldest written literature containing such information. In them, as in the oral literature, time was structured in terms of generations, which are, together with other natural time cycles, basic units of time in which social and political order can be expressed.

As the epigraphical tradition develops, the first simple statement about time, the regnal year, grows into a long formula consisting of information on various time-related social and astronomical phenomena. First, the names of father and grandfather are added to the regnal years (first century CE); then longer lists of ancestors leading back to the founder of the dynasty are given. The next layer is added when a king starts to count the years from the beginning of his family's rule besides or instead of his regnal years, and thus introduces the beginning of a dynastic era (second century CE). During the time of the Gupta empire (320 CE–510 CE) the political symbolism of the dynastic era becomes evident when little kings take over the era of their overlords and their system of temporal reference as one of the symbols of their lower status. Not all dynastic eras reflect a family's establishment as independent rulers; there were also eras connected with cosmological dates, events of regional importance, or heroic figures.

During the middle ages little kings sometimes adopted the dynastic eras of the great kings for a longer time than these existed, especially when the connection of the dating system with the origin of the former overlords was not too obvious. After its downfall, the great family's dating system was then retained by the new overlords and their little kings, and contributed in this way to the establishment of regional eras and styles of dating. One of these former dynastic eras, the "Scythian" or *Śaka*-era of 78 CE, was eventually spread from its region of origin in Gujarat over virtually the whole subcontinent and is today an official era in Indian and Nepal.

Western India, the region of Gujarat, Sindh, Malwa, and Rajasthan, might well have been the point of origin from which the idea of dating records in family eras spread to other areas of the subcontinent. Here it was closely linked with the dominant regional era, the Śaka era. The emergence and development of this important era shall be used as an example of an Indian inscriptional dating system. Even though it is clear that the date of the era is the Christian year 78 CE, the identity of the founder of the era is still a matter of debate.

The mainstream opinion among historians is—despite the fact that the era is called the "Scythian era"—that it goes back to the greatest of North Indian kings of the first century, the emperor Kaniṣka I of the Kuṣāṇa dynasty.[16] The Kuṣāṇas were a family of Central Asian kings who became rulers over a large northwest and north part of the Indian subcontinent. Kaniṣka I successfully extended their sphere of influence into the Śaka territory in present Gujarat. From this it has been concluded that the term "Śaka" was taken after some time as a generic term for any great king of foreign origin who ruled in Northwestern India. But the date 78 CE of the Śaka era tallies much better with the beginning of the rule of a small family of actual Śaka origin: it is the first year of the rule of the Indo-Scythian satrap (*kṣatrapa*) Caṣṭana.[17] His era continues up to the end of Śaka rule in 409 CE within his family, and as a regional era has survived both the downfall of the great Kuṣāṇas as well as the suzerainty of the Gupta empire (320–510 CE). Hindu temples and Jaina monasteries of the middle ages also used that era, a fact that allows us to assume that they, as well as astronomers and astrologers, further helped the Śaka era to persist through the time after the end of Śaka rule until the era was again used by a royal dynasty. From the seventh century onward the Śaka era can be found in the inscriptions of the Cālukyas of Badami in Karnataka. With the conquests of the Cālukyas and the foundation of numerous smaller Cālukya principalities in South India, the era spread over sizable parts of the Dekkan Peninsula and became a standard of time reckoning in virtually all of Hindu India from the eleventh century onward.[18]

The example of the Śaka era shows that an era can, once it has been established as a part of a regional historical tradition, display a rather persevering character. Especially in circles that were not directly influenced by a change at the top of a power pyramid, such as peripheral political and religious administrations and little kings, the inclination may have been to prefer a new interpretation of an already existing era rather than to adopt an entirely new dating system. To establish a new dynastic reckoning against the interests of conservative regional elites may be considered a substantial political and also symbolic task. Once it was successfully put into practice, the founding of the new era itself could then become a topic of state propaganda.

Future Kings

In order to justify and legitimize actions in their present, the medieval kings of India related themselves to figures of a legendary or historic past that they took as role models. But there was also the question of the fate of their political, and even more their religious, activities in the future. Even though the many donations for the spiritual benefit of mother and father or for the donor himself were been made in a religious context, it is not common for inscriptions to explicitly relate a king's actions to philosophical or religious traditions, such as the time-related questions of rebirth and the fate of the king as an individual after his death. The central question of concern is the future of the donation itself, which is intended as a perpetual gift-giving ritual that should last "as long as sun and moon exist," that is, until the next re-creation of the universe according to the Purāṇic ideas about space and time.

Even the oldest epigraphical sources, Aśoka's inscriptions, contain passages in which the doings of past kings are denounced and the future is seen as a brighter place as long as the Buddhist dharma is followed by everybody. But most of his concerns about the effects of time deal with his own family. Most important is the fact that Aśoka never mentions any future king outside of his own family.[19] Indian kings of later times also made a close connection between the future of the earth and the future of their family by avoiding mention of any king of equal or greater rank in their eulogistic genealogies and by constructing a direct link between the gods and the earliest members of their dynasty. These inherent godlike qualities of the royal family[20] may also be seen to imply a claim to a very long, if not virtually eternal, reign that will last to end of the present creation.[21]

The future is a timescape that is endangered by the continuous decay of the human existence on earth as it is prophesied in the Purāṇas. The king, whose task it is to uphold the dharma and to protect the earth and all human beings thereon, is the only institution that can slow this moral and physical decay, at least for his lifetime. In this way, a dynasty of dharmic kings is able to reverse the course of time and reestablish a golden age as long as the royal family plays its role as protectors of the Brahmins and keepers of the law. In such a world, misdeeds of individuals have to be punished, but there is no threat of a general crisis caused by the decay of the dharma.

But this rather optimistic propaganda, which was more intended to impress the subjects in the respective present time of the king than to express a concrete belief in the state of the future, did not prevent the kings from expressing more realistic concerns about the future fate of their donations in less conspicuous parts of the inscriptions after the pompous *praśasti*. At the end of the texts, after the details of the deed are stated, future kings are routinely admonished to take care about the fate of the donations of their predecessors and to recognize the rights to the donated lands. There exists a corpus of a maybe a hundred verses (*Pönformeln*) attributed to the Vedic sage Vayāsa that take up the appropriate injunctions from oral traditions, epics, and the law manuals (*Dharmaśāstras*). In them, the future kings and other humans are strongly advised that it is more fruitful to protect the donation of another than to donate oneself, and horrible consequences are predicted for those who infringe upon the rights of the donees or even take away their land.[22]

These almost omnipresent *pönformeln* make sense only if one assumes that the donating kings and the receiving Brahmins really believed in a continuing rule by the king's line "as long as sun and moon exist." Yet, they were obviously well aware of the facts that shifts of borders and loss of territory were frequent, that a successor could prefer a new religious affiliation, and that a changing political climate or an empty exchequer could cause a donation to be revoked entirely or its terms changed so that the privileged landlord had to pay taxes like other land owners. In some instances donated land was taken back by a new dynasty for distribution to its own Brahmin priests and administrators.

Zeitdenken and Historical Consciousness

I have used the emergence of genealogies and eras to show that in the texts produced by Indian kings, time is a construct that combines the necessities of political praxis and religious tradition, sometimes harmoniously, but very often in a conflicting way.

After the epigraphic tradition came into being in the time of Aśoka Maurya in the third century BCE no other epochal innovation was introduced by a single great ruler. Changes came about slowly over generations through the given forms' adaptation to the needs of a ruling family, by adopting regional conventions and reactions to general trends in all-Indian cultural changes, especially the incorporation of the Purāṇic world-view and classical Sanskrit literary styles into the inscriptions. Writing inscriptions in India had never been merely an administrative act; it always had its religious and propagandistic functions. In medieval India, the big royal inscriptions with their different parts could very well attribute different notions of time to different parts of the text, ranging from the eulogy at the beginning, which contained cosmological ideas and mythical genealogies, to astronomical details, which determined the dates of the religious festivals that were the settings of royal gift-giving rituals, and lastly to details of the agrarian cycle further back in the text, often in the spoken language of the region. These sublime differences show that inscriptions had become a part of the literary production at the royal courts, a process through which they appear, rather uniformly in style all over India, by approximately the year 1000 CE and also dealt with notions of time in very similar ways, even though the preference for stone inscriptions is still to be seen in West India while copper plates are more frequent in the east. Slightly later, with the great state temples, the stylistic differences between the elaborate royal court documents and the more matter-of-fact temple inscriptions become more pronounced. All these types of epigraphic documents, however, document not only the typical Indian *Zeitdenken* in the form of cosmology, calendrical system, and eras, but also time as a part of practical reason (*Zeitpraxis*) in the field of state ritual and politics.

Inscriptions give us an impression of how great kings used, among other resources, the time of their subjects and little kings to enhance their power and to display their supremacy through symbolic acts. In the system of interdynastic communications in medieval India some modes of contact such as war, marriage and tribute relations, temple patronage in border regions and foreign territory, etc. are connected with the use of individual and collective time. It has already been mentioned that little kings were expected to use the era of their overlord. For the losing party, wars may end in public rituals of surrender that are connected with prolonged periods of waiting at court to be let into the presence of the victor, and they often establish obligations in the cycle of the year, such as traveling to the court of an overlord at his command and public displays of submission at certain times in the festival calendar of the stronger king. Another exercise of symbolic command over the time of another noble was to oblige the weaker party to take part in the wars of the overlord. In such ways, a little king and members of his family, e.g., hostages or women given away in political marriages, were be obliged to dedicate parts of their ritual time or even their life time to the overlord. Since these contacts went on over generations, they also served as channels of information between courts of regional powers.[23]

Under such circumstances the reproduction of already known conventional formulae and metaphoric expressions of time was the norm. Inscriptions were rarely a place of innovation and were no platform to reflect about the character of time in general or the finite nature of human existence. Administrative routine did not encourage stylistic creativity or rapid changes in the content of inscriptional sources. Strong division of labor is one of the characteristics of traditional Indian society, which leads to the assumption that the composers of medieval royal inscriptions were often to be found among the court poets and other learned Brahmins, as the names of composers suggest, while scribes and other specialists in the field of administration who were the managers of practical time merely provided the raw materials.

Inscriptions are only loosely associated with the world of the ascetics and philosophers whose predecessors and teachers had written the treatises about cosmology and cyclic time. There, time (*kāla*) plays both the role of fate and of the *śakti,* the personified divine energy and female companion of the god Śiva. There is no evidence in epigraphical sources of their direct involvement in the making of epigraphical records. But this world of intellectual speculation surely had its subtle influences on the motives of royal policy and may also have played a role in the usage of symbols and narrative abbreviations, even though they are not mentioned directly in the group of texts discussed here.

There is also no expressed necessity in the texts to make one's place in history. Therefore, no elaborate constructions of time systems were ever created and maintained as repositories of knowledge about the historic past, as in China or in Europe. Time was never conceived as something absolute, existing outside and above the world of men and the gods. Linear time was predominantly time structured by the succession of the generations; history is therefore a narration in the form of the *vaṃśāvali,* the tale of the royal line (*vaṃśa*). This unilinear form of dynastic history is sometimes transcended in other literary forms of more ambitious character, such as the poetic chronicles and biographies or the Purāṇas. Here the single strands are interwoven into a network of tales, augmented by court poets with artful embellishments and by priests and ascetics with their symbols of religious devotion.

The writers of inscriptions, however, are not interested in telling the historically "correct" events or in dating them. It is therefore often hard to determine the position in time of a document relative to another one, sometimes even if it is known that two inscriptions belong to one family—more often when the historian tries to determine the temporal sequence of documents from different dynasties. The dating in the Śaka era and other less popular systems is therefore a very helpful, if often late, introduction into the epigraphical tradition of a region. But even in cases when eras like Śaka appear only after 1000 CE, this does not justify assuming that there had previously existed no historical consciousness. The "neutral date" of a nondynastic era is a late invention in most parts of the subcontinent and even then is often accompanied by popular legend.[24] An absolute time, reckoned from a specific date outside the context of ordinary history, is—other than in the Jewish, Muslim, and Christian traditions—rarely used.[25] Time does not occupy an axiomatic position in the world-view of the medieval inscriptions, but is on the contrary dependent from or even identical with fate (*daiva*) and thus an aspect of the highest divine being.[26] In the popular Purāṇic myth as well as in medieval *bhakti*-religion God is the lord of fate and time, and enables devotees to overcome their bonds and be freed from the cycle of rebirth. But this religious belief is no proof of an entire absence of historical consciousness, such as has been attributed to India, especially by nineteenth century thinkers like Macaulay and Hegel.

In the medieval Hindu world the king is bound to fulfill his duty as the protector of creation and dharma. In this role he is in no position to express doubts about or formulate explanations of the historical preconditions and present circumstances that constitute his historical present. With very rare exceptions, activities are justified within the context of the conventional framework of dharma and power politics.[27]

Indian inscriptions therefore present no narratives of events, seen as links in a chain interwoven by cause and effect. But the records display webs of relations in which the past is—in Marshall Sahlins's words—encoded in a "meaningful order" different from ours.[28] The past is structured with local and regional relations in view,[29] it is multi-dimensional[30] and encoded in the language of myth and poetry. Keeping this in mind, it is neither astonishing nor a relapse into "primitivism" when around the year 1000 CE the texts begin to tell elaborate tales of a mythical nature that overgrow the previously shorter and more straightforward genealogies. This change correlates closely with the use of the Śaka era as the common dating system, which freed the genealogies, so far the most important way of dynastic time-keeping in public records, to serve a new purpose. A third change might have accelerated this development: the new monumental state temples attracted masses of pilgrims and displayed for them an image of the king and his dynasty that was appropriate to the popular beliefs. Temples and inscriptions use similar Purāṇic imagery to represent the king as the highest devotee of the state deity.

In the encoded narrative of the past, "events" are meaningful in the cultural order of society.[31] More important than a description of the exact circumstances or the astronomical time of the occurrence was the relation between the actors, which is mentioned in terms of: (1) status: e. g., as titles[32] that abbreviate narrations about victories over neighbors, performative acts such as a horse sacrifice, one's own dynastic era, mention of marriage alliances with dynasties of high status, an epic genealogy; (2) style: using the literary style of the eulogy (*praśasti*), a distinguishing feature of high poetry (*kāvya*), the amount of space a temple inscription occupies at a prestigious location increases in relation to texts of competitors of similar status; (3) context: the prestige of Brahmin donees, number and location of inscriptions at a particular temple, number and location of patronized temples, value of the grants, etc.

All this shows that in inscriptions there is no need for the detailed reconstruction of historic events and their precise times, as would be necessary in, for instance, a royal apologia in the face of some supernatural entity. This kind of *Zeitdenken* is present in the Near East and in Western religions, but not in South Asia. No Indian king of the middle ages ever issued statements of justification or explanation, but rather let his court poets write hymns of praise about him as a successful protector of the dharma. This emphasis on the king's role as the protector of the social standard may also be a reason for the uniformity of medieval inscriptions in terms of themes and style throughout the subcontinent.

We can also assume that the absence of "objective" historical writing in India as an independent counterpart of royal propaganda plays an important role in the development of *Zeitdenken* in Indian epigraphy. Except the historian Kalhaṇa, the author of the *Rājataranganī* of 1148,[33] no writer has had both insider knowledge of the political affairs and administrative proceedings at the regional royal court, and sufficient ability as a critically and independently minded writer, to compose a work acclaimed by both his contemporaries and later generations of critics and scholars. The critic was most likely a religious person with no interest in writing literature, the poet was employed to produce the standard eulogies, and the administrator in charge of the data had no voice outside the court itself. Thus Kalhaṇa's biography, which combines these roles, is unique and thereby shows that there was no traditional role model in which the separate worlds of the court officials combined to bring forth the type of the chronicler or historian who writes for a larger community.

Notes

1. Since information of a more general nature about the history of South Asia is not included here, the reader may refer to H. Kulke and D. Rothermund, *A History of India* (London and Sydney 1986). The following abbreviations are used in the text: *CII (Corpus Inscriptionum Indicarum), EI (Epigraphia Indica), IO (Inscriptions of Orissa).*

2. I am using the term "premodern" to avoid the tricky question of periodization in South Asia. See H. Kulke, "Gibt es ein indisches Mittelalter?," *Saeculum* 33 (1982): 221–239.

3. See G. Berkemer, "The 'Centre out There' as State Archive: The Temple of Siṃhācalam," in H. Bakker, ed., *The Sacred Centre as the Focus of Political Interest* (Groningen 1992), 119–130.

4. See Arthur L. Basham, *The Wonder That Was India* (New York 1959) (and many other editions), Appendix II, "Astronomy," 489ff., and L. D. Swamikannu Pillai, *Indian Chronology (Solar, Lunar and Planetary): A Practical Guide to the Interpretation and Verification of Tithis, Nakshatras, Horoscopes and Other Indian Time-Records, B.C. 1 to A.D. 2000* (Madras 1911 [Delhi 1982]). The words for

year in Sanskrit are mostly derived from seasonal or yearly natural cycles. The term *varṣa* means literally "rain" and reminds one in the same way as *abda* (*ab-da:* giving water) of the monsoon rains in the cycle of the year. Only the very common term *saṃvatsara* and its abbreviation *saṃvat* are derived from a cycle whose basis is less suggested by natural phenomena: the *yuga,* a cycle of five or sometimes six years, in which the *vatsara* is the last year, and the *saṃvatsara* the first. From there *saṃvatsara* developed the meaning of "year" in general. The following date is an example: *mīnamāsa-navame somavāre … śakābda 967* – on the ninth (lunar day) of the (solar) month (*māsa*) Mīna (the sign Pisces of the zodiac), on the solar day Monday (*soma:* moon, *vāra:* day) … *Śaka*-year 967 (Śaka: the Scythian era of 78 CE); the date corresponds to 3 March, 1046 CE (from: "Narsapatam Plates of Vajrahastadeva III.; Saka Samvat 967", ed. Sten Konow, *EI* 11 (1911): 147–153); conversion of date according to *IO* 3/1, 1, fn. 3.

5. This is most likely also the time when writing was introduced into India from the West by the Maurya dynasty of North India; see Harry Falk, *Schrift in alten Indien* (Tübingen 1993).

6. Most important was the *aśvamedha* (horse sacrifice), a ritual expression of political independence. This public rite was later gradually replaced by other symbols such as the dynastic era and disappeared almost completely after 500 CE. See Dines Candra Sircar, "Some Performers of the Asvamedha," in K. K. Pillai, A. S. Narayana Pillai, and V. I. Subrahmaniam, eds., *Professor P. Sundaram Pillai Commemoration Volume* (Tinnevelly 1957), 93–97.

7. See Erich Hultzsch, ed., *Corpus Inscriptionum Indicarum,* vol. 1: *Inscriptions of Asoka* (new edition Oxford 1925); Ulrich Schneider, *Die großen Felsenedikte Ashokas* (Wiesbaden 1978).

8. See for the first century CE, Vasudev V. Mirashi's edition of the "Daulatpur Inscription of the Reign of Caṇḍana: Year 6," *Journal of the Oriental Institute, Baroda* 28, no. 2 (1978): 34–37 of 84 CE and "Andhau Inscriptions of the Time of Rudradāman," R. D. Banerjee, ed., *EI* 16 (1921): 19–25 dated ca. 150 CE. For South Asian hero stones see S. Settar and G. D. Sontheimer, *Memorial Stones* (Dharwad and Heidelberg 1982), and G. Berkemer, "No Heroes in Kalinga? On Death in Kalinga Inscriptions," in Elisabeth Schömbucher, Claus-Peter Zoller, eds., Ways of Dying: Death and Its Meanings in South Asia (New Delhi 1999), 179–189.

9. Kashi Prasad Jayaswal and R. D. Banerji, eds., "The Hāthīgumphā Inscription of Khāravela," *EI* 20 (1929) 72–89; S. Kant, *The Hāthīgumphā Inscription of Khāravela and the Bhabru Edict of Aśoka: A Critical Study* (Delhi 1971 [2000]).

10. We must note that the Mauryan Empire was contemporary with the Seleucid state in Iran, and that it was Seleukos Nikator who was the first king to introduce a dynastic era, the "Alexander-era" of 312 BCE. This era served to legitimate him and his successors as heirs of Alexander the Great. The first Indian era is the already mentioned Scythian era of 78 CE.

11. Since giving gifts to Brahmins was one of the most important duties in the then dominant Hindu *dharma,* the theme of gift-giving incorporates the *dharma* theme. Indeed, both are virtually identical. The memory of the dead—a religious activity by no means confined to epigraphy—remains a topic in many temple inscriptions. From the time when such documents are found in increasing numbers, i.e., from about the eleventh century onward, it is clear that this theme survives as a "private," if very popular, form of religious ceremony. The fourth theme, of royalty rendering account, is replaced by the eulogies in which the king is styled as a conventional figure whose deeds are always in accordance with the royal *dharma.*

12. See V. Naryana Rao, "Purāṇa as Brahminic Ideology," in W. Doniger, ed., *Purāṇa Perennis* (Albany 1993), 88f.; K. Mylius, *Geschichte der altindischen Literatur* (Bern 1988), 133–134.

13. So far, there is no critical edition of the Ganga *praśasti* available. A translation into English of one version is included in Nagendra Nath Vasu, ed., "Copper-plate inscription of Nṛsiṃha-dēva II of Orissa, dated 1217 Śaka," *Journal of the Royal Asiatic Society of Bengal* (1896): 229–271.

14. Now called Mukhalingam, a village and archaeological site in Srikakulam District, Andhra Pradesh.

15. This passage results from the reading of A. Tenenti's description of the differences between medieval thinking and early humanistic and renaissance ideas in *Die Grundlegung der modernen Welt* (Foundation of the Modern World). *Fischer Weltgeschichte,* vol. 12 (Frankfurt/ M. 1981),

135; and Bernard Cohn's distinction between historic past and traditional past in his article "The Past of an Indian Village," in B. Cohn, *An Anthropologist among the Historians and Other Essays* (Delhi 1987), 88–99. The best example of this kind is the Mādaḷā Pāñji, the chronicle of the Jagannāth-Temple of Puri (most accessible in the Sanskrit versien of 1820: G. C. Tripathi and H. Kulke, eds., *Kaṭakarājavaṃśāvaliḥ: A Traditional History of Orissa,* vol. 1 (Allahabad 1987); see also H. Kulke, "The Chronicles and Temple Records of the *Mādalā Pāñji* of Puri: A Reassessment of the Evidence," in H. Kulke, *Kings and Cults: State Formation and Legitimation in India and Southeast Asia* (New Delhi 1993), 137–158; and H. Kulke, "Reflections on the Sources of the Temple Chronicles of the Mādaḷā Pāñji of Puri," in H. Kulke, *Kings and Cults,* 159–191. In the first of the mentioned articles Kulke is looking into the problem of the compilation of the *Mādalā Pāñji* and shows how the authors constructed a "very strict system of highly symbolical numbers" (H. Kulke, "The Chronicles and Temple Records of the *Mādalā Pāñji* of Puri," 142) out of the symbolically insignificant number of kings who actually had ruled Orissa. The documents for the reconstruction of "historical facts" in the form of copper plates and other title deeds must have been accessible to the authors. In this way, the 15 historical kings of the Ganga dynasty have been made into 18, while other dynasties are said to have had 18 or 36 kings. Altogether there are—including the king who ruled at the time of the writing of the *Kaṭakarājavaṃśāvaliḥ*—108 kings. This is a perfect number according to Hindu tradition.

16. Widely acknowledged among historians of India is the date c. 78–101 CE, while specialists on Central Asia prefer the second century (e.g., 144–ca. 172). A recent paper on the issue advocates approximate dates of either 100–126 or 120–146 (Nicholas Sims-Williams and Joe Cribb, "A New Bactrian Inscription of Kanishka the Great," *Silk Road Art and Archaeology* 4 (1995/96): 75–142; see also the critique by Gerard Fussmann, "L'Inscription de Rabatak et l'Origine de l'Ere Śaka," Journal *Asiatique* 286, no. 2 (1998): 571–651; and Nicholas Sims-Williams, "Further Notes on the Bactrian Inscription of Rabatak," *Beiträge zur Iranistik* 17, no. 1 (1998): 79–92.

17. See Vasudev V. Mirashi's "Daulatpur Inscription of the Reign of Caṇḍana: Year 6," 34–37, and Mirashi, "The Riddle of the Mewāsā Stone Inscription of the Western Kṣatrapas," *Journal of the Oriental Institute, Baroda* 28, no. 1 (1978): 56–62.

18. Dines Chandra Sircar, *Indian Epigraphy* (Delhi 1965), 259 points out that only the Cālukyas of Badami began to call the era the "*Śaka*-year" (*śaka-varṣa*). Previously it was just called *saṃvatsara* ("year") or abbreviated *śaṃvat.* For other Indian eras see D. C. Sircar, *Indian Epigraphy,* chap. 7, 219–326.

19. See Aśoka's *4th Mayor Rock Edict,* line (F): "And the sons, grandsons, and great-grandsons of king Devānāṃpriya Priyadarśin will promote this practice of morality until the aeon of destruction [of the world], [and] will instruct [people] in morality, abiding by morality [and] by good conduct." (Transl. by E. Hultzsch, ed., in *Corpus Inscriptionum Indicarum,* vol. 1: *Inscriptions of Asoka,* 8).

20. But there was never a god-king in India. Kings are said to carry fragments of a godlike nature that comes with kingship, but they are never worshiped or deified.

21. See especially the *praśastis* of the great dynasties after 1000 CE, especially the Cālukyas, the Cōḷas and the Imperial Gangas. Their less prominent successors use the same genealogical topoi. There exists, for instance, a copper plate inscription of the Nāyakas of Madurai dated 1634 in which the overlords, the kings of the third dynasty of Vijayanagar, are depicted as having a divine pedigree while the Nāyakas who issued the grant claim only human ancestors. This example shows how such a myth can be used as a symbol of power and hierarchy (H. Krishna Shastri, "Kuniyur Plates of Venkata II," *EI* 3 (1894): 236–258).

22. E.g., verse 14 of the Buddhist "Nalanda Stone Inscription of the Reign of Yasovarmmadeva" (6th cent. CE), Hirananda Sastri, ed., *EI* 20 (1929): 37–46: "Whoever interferes with this gift, which has to last as long as the world endures, will, void of virtue as he is, have the dire fate of one who commits the five [deadly] sins," and a line from a tenth-century hero stone states in Kanarese: "He who destroys this shall incur the guilt of destroying Vāraṇāsi and a thousand brown cows and a thousand Brāhmins! He who protects this is a protector of just so much!" (John F.

Fleet, ed., "An Inscription at Devageri," *EI* 11 (1911): 1–7). See Sircar, *Indian Epigraphy,* 176–201 for a selection of these verses. No systematic study of these verses has been done so far.

23. An interesting example is the rivalry between the two pretenders for the throne of the Cōḷas king of Tanjore (Tamil Nadu) in 1071, when the ruling line had died out. The usurper Kulottunga of the family of the Eastern Cālukyas (ruled as Cōḷas King 1071–1122) and his rival, later the Ganga king Coḍaganga Anantavarman (r. 1077–1147), who was still a minor in 1071, led propagandistic and military campaigns against each other over decades. One result was the adaptation of the Cōḷas chancery style, calendrical system, legends of legitimation, religious politics, organization of the army, etc. by the Gangas (see G. Berkemer, *Little Kingdoms in Kalinga,* [Stuttgart 1993], 154–168).

24. See for instance the fight between the kings Wurmagoo and Shalavahum (Vikramārka and Śalivāha) which reminds of the names of the Vikrama and Śaka-Śalivāhana eras (*Fragment of an Historical Account of the Vizianagram Family: Translated from the Persian;* British Library, Oriental and India Office, MSS EUR/Mack Gen 7, text no. 15, 389–393).

25. The beginning of the Kali-yuga of 3104 BCE is sometimes taken as a nondynastic, absolute date. But it is at best a supplement to the dynastic eras.

26. While time (*kāla*) is considered as an agent in old texts such as *Atharvaveda* XIX, 53), in the *Mahābhārata* it becomes a tool of Karman, the result of action personified. In Indian philosophy *kāla* is explained by different schools in different ways, as an inherent quality of matter, as a sequence of infinitesimal moments, or even as pure illusion. See J. Scheftelowitz, *Die Zeit als Schicksalsgottheit in der Indischen und Iranischen Religion (Kāla und Zruvan)* (Stuttgart 1929), where *kāla* appears also as the god of death, Yama, and H. S. Prasad's collection of articles in *Time in Indian Philosophy* (Delhi 1992).

27. Such rare exceptions come from the sphere of personal religiosity (*bhakti*) and may in some cases be documents of extreme desperation. There exists, for instance, a prayer in the Oriya language inscribed at the Jagannath temple of Puri. The text, by king Kapilendra (r. 1434–1467), was written in his twenty-ninth regnal year. In the fifteenth century, the great kings of Orissa did not include genealogies in their inscriptions any more because annals and chronicles were now being used instead of the older inscriptional forms of collective memory. Perhaps it was in this context of change that the king chose to express his feelings of disappointment and rage over his unfaithful little kings in a more personal way than former conventions allowed. He says: "Oh Jagannath! Thus prayeth Thy servant. Throughout the kingdom, I maintained from childhood, these [feudal] lords [or nobles] including the infantry and cavalry and gave them wealth. All of them have forsaken me. I shall deal with them [and] punish them each according to his desert" (Inscription no. 24 in K. B. Tripathi, *Evolution of Oriya Language and Script* (Cuttack 1962).

28. See the "Introduction" in Marshall Sahlins, *Islands of History* (Chicago 1985).

29. In relation to ruling families and dynasties, dominant local kin-groups, Buddhist or Jaina monasteries and orders, temples, important towns (e.g., Benares, Mathura) or in the late middle ages also common regional languages and beliefs.

30. In both the generation-long series of temple inscriptions and the genealogical parts of copper plate inscriptions of a family, which often were used and augmented over generations, forms of "challenge and response" between dynasties occur. They hint at information on topics such as diplomacy, war, marriage, pilgrimage, land grants, etc. in the form of hidden dialogues between competing families, often without giving details about the addressees of the challenges.

31. Examples are the inscriptions issued by Pulakeśin II (ca. 609–ca. 642) ("Aihole Inscription of Pulakeśin II," *EI* 7 (1900): 1–12,) and Arjuna I (1252–1283) ("Dibbida Plates of Arjuna of the Matsya Familiy; Saka-Samvat 1191," *EI* 5 (1898): 106–112). The author of the first examples alludes frequently to the works of the most famous of all Indian poets, Kalidāsa. This imitation was supposed to set both author and king close to their North Indian models and thus far above their contemporary, local competitors. In the second case the author attempts to connect the genealogy of the Matsyas, a family of mere local standing in the hills of upland Kalinga, with the traditional canon of dynastic foundation myths of the epic and Purāṇic tales. In this way, the king seeks "horizontal legitimation" (Kulke) of his rule among neighboring little kings and his

own overlord, the imperial Gangas. The literary technique is similar to the first example, but both model and target group are oriented toward the popular myths.

32. There has not yet been a systematic study of Indian royal titles. It is clear that they—as well as other political symbols—tend toward ever more grandiose forms over the centuries. A rough typology may distinguish between titles denoting rank and function (e.g., *mahāsāmanta:* great vassal, little king; *mahārājādhirāja:* overlord of the great kings; *kaisara* or *cakravartin:* lord of the world, emperor), religious titles (e.g., *paramavaiṣṇava:* highest devotee of Viṣṇus; *mātāpité-pādānudhyāta:* the one who pays homage to the feet of mother and father), legitimatory titles (e.g., *licchavidauhitra:* son [by marriage of the father to] a daughter of the Licchavis), and honorific titles (e.g., *gulbargeśvara:* lord of Gulbarga [by conquest]).

33. The most important edition with a translation is Marc Aurel Stein, *Kalhaṇa's Rājataranginī or Chronicle of the Kings of Kashmir: Sanskrit Text with Critical Notes* (Bombay 1882, [Delhi 1960]) and Stein, *Translation with Introduction, Commentary, and Appendices* (Westminster 1900 [Delhi 1961]). See also C. Schnellenbach, *Geschichte als 'Gegengeschichte'? Historiographie in Kalhaṇa's Rājataranginī* (Marburg 1996).

CHAPTER 6

Interpretations of Time in Islam

Otfried Weintritt

Attempting to define the Islamic concept of time, one is confronted with two problem complexes: the relevant theological foundations on the one hand, and on the other the socioreligious strategies that have developed in the process of adhering to the regulations and obligations of the Islamic life order. A consideration of both these areas will contribute to this attempt to define the category of time in relation to its significance for historical consciousness.

An analysis of the concept of time in Islam must be based on the foundations of its transcendent monotheism, out of which have emerged specific concepts of time and perhaps also a conception of history typical for Islam.[1] To this end, Islam is understood as a historical phenomenon; the development of its religious foundation and concepts of time is complete. Regional variants of Islam with possibly divergent interpretations of time will not be taken into consideration here. But they too are marked by Islamic concepts.

There are numerous terms for time in Arabic. They appear in the Koran and were defined in later times by lexicographers. The differences in meanings, however, are not particularly relevant to the following reflections as they neither connote a specific concept of time nor are semantically unequivocal. Louis Gardet has defined the terms in question approximately as "zamān, time (sometimes a synonym of al-ayyām, <the days>); ān, present instant, <now>; ḥīn, fixed moment or period; waqt, a particular point in time; ṃīqāt, a memorable encounter; dahr, duration."[2]

In the theology of Islamic orthodoxy, the *sunna,* the idea of a temporal atomism was developed according to which God, at every moment, in a succession of commands, creates the world, i.e., everything that exists in time and space.[3] Theologists were bound to the Koranic idea of the contingency of

Notes for this section begin on page 91.

nature. This led them to maintain that God created the world in the beginning and that the world has a temporal beginning. In contrast to Islamic philosophers of the neo-Platonic Aristotelian tradition, the theologists rejected the idea of the eternity and necessity of the world and based the continuation of the world on the creatio continua.[4]

In order to be able to conceive of God's creation as continuous, theologists developed the concept of divine habit, which guarantees the continuity of the natural order. God's creation, however, has momentary character. In order that it be permanent, at every moment an act of creation is required to repeatedly create the world anew. A possible future—as thought or even planned—is dependent on God's will, which eludes mankind. This theological framework explains Muslim reservations regarding the temporal dimension of the future, which are expressed often in everyday life in the expression "*in šā'a llāh*," "If God is willing." However, the conception of the continuous activity of God does not result in a rigid determinism. Theologists, in occasionally complicated lines of thought, were able to retain divine omnipotence while at the same time acknowledging man's freedom of will. Thus it is the human being whose decisions determine the measure of reward and punishment in the hereafter.

Medieval Islamic philosophers responded to the atomistic occasionalistic concept by developing a principle of causality for all temporal and spatial beings based on Aristotelian natural laws. In this, however, they were largely unsuccessful. Other formulations of the time concept (for example of the Shi'a) would have to be described in detail. However, as isolated cases they are insignificant for the Islamic perception of time in relation to historical consciousness.

A concept of time that appears to be occasionalist was ascertained by Pierre Bourdieu in his investigation of the inhabitants of the Kabylei. "Time experienced, just as space perceived, is discontinuous, made up of a series of heterogeneous islets of differing duration."[5] Clifford Geertz writes similarly of the Balinese, that for them time dissolves into single particles.[6] Neither ethnologist refers to a possible theological basis for such temporal concepts.

In the Sunni systematics of Islamic faith, spatial and temporal concepts interpenetrate. Temporality is a characteristic of life on earth, whereas the hereafter, which is conceived as timeless, has a spatial dimension. As such, it symbolizes duration. For this reason a martyr can enter Paradise directly, without having to wait for the resurrection or the Last Judgment. The temporality of life on earth is relevant for the Muslim in regard to fulfilling his religious duties.[7] Every cosmic day, caused by the movement of the planets, represents a challenge and an opportunity for a Muslim to fulfill the commandments of the *šarī'a,* the divine law in general, and in particular the duties that have been imposed on the individual. The temporality of life on earth provides the conditions under which man can fulfill the commandments of divine law. The obsession with time in Sunnitic Islam (for example, calculating the calendar) arises from the concept of faith as orthopraxis. Prayer times, though are determined, according to the position of the sun and the length of shadows and not according to abstract time. Time is a means of carrying out divine law that promises the faithful that they will enter eternity after the Last Judgment.

Muslim scientists' efforts to come to terms with astronomical problems, from which European medieval scholars heavily borrowed, should be seen in this context, as science in the service of religion. Technical-manual applications of mathematical-astronomical knowledge in determining the calendar and prayer time, are dictated by religious requirements.[8]

Actions that comply with the requirements of the šarī'a transcend the time of human earthly life, because each action taking place in this period will be completed only when it is assessed at the Last Judgment. Only then can past happenings draw to a close; until then they will persist in a state of unfinished present. Thus the past is not the past in a linear concept of time that becomes ever more remote. The moments brought about by divine commands "are represented as a number of tangential points—points of contact rather than intersection—between human time and the eternity of the Most High."[9] They complete one another to become the events in this life, i.e., individual and collective time. At the same time, the moments aspire to overcome the separation from God's eternity.[10] Koranic revelation, which puts an end to the time of life on earth, is the guarantee that this will happen.

Temporal atomicism, which presupposes discontinuity, shapes the perception of time and the scholarly concept of history. History, in as much as this term is applicable to Islamic conditions, is conceived as a discontinuous succession of individual events.[11] They are recorded and chronologically ordered but not examined in terms of causality. What appears from the outside to be a shortcoming might be a result of the above-mentioned occasionalism. Historiographic works demonstrate also, however, despite the descriptive mode of portrayal required by the segmentation of reported events, that their authors have attempted to find ways to point out the progressive character of historical events.[12] Thus it is legitimate to interpret chronologically reported historical time as continuous, and to assume that the theologians' atomistic conception of time did not have a far-reaching impact on history-writing.[13] This does not mean that Islamic historiography is not strictly orientated towards calendar chronology. The spectrum of order schemes to be found in history-writing, such as chronicles, annals, or biographical lexica,[14] ranges from the temporal unit "day" to varyingly long periods of time, measured in years.

In Islamic historical theology, according to Koranic revelation, time and history signify a delaying of the prophesied final condition that will be entered on after the events of the Last Judgment. Koranic revelation ushered in the last phase of the history of salvation, which lasts from creation to the judgment day. Muslim theologists have specified two aspects of the course of the last phase of the history of salvation. The community of the faithful as such will experience the Last Judgment. At the same time, however, they will be confronted with the inexorable decline of the faith. It is in this context that we should view the idea of periodical renewal of the faith by an outstanding personality. This can be seen as a strategy to come to terms with historical experience by attempting to reconcile pessimistic observation with the certainty of salvation.

In the Islamic view, history begins with creation. For the Muslim, time is a phenomenon that begins with the creation of day and night and ends with the Last Judgment. Early Islamic historians usually preface their works with the definition of time as the hourly unit of day and night. "Time is the hours of night and day. As for the hours, they are the crossing through the degrees of the celestial spheres by the sun and the moon."[15]

Because Islamic history takes place throughout this period of time, it was necessary to establish a temporal measure for the period prior to the final divine revelation. In addition to defining time, early Islamic historians presented doctrines about the duration of the world, because from this the end of the world could be calculated. There are varying traditions, the majority of them stating a period of 7,000 years, of which the time after Koranic revelation is only one part (for example, 800 years).[16] Koranic revelation forced Muslims to reflect on time, in order to establish the temporality and content of the history of salvation.

Universal history as a historiographical type was developed in the early Islamic period. Historical works of this kind begin with creation and lead up to the time of their authors. In later periods the cosmological-cosmographic part and the world history (from the creation of Adam to the appearance of Muhammad, i.e., the history of the prophets and large dynasties) are left out. This conception of history represents the original Islamic attempt to order knowledge of the past from the perspective of the last divine revelation to mankind. Such endeavors, which began in the early nineth century, or 200 years after Muhammad's emergence, were hardly influenced by ancient traditions. The pre-Islamic period was not put into chronological order but rather divided up into periods. Neither were there attempts in later periods or in modern times to achieve a temporal structuring of the past. Whenever it was necessary to fix dates for pre-Muhammadan history, other time calculations were employed.

In the Islamic view, the history of creation is the history of salvation. As such, it is meaningful and has a divinely ordained goal. Divine order in creation makes it possible for man to fulfill his mission and destiny. At the moment of Muhammad's prophecy, history and time reached their zenith. The Islamic concept of time is shaped by the notion that human time and divine non-time represent different forms of existence. Through revelation, history was concluded. In its last phase, history takes place as the carrying out of the mission to fulfill divine law. Such history is ahistorical, in the sense of Western developmental thought. Time and human history form a unity that encompasses creation and the Last Judgment. In the Koranic stage of the history of salvation, time is conceived as a condition for fulfilling divine law. For this reason, the discipline of time measurement, a noteworthy area of Islamic science, was developed to ever more perfection in the service of the faith.

The concept of Islamic history-writing demonstrates clearly the Muslim understanding of history. Islamic dynasties are responsible for preserving and upholding divine law. Historical memory thus sees them as manifestations of divine will. In this regard, Ibn Haldūn, the famous historian of the fourteenth century, is no exception. He attributed the course of organized rule to the conflict between urban and nomadic lifestyles. Dynasties are subordinated to the primacy of divine law. In this function they are the object of historical memory reflected in this genre and guarantee the continuity of the postrevelation period by fulfilling divine law.

The early Islamic period has the quality of a primordial era to which all later history refers back. After the ideal early period, history, which is understood as the history of the Islamic community, follows a downward movement. Historical meaning is construed from the perspective of a specific time. Each phase of Islamic history that emerges in historical memory acquires meaning in relation to this base period. History and its interpretation are linked to it.

The history of salvation in Judaism is conceived as an amelioration of fate on earth. Christianity has a two-step salvation plan in which the coming of Christ—redemption—is prepared in the history of Israel. The Christian history of salvation begins with the birth of Christ, through which the achievement of salvation—a period of growth—becomes possible. Through the act of redemption time acquires a new quality, as the history of salvation can take place in it. The Islamic view of the history of salvation differs from both these conceptions. The history of salvation commences with creation and ends with the Last Judgment. As Muhammad did no more than to proclaim, once again and finally, divine law, thereby concluding the series of prophets sent by God, the history of salvation merely enters its final phase with Muhammad. There is no qualitative change in the timeless relation of the divine and the human orders—the here and now and the hereafter. Time and history acquire neither a new quality nor the capacity for development. However, in the Koranic phase of history a decline in faith takes place that can no longer be corrected by revelation.[17]

The past became relevant for Muslims as soon as the Islamic community was no longer able to order its affairs according to the practice of the prophet, or according to certain memories thereof that were unquestioned because they were rooted in sufficient temporal proximity. In this situation the need arose for authentic knowledge about the life of the prophet. The chain of transmission was the criterion for assessing authenticity, a method that acquired significance in every area in which knowledge of the past was handed down.

Due to its genesis, Islam developed a two-phase model of history. The past became synonymous with the time of the prophet, his successors, and the period of expansion lasting some 100 years in which the empire came into existence. The later course of Islam is directly linked in identificatory fashion with this phase: imitation was recommended. This period, which also has the greatest sentimental value, was in a very concrete sense an interpretive model for later times. History, as viewed from the perspective of this era, has significance because the early period has a moral dimension and is replete with models and good examples that have never lost their validity or authority. As historical time it has a timeless status.

It was of consequence that the revealed divine law, which renders every human activity sacral, and faith reached their zenith in the period of the first four successors of the prophet Muhammad.[18] The ideal Islamic community does not have a utopian quality, but is, rather, a historically verifiable presence. The historicity of the Islamic community derives from the (undesired) distance from its initial situation, the details of which are projected into the present by virtue of the canonized knowledge of the life of the prophet and the prophets' companions. As a historical constellation, the early period calls for imitation if not repetition. Being well aware that changes cannot be averted,[19] Muslims came to regard the early practice as normative. The value of faith in a given period, that is the Islamic character of the community of the faithful, is measured on the basis of the proximity to or distance from the ideal past. One theme of Islamic history-writing is to account for this ever changing disparity, i.e., to report on adherence to the law and the failures to do so that jeopardize the salvation of the community. History thus gives information about the loss and the guarantee of salvation. The concept of salvation is grounded in the tension between the model of a past, ideal community and the distance from it at any given time. This results in time being experienced as injury, since it leads away from the ideal.[20]

The Muslims' relation to their initial state gave rise to a historical mentality that generated a longing for the origins. At the same time, revelation initiated the last phase of history and salvation is certain. The path taken by Islam led to intense retrospection of the early period, which, despite the linear concept of time underlying eschatology, appears as time at a standstill. Historical consciousness and time concepts have, therefore, a regressive character.

Remembering ideal conditions as a form of referring to the past, appears also in other areas. The Arabic language of the Koran is considered to be of divine origin and therefore inimitable. The so-called dogma of inimitability of the Koran has resulted in a language ideal that cannot be changed, having a normative influence on literature. The developing Arabic literature was deeply influenced by being defined in relation to the Koran. The norm of standard Arabic—the divine language of the Koran—has held its ground up to our times. The numerous dialects, on the contrary, are seen as deviations from this norm.[21] The experience of the present is, in this case too, shaped by the reference to a past fact whose significance cannot be apprehended.

Another form of permanent reference to the Islamic early period arose during the fixation on the model of Muhammad as a source of legal authority. From the middle of the eighth to the ninth century, a corpus of judgments and reports was built up that historically objectified Muhammad's activities by providing proof of authenticity via the oral transmitter. Thus a method was developed to verify a statement of the prophet based exclusively on the credibility of the transmitter and not on the contents of the statement—a method by which one hoped to come as close as possible to the past occurrence. These procedures for producing proof are fundamental to the Islamic understanding of science. Historical phenomena from an ideal past time, which are not modified in the process of transmission, regulate the experience of the present.

Creation and the Last Judgment, as the perimeters of human history, belong to a theology in which divine and human order relate to one another. The connection of both is established by divine law, which was revealed in the Arabic language. This conception has a temporal and historical character, as divine law represents an eternal order that not only regulates human existence but also relates it, by guaranteeing the certainty of its outcome, to divine order. The law may get watered down in this world. Before Muhammad, this had led again and again to the dispatching of prophets to reveal the same divine law. However, according to Islamic doctrine, this cannot happen again, since with Muhammad's final revelation the line of prophets draws to a close. The perceived demise of faith is compensated by the certainty that the Islamic community will not completely break up before the Last Judgment—the end of time and history—comes.

This certainty is reflected in the concept of the regular renewal of the faith, which manifests the effort to repeatedly make the Islamic faith plausible. This renewal becomes a dynamic element in the attempt to revert the downward tendency of the Islamic community by trying to approximate the idealized early period.

Notes

1. The following presentation focusses on the concept of time of orthodox theologians as it was developed in the classical Islamic period.

2. See Louis Gardet, "Muslim Views on Time and History (An essay in cultural typology)" in UNESCO 1976, ed., *Cultures and Time,* (Paris 1976), 197–227, here p. 199.

3. In orthodox Sunni doctrine movement, and time as its measure, are elements of creation that God transcends: "Time (al-zaman) does not affect Him," al-Nasafi (d. 1142); quoted in Lenn Evan Goodman, "Time in Islam," *Asian Philosophy* 2 (1992): 3–19, here p. 17.

4. See Goodman, "Time in Islam," 4.

5. Quoted in C.A.O. van Nieuwenhuize, "Muslim Attitudes towards Planning," *Internationales Jahrbuch für Religionssoziologie* 5, Special Issue: *Religion, Kultur und sozialer Wandel* (1969): 89–117, here 96.

6. See Clifford Geertz, *Dichte Beschreibung: Beiträge zum Verstehen kultureller Systeme* (Frankfurt/ M. 1989), 177, n. 31.

7. The this world-other world concept presented here is based mainly on Jane I. Smith andYvonne Y. Haddad, *The Islamic Understanding of Death and Resurrection* (Albany 1981).

8. On Islamic natural sciences and methods and instruments for measuring time in medieval Islam see David A. King, "Science in the Service of Religion: The Case of Islam," *Impact of Science on Society* (UNESCO newspaper) 40, vol. 3, no. 159 (1990): 245–261.

9. Gardet, "Muslim Views," 204.

10. Ibid.: "Consequently, each instant, by a motionless reversion to itself, tends to re-enact the original 'break' accomplished by the divine kalima, the creative Word which brings into being what must be, irrespective of any circumstances of time and space." See Smith and Haddad, *The Islamic Understanding,* 5f.

11. In a narrower sense history is nothing more than the temporal course of this world, which extends to the day of the Last Judgment.

12. See Stephen Humphreys, "Islamic Historiography," in Joseph R. Strayer, ed., *Dictionary of the Middle Ages* (DMA) vol. 6 (New York 1985), 249–255, here 253f.

13. See Jean-Paul Charnay, "Temps Sociaux et interprétation historique en Islam," *Studia Islamica* 28 (1968): 5–27, here p. 8.

14. Aside from the alphabetically ordered biographical works there are also some in which biographies are ordered chronologically according to the dates of death.

15. Tabari (d. 923), *Tārīh ar-rusul wal-mulūk,* ed. Muhammad Abu l-Faḍl Ibrāhīm, 8 vols. (Cairo 1960ff.) The chapter is entitled: "Had God, before he created time, night and day, already created something else?" Vol. 1, 22.

16. Ibid., 9ff.

17. The decline is rooted in the beginning. Its inevitability is illustrated by the following statement by the prophet: "As the prophet said: today I have completed for you the object of your faith and given you Islam as a religion, everyone was enthusiastic. Only the great Abu Bakr cried. Why are you crying, asked the prophet? Because, envoy of God, answered the later first caliph: because every time a thing is completed, it begins to lose." Quoted in Charnay, "Temps sociaux," 5.

18. The declining movement is clarified in a statement by the prophet Muhammad: "The best people are my generation; then, those who immediately follow (that second generation)." Franz Rosenthal, *History of Muslim Historiography* (Leiden 1968), 326.

19. Discussion about new things, which might have led the community astray, left distinctive traces in the spiritual-religious development. Innovations (arab. *bid'a* = forbidden innovation) were confronted with an evaluation scheme. Since change could not be completely averted, it was agreed that some innovations were good; others, on the contrary, would lead to damnation.

20. The Ottoman historian Katib Çelebi (d. 1657) expressed the following view on the matter: " if the people of any age after that of the prophet were to scrutinize their own mode of life and compare it with the Sunna [the prophet's practice], they would find a wide discrepancy … . Scarcely any of the sayings or doings of any age are untainted by innovation." Katib Çelebi, *The Balance of Truth,* transl. Geoffrey L. Lewis (London 1957), 90.

21. That the difference between the standard language and the dialects is hard to imagine and very large, is made clear in the following description: "The difference between spoken and colloquial Arabic and classical or literary Arabic involves something close to the difference between, say, modern French and medieval Latin." Hisham Sharabi, "The Neopatriarchal Discourse: Language and Discourse in Contemporary Arab Society," in George N. Atiyeh and Ibrahim M. Oweiss, eds., *Arab Civilization: Studies in Honor of Constantine K. Zurayk* (New York and Albany 1988), 140–165, here 149.

CHAPTER 7

Constructions of Time in the Literature of Modernity

HARRO MÜLLER

My reflections on constructions of time in the literature of modernity begin with an ironic quotation from Nietzsche that I myself do not quote without irony: "He who has once contracted Hegelism and Schleiermacherism, is never quite cured of them."[1] After an introductory section I will start with "Hegelism"; by the end of the text I will turn to the "Schleiermacherism."

Introduction

In the extensive literature on the problems with time, Aurelius Augustinus is often named as the founder of the discourse for introducing the distinction between internal and external time.[2] His whole concept of time, which offers, among other things, an explicit theory of the modalities of time, can be derived from the central distinction between time and eternity. Eternity is a precondition for time; time is the creation of God. Man, who is temporal and finite, is the creation of God, who is atemporal and infinite. Thus, this strongly metaphysical conception of time proceeds from an atemporal, infinite center, that is, the source of the beginning and end of time: from eternity to eternity, amen.

Augustine's semantic of time can be correlated well with stratificatory models of society, which proceed from the idea that society has a center, a pinnacle. Modern society, if we follow Niklas Luhmann, is a functionally differentiated society in which there is no longer a central authority or pinnacle, because neither can one subsystem take over the function of another, nor can one central distinction that encompasses all subsystems be supposed. Thus, the

Notes for this section begin on page 103.

formerly all-encompassing central distinction between finite and infinite retreated into the system of religion and in this way has become irreducibly particular. That means that the claims to universality made by the distinction finite/infinite are only valid in the system of religion. Therefore substantial semantic transcriptions are introduced by the conversion from stratification to functional differentiation. They affect conceptions of time: the way that time is observed, and how, in turn, observations of those observations are made.

It is the thesis of this chapter that, after the heroically exuberant first phase with its philosophies of history (e.g., Schiller), in the modern system of art, ways of observing time develop beginning with assumptions of paradox or difference. In the system of art, observations of time are tested that are capable of considerably irritating other subsystems, such as the systems of science, politics, or education. This presupposes not only that in the functionally differentiated system of art, with its guiding code and various programs, the communicative aspect of information becomes information, but also that, in the framework of the medium/form relationship, the construction as well as destruction of concepts belongs to the formation achievements of the system of art. The texts selected for this article will therefore be read so as to highlight their respective conceptions of time. Not taken into account are the connections of these concepts with, for example, anthropological assumptions, or their contradiction by textually produced undecidabilities. We are, then, dealing with reductive readings, which could offer aid in understanding the capabilities of the extraordinarily flexible modern system of art with its preadaptive advances.

Hegel: The Dialectics of Time

I begin with a concise reading of Georg Wilhelm Friedrich Hegel's *Lectures on the Philosophy of History,*[3] analyzing it as a historical novel. If questions as to empirical validity or precise argumentative conclusiveness are suspended, Hegel proves to be an author of great constructive force. His novel begins with a metanarrative reflection on the writing of historical novels and portrays world history as the process of the world-spirit coming to itself: "World history is the progress of the consciousness of freedom"[4] and "the rationally necessary course of the world-spirit.[5]"

In his construction of world history, Hegel combines achronic, synchronic, and diachronic elements. The world-spirit is timeless, and at the same time realizes itself in time, within a conception of stages. World history is teleological history; as "process history," it is linear, dialectical, and cyclical all at once. It is linear because the spirit is always the spirit. It is dialectical because the dialectic makes progress possible. It is cyclical because the spirit, having come to itself at the end of history, returns to the origin. Particularly the dialectic, with its threefold definition of *negare, constare, elevare,* is capable of presenting history as a dynamic process history in which the agents of the world spirit—the spirits of the peoples, and individual movers of world history (Alexander, Caesar, Napoleon)—act as heroes who, by passionately pursuing particular goals, serve general progress as "agents of world history."[6] Hegel starts, then, from a metaphysically and materially loaded assumption of identity, which, within world history as the play of identity and difference, leads to synthesis as identity at level of completion. History may be a slaughterhouse, and pages of happiness in the book of history may be blank sheets;[7] nevertheless, all events, actions, and processes lead to the being, in and for itself, of the world-spirit, and to that extent are necessary, rational, and progressive. World history is theodicy, and therefore Hegel can end his conflation of narrative and metanarrative discourse on this note:

> That world history, with all the changing scenes which is annals present, is His process of development and the realization of Spirit: this is the only true theodicy, the justification of God in History. Only this insight can reconcile Spirit with world history—namely, that what has happened, and is happening every day, is not only not "without God," but is essentially His work.[8]

How does time present itself in this historical novel, with its tragicomical characteristics and legend-like ending "Thus the state is rational and self-conscious freedom, objectively knowing itself?"[9] Time is process time, with beginning and end, with arché and telos; and epochs and events/actions are to be situated within the processes. Consequently, for this conception of time it is central that in the history of the world-spirit calendar time and historical time along with, "objective time," individual time, and social time, are sublated in a grand metaphysical comprehensive premise, because the spirit is present in all three forms of time, and therefore also natural time as calendar time is the being-other of spirit.

In literary modernity, this identity of identity and non-identity between chronological and historical time is hollowed out, annulled, and replaced with assumptions about paradox and difference of varying origins and range. There are numerous procedures for this. I would like to outline four:

1. Dialectic is given over to paradox.
2. De-teleologization enables an emphasis on the moment.
3. Indicative world history is switched to the subjunctive and contingent.
4. An assumption of difference is inserted between the present and the past, and between the present and the future, so that each present must always be redefined in its difference from the past and future.

Büchner: The Paradox of Time

One of Hegel's favorite metaphors for world history is theater. Georg Büchner's drama *Danton's Death* also deals with the theater of history, and with the spectacle of revolution. Because the world-spirit has not yet come to itself in this play, the completion of history—the rejuvenation, the reincarnation of humankind—must be put off for the future. The Jacobins Saint-Juste and Robespierre organize the Revolution to this end, using misery, above all shock and terror as their instruments. If there is a hero or heroine in this prose drama about the prosaic relations in year five of the French Revolution, then it is the guillotine sanctified by the Jacobins. The guillotine functions simultaneously as metaphor and metonymy: as metaphor, because in the moment of decapitation the people's sovereignty is shown in action; as metonymy, because this moment cannot be captured, but constantly demands new moments, new decapitations.

If, in their unflinching, unrelenting inability to learn, Saint-Juste and Robespierre bet on the feasibility and producibility of history and stylize themselves as the authors, directors, and actors of this historical tragedy, with a virtuous and good final goal to be brought about, then the basic assumption for Dantonians—particularly for Danton himself—is that events, actions, and the historical process can no longer be synchronized toward a single goal: "These days everything is fashioned out of human flesh,"[10] but "human flesh" (*Menschenfleisch*)—the corpses, the victims, the murdered of the Revolution—cannot be defined as a means to the end of reincarnating humankind. To Danton history appears as a fatal, subjectless process history whose origin is unknown, whose mechanisms make the actors into marionettes in a play not interwoven with the cunning of reason, and whose overseers—if they exist—are not endowed with ultimate benevolence. Rather they are probably sadistic, misanthropic gods who take great pleasure in the spectacle of tormented and self-tormenting humans, and who rejoice with great cheerfulness at the perpetual human history of suffering and dying: "Are the heavens with their winking eyes of gold bowl of golden carp that stands on the table of the blessed gods, and the blessed gods laugh forever and the fish die forever and the gods delight forever in the dancing colors of their dancing agony."[11]

In this play on the drama of the French Revolution, which functions as the historical sign of modernity, historical time appears as a mixture of qualitative and quantitative time that can no longer be ordered or finalized through a central transcendental signified such as the collective singular spirit (Hegel), history (historicism), or human beings (anthropological discourse). This becomes evident not least in that the revolutionaries acting in the political system are completely helpless in the face of the disastrous state of the economic system. There, neither emphasis on virtue (Robespierre), nor material philosophy of history (Saint Juste), nor the utopia of pleasure (Dantonians) are of any help, and even the people's program for the utilization of murder and corpses leads astray: "Let's tear the skin from their thighs and turn it into trousers, let's melt the fat of their bellies to lard our soup with. Come on! Death to all them with no holes in their coats!"[12]

From today's perspective, one could say that the systemic individual times of the subsystems render the realization of a general semantic conception such as virtue, happiness, or progress impossible and point out that history is (histories are) fundamentally laden with risk, for which there is no general guiding authority to guarantee the rationality, meaning, or goal of authority. In other words, the play between intentional moments (history as subject-history or spirit-history carrying exuberant, effusive goal projections) and non-intentional moments (history as subjectless natural history with its associated resignative, melancholic imputation of fatalism) is without synthesis, without reconciliation. There are only precarious, risky de-paradoxifications, which lack ontological safeguards. History as a linear progressive history with a goal, history as cyclical suffering, and history of decay without a goal prove, when seen from the perspective of spectator, to be interpretive ascriptions constructions that could turn out either one way or another.

The keywords "contingency" and "interpretation" and the continually new necessity of interpretations of history (-ies) in each present (in contrast to past and future), appear to me to be leitmotifs of literary modernity. I thus come to my second topic: conceptions of the moment.

Nietzsche: The Moment of Time

There are numerous examples of accentuated conceptions of the moment in the literature of modernity, from Baudelaire's *A une Passante,* to Proust and James Joyce, to Virginia Woolf, to Gottfried Benn's search for moments of mystical participation, and not least, to Peter Handke's *Die Stunde der wahren Empfindung.* These range from discontinuous, shocklike transformations of perception, to kairological moments, to epiphany, which collects the complete meaning of the world into one moment, whereas authors such as Gert Jonke, in his novel *Der ferne Klang,* describe the concept of fulfilled present time only as a theoretical model in the subjunctive.[13] I will concentrate on one of the founders of the discourse on the concept of the moment, about whom Gottfried Benn claimed:

> In fact, everything my generation discussed, everything we mentally grappled with, everything we *underwent,* so to speak, everything we *entered into*—all of that had already been expressed and exhausted by Nietzsche. He had already given it definitive formulation, the rest was just exegesis.[14]

Hegel assumed that time and occurrence, event and meaning could be strictly finalized. Büchner presents a paradoxical mixture of qualitative and quantitative conceptions of time, without synthesis. Friedrich Nietzsche proceeds from a dichotomy of qualitative time full of meaning and quantitative time immune to meaning, as Armin Nassehi has put it.[15] Nietzsche's oeuvre is set up so that his consideration—that interpretation is a form of the will to power[16]—can be attributed to a self-fulfilling prophecy. Nonetheless, the following interpretation of Nietzsche's theory of the eternal return seems plausible to me. Counter to all forms of Hegelianism and with great radicality, Nietzsche bets on the discontinuous moment, and not on continuous development or dialectical process. Thus, the statement "The origin is the goal," with its various interpretive possibilities, is strictly denied, and the de-teleologization of history is decisively sharpened. Being left over as historico-philosophical minimum or perhaps as intensified maximum, the moment passes away and, in contrast to the senseless sequencing of chronological time, contains the possibility of a momentary ecstatic upswing, the possibility of a Dionysian fulfillment: "All joy wants eternity/ wants deep, deep eternity."[17]

This moment, a moment of abundant meaning, is always just a single moment. Always inscribed in this moment is its passing away, and therefore each moment must be newly and differently filled constantly. This project of knowledge about the passing away and, at the same time, the decisive affirmation, which even in hindsight says "what I have wanted in this moment, I will always do," is the project of the *Übermensch*, presented by Nietzsche in his book for everyone and nobody, *Thus Spoke Zarathustra,* as "the lightning from the dark cloud Man"[18] and declared there to be "the meaning of the earth."[19] As radical apologist of the discontinuous moment, Nietzsche bets on the cyclical inexorability of the permanent beginning—"*Anfang ist immer*"—and rejects all conceptions that work with linear, nonlinear, or dialectical models of process, apologist it is irrelevant whether the accent is on "progress" or "decline." Nietzsche, then, for example, would pour scorn and mockery on Habermas's project of the incomplete modernity that needs completion. He would subject it to a symptomatic reading and denounce it as an expression of resentful thinking that is philistine, and vulgarly conformist.[20]

In modernity, in all subsystems, (particularly in the system of art and least in the religious system) there is a turn toward innovation. Art must always be new and different. It is no wonder that Nietzsche's aestheticization of the world—"for only as an aesthetic phenomenon is existence and the world eternally justified"[21]—has found many followers up to the present day, for example, in literary criticism. In any case, I would like to offer three counterarguments for consideration:

1. All those who repeat Nietzsche's conception and emphasize the moment as "hectics of the mind"[22] are from his perspective mere recidivists, intellectual causes for mourning, who make discontinuity impossible through their repetition and who, against their own elitist self-understanding with their flights of sublimity, constantly work toward common, chronological time.
2. The Nietzschean emphasis on discontinuity seems excessive to me. Somewhat more complicated historical, autopoietic systems, with their

constant play between self-reference and reference to the other, must hold some of their elements constant precisely when they produce something innovative. They can hardly bet on totaliter aliter or on Saul/Paul. To that extent, it seems reasonable to proceed from the relational assumption of repetition/variation, and to take leave of all absolutizations of discontinuity, as of continuity.

3. The initial dichotomy between qualitative time full of meaning and quantitative time immune to meaning has questionable elements. Meaningful differentiations between earlier and later, between before and after, between past and future, are made possible by chronology. If one is to work at all with the differentiation of qualitative/quantitative conceptions of time—and it is not necessary to do so—then there are complementary relations between the qualitative and quantitative conceptions: they mutually presuppose, complete, and exclude one another, without it being possible to derive the one from the other.

Assumptions of complementarity and binary categorizations that proceed hierarchically cannot be reconciled theoretically. In particular, models that work with original, primary, qualitative, heterogeneous time and with derivative, secondary, quantitative, homogenous time have lost much of their cognitive dignity. Thus, for example, whoever prefers the asymmetrical distinction original/derivative must bear in mind that the original presupposes the derivative. Therefore the asserted unmediated originality is always mediated, that is, in no way original. The assumed hierarchical or asymmetric relation between original term and derivative term falls into a hardly solvable dilemma.

Musil: The Modality of Time

With the rejection of expansive historico-philosophical models, not only the model of the moment becomes more attractive, but also the category of possibility. Each moment contains an unlimited number of possibilities, and we do not have universally valid rules ready for the ever necessary reduction of complexity. Robert Musil's *The Man without Qualities*[23] no longer succeeds in producing a "closed" horizon such as Nietzsche presents in his *Untimely Meditations* "On the uses and disadvantages of history for life": [T]his is a universal law: a living thing can be healthy, strong and fruitful only when bounded by a horizon."[24] Musil writes a modal novel about the modal relation actuality/possibility, for:

> If there is a sense of reality … then there must also be a sense of possibility. … The sense of possibility could be defined outright as the ability to conceive of everything there might be just as well, and to attach no more importance to what is than to what is not. … Possibilists are said to inhabit [a more delicate medium, a

> hazy medium of mist, fantasy, daydreams and] the subjective mood. … They [possible experiences, truths] have a conscious utopianism that does not shrink from reality but sees it as a project, something yet to be invented.[25]

The accentuation of the category of possibility, of the subjunctive, and of possibilities not lived makes discontinuous innovative connection and experiences of the other condition possible. But utopia is—as Musil points out—a direction, not a goal. There is always, of course, the problem of cloudy possibilities, of too many, of self-contradictory possibilities, and furthermore we are not able to divorce possibility from impossibility, possibility from necessity. The emphasis on the modality of possibility means particularly a downgrading of the modalities of necessity and impossibility—which stand for unavailability without surprises. Along with the emphasis on the modal category of possibility, not only is the border between possibility and necessity/impossibility historicized, but also the space of possibility in modernity is maximized in such a way that modernity, which is full of risks and surprises, can less and less be grasped with a simple narrative model that, with its "when," "before," and "after," produces a "tried-and-true 'foreshortening of the mind´s perspective'"[26] by suggesting a diachronically organized order of time and sense. Ulrich, for example, cannot reduce the plurality of possible interior decorating styles to one style, which in turn might be plural itself. For this reason, he leaves the construction of the house to the "genius of his suppliers."[27] If each thing is just one congealed instance of its possibilities,[28] and if the field of possibilities cannot be closed off, then there follows a pluralization of conceptions of time that must be recursively formulated. I am aware of the unending play of actuality and possibility, and at the same time, I must constantly play it again without reaching the goal or the end:

> The train of events is a train that lays down is own tracks as it goes along. The river of time carries its own banks along with it. The traveler moves on a solid floor between solid walls, but the floor and the walls are strongly influenced by the movements of the travelers, though they do not notice it.[29]

Kluge: The Difference of Time

After these remarks on the fragmentary novel *The Man without Qualities,* I come to my last example, Alexander Kluge, who likewise has a lively interest in the modal category of possibility and whose documentary techniques, methods of montage, and mixtures of discourse and media similarly reject the homogeneous effects of traditional narrative methods.

> Early: still warm. Blood pulses as normal through the body. Ten officers force the soldiers into position. Around 13:00, Reinbrecht tests if it would be better to stay in the foxholes till evening. After a few strides of his run he is heavily wounded. A few minutes earlier, he had tested whether this was the right way; if he had

> decided otherwise, he would still be sitting safety. Hope flees. The body gets colder. On the transport back to central Stalingrad, he suffers severe frostbite. He lies untended for a few days; then he dies.[30]

In *Schlachtbeschreibung,* a polyperspectively structured "*Vielheitsroman*" (Fontane), a novel that places theoretical and narrative sections on top of each other in a variegated montage, causes the border between found and invented texts to oscillate, and integrates passages of science fiction, Alexander Kluge shows that there are cases in history where there is no way out: "the breakout could then have been carried out at any point until November 28th."[31] Time can stretch; time can contract: "on the contrary, to know [time] means to move forwards and backwards within it and thus always to go beyond the actual course of events."[32]

The historical novel *Schlachtbeschreibung* presents an extreme example for the constriction of time, radically minimizing the dimension of the future for most participants: "G. clawed at the hard-frozen earth. A nail broke. The tank looked on, then rolled over G."[33] Kluge understands himself to be the guardian of the difference between past and present, between present and future. As the guardian of difference, Kluge sets off on a search for ways out, knowing full well that for this search for ways out in modernity there is no Archimedian point that could provide the key: "There is no place, no standpoint, no production process, no collective in the production of consciousness which reliably separates right from wrong and which would, so to speak, build up a vast stock of rightness."[34] In the complexity and polycontextuality of modernity, the social, material, and chronological dimensions can no longer be coordinated by a transcendental signifier: "The legacy of magic ('comprehending the world from a single point') left behind in our rational concepts (and in our consciences) is violent: it carries out a cruel distortion of and in reality."[35] There is no string left by Ariadne, by which Theseus could wind his way to the exit as on a ball of yarn, not for society in modernity.[36]

Therefore, in view of the asymmetrization of time-modalities, search movements that continually begin anew are needed to be able to act in the present concerning the present, with an eye to an open future, open futures, an open past—and open pasts, with their different fields of contingency, so that history/histories can purified of the tragic and hopeless situations like Stalingrad can be avoided. Kluge's risky attempt, aided by an experimental aesthetic of ruptures and gaps that offer no solutions, to excite the powers of imagination of his readers so that they start to search for ways out on their own, also makes use of the metaphor of reading:

> The complexity of reading is striking. Books make it easy for us, since we don't have to worry about what's a vowel and what's a consonant, or about where the words are separated from each other, or where paragraphs begin and end. But imagine if books were printed so that the words were unpunctuated and unseparated. In addition, all the vowels would be taken away, so you'd have to figure

> them out by association, just from the context. Worst of all, everything is written on top of everything else, as if a single typewritten page had been typed over four thousand times. This writing-on-top of corresponds exactly to the way historical relations work, to the workings of generations and their uses of language. As well as all of this, every one of these elements, these 'not-letters', is in living motion. It turns out that each element is not only written on top of others, but it changes as a result of being written on itself. All texts change while you're working with them. But even all this is only a superficial representation, since we must consider the relation *between* texts, in other words, the living reaction of one to another. Thus every change in perspective results in a new combination of texts. What I am trying to get at is the richness of social relations: it is in this wealth that we find the escape routes of history, i.e. the texts that are worth reading.[37]

Conclusion

After this Kluge citation, we are led to the concluding coda: "Schleiermacherism." Pluralization, modalization, asymmetrization, individual times of subsystems that cannot be synchronized from an external point; past present and present past, present future and future present, are all starkly disjunctive. The future is known to be unknown (Koselleck); the future never begins (Luhmann). All this may be true, indicating, among other things, that in modernity the social system and psychic system, social time and individual time, can no longer be synchronized. They remain, in spite of structural ties—for example through the medium of language—one another's environments, while the construct of "objective" chronological time in functionally differentiated society serves to produce the operative connections between the individual subsystems as a world clock. For social time (in so far as it exists) the sentence uttered by the executioner of Versailles and Avignon Jouve in Christian Dietrich Grabbe's historical drama *Napoleon oder die hundert Tage* is valid: "At the end … there always follows another beginning."[38]

This is not valid for individual psychical time. Individual psychical time is in accord with the laconic old Roman maxim, *Mors certa, hora incerta,* which in systems-theoretical terms reads: "We can imagine our own death as the end of life, but not as the end of consciousness."[39] Death, which one cannot experience, is then a certainty in the future. The uncertainty in the future is the moment, the point in time, of death. In respect to death Georg Wilhelm Friedrich Hegel puts his faith in the dialectic of individual and species that makes progress possible. In the preface to his *Phänomenologie des Geistes* he writes with the certainty of the substance-subject dialectitian the following powerful sentence: "But the life of Spirit is not the life that shrinks from death and keeps itself untouched by devastation but rather the life that endures it and maintains itself in it."[40] Danton, the melancholic, many-sided character actor in Büchner's historical drama *Danton's Death,* reflects: "There is no hope in death: death is putrefaction plain and simple, in life we putrefy with more sophistication, more subtlety, that's the only difference!"[41] In the work of Friedrich Nietzsche, who was characterized by Gottfried Benn as a "a positive youth,"[42] I find the following observation: "There is among men no greater banality than death, second in lines stands birth …; then follows marriage."[43] A relevant passage from Robert Musil's diary reads: "It is distressing to reflect that we hurry like little hunted dots along the line that is our life and finally disappear down some unforeseen hole. And that, in front of us and behind, at intervals that nothing can reduce, other similar dots go racing along which have some kind of temporary link with us, like the next links in the chain of a paternoster lift that goes racing round."[44] Some sentences in Alexander Kluge's historical novel *Schlachtbeschreibung* were written in an expressively restrained, laconic style: "An exhausted man lies down with a group of the dead." "The last company of an artillery regiment is killed off with grenade launchers." "The crows set to work on the eyes of frozen corpses."[45]

In modernity, social systems and psychical systems, social time and psychical-individual time, can no longer be homogenized by a fundamental final category, a "god term" (Kenneth Burke). Every person may come to terms with this difference in his or her own way. It is not necessary to perpetuate the edifying pathos of Hegel, who as metaphysician of spirit lives on the prerequisite that "Reason is the comprehension of the divine work;"[46] and it is necessary neither to perpetuate those forms of fatalist and tragic-heroic writing nor to fall into forms of sophomoric insolence (Nietzsche) that take pleasure in inverse turns, in which the experience of the loss of the transcendental signifier, the death of God, resonates all too clearly. For modernity the following state of affairs must be diagnosed: the comforts of the religious system are limited to the religious system. Everything else would be "Schleiermacherism."

Translated by J. W. Compton

Notes

1. Friedrich Nietzsche, "David Strauss, the Confessor and the Writer," in Daniel Breazeale, ed., *Untimeley Meditations,* transl. R. J. Hollingdale (Cambridge, MA, 1997), 27.

2. The literature on the time problematic is "boundless." Nevertheless, I want to call attention to the following books: Norbert Elias, *Über die Zeit* (Frankfurt/ M. 1984); David Carr, *Time, Narrative, and History* (Bloomington 1986); Günter Dux, *Die Zeit in der Geschichte: Ihre Entwicklungslogik vom Mythos zur Weltzeit* (Frankfurt/ M. 1989); Anthony Giddens, *The Consequences of Modernity* (Stanford 1990); Jörn Rüsen, *Zeit und Sinn: Strategien historischen Denkens* (Frankfurt/ M. 1990); John Bender and David E. Wellbery, eds., *Chronotypes: The Construction of Time* (Stanford 1991); Roger Friedland and Deidre Boden, eds., *Now Here: Space, Time, and Modernity* (Berkeley 1994); Peter Osborne, *The Politics of Time: Modernity and Avant-Garde* (London 1995). For my reflections the following were particularly helpful: Niklas Luhmann, "Gleichzeitigkeit und Synchronisation," in Luhmann, *Soziologische Aufklärung 5*: *Konstruktivistische Perspektiven* (Opladen 1990), 95–130; and the rich study by Armin Nassehi, *Die Zeit der Gesellschaft* (Opladen 1993). Cf. also my essay "Identität, Paradoxie, Differenz: An Essay on Conceptions of Time in

the Literature of Modernity," in Jürgen Fohrmann and Harro Müller, ed., *Systemtheorie der Literatur* (Munich 1996), 225–237.

3. Georg Wilhelm Friedrich Hegel, *Reason in History,* transl. Robert S. Hartmann (Indianapolis 1953).

4. Hegel, *Reason in History,* 24.

5. Ibid., 12.

6. Ibid., 41.

7. Ibid., 35, 42.

8. G. W. F. Hegel, *Lectures on the Philosophy of History,* transl. J. Sibree (London 1890), 477. "Daß die Weltgeschichte dieser Entwicklungsgang und das wirkliche Werden des Geistes ist, unter dem wechselnden Schauspiel ihrer Geschichten—dies ist die wahrhafte Theodizee, die Rechtfertigung Gottes in der Geschichte. Nur die Einsicht kann den Geist mit der Weltgeschichte und der Wirklichkeit versöhnen, daß das, was geschehen ist, und alle Tage geschieht, nicht nur nicht ohne Gott, sondern wesentlich das Werk seiner selbst ist." Hegel, *Vorlesungen über die Philosophie der Geschichte: Werke,* vol. 12, ed. E. Moldenhauer and K. M. Michel (Frankfurt/ M. 1970), 540.

9. Hegel, *Reason in History,* 60.

10. *Danton´s Death,* 3.3. Georg Büchner, *Complete Plays, Lenz and Other Writings* (London 1993), 49. See my article "Theater als Geschichte—Geschichte als Theater: Büchners Danton´s Tod," in Harro Müller, *Giftpfeile: Zu Theorie und Literatur der Moderne* (Bielefeld 1994), 169–183.

11. *Danton´s Death,* 4.5. Büchner, *Complete Plays,* 69,

12. *Danton´s Death,* 1.2., 19, Ibid., 10.

13. Cf. Ulrich Schönherr, *Das Unendliche Altern der Moderne: Untersuchungen zum Romantrilologie Gert Jonkes* (Vienna 1994), 133ff.

14. Author's translation. "Eigentlich hat alles, was meine Generation diskutierte, innerlich sich auseinanderdachte, man kann sagen: erlitt, man kann auch sagen: breittrat—alles das hatte sich bereits bei Nietzsche ausgesprochen und erschöpft, definitive Formulierung gefunden, alles weitere war Exegese." Gottfried Benn, *Essays und Reden in der Fassung der Erstdrucke,* ed. B. Hillebrand (Frankfurt/ M. 1989), 495.

15. For this and the following see Nassehi, *Die Zeit der Gesellschaft,* 360ff.

16. Nietzsche, *Werke,* ed. K. Schlechta, 3 vols. (Munich 1966), vol. 3, 489.

17. Friedrich Nietzsche, *Thus Spoke Zarathustra,* transl. R. J. Hollingdale (London 1961), 333.

18. Nietzsche, *Thus Spoke Zarathustra,* 49.

19. Ibid., 42.

20. Cf. in contrast the symptomatic reading Habermas gives Nietzsche: Jürgen Habermas, *Der Philosophische Diskurs der Moderne* (Frankfurt/ M. 1985), 104–129.

21. Friedrich Nietzsche, *The Birth of Tragedy and Other Writings,* ed. Raymond Geuss and ed., transl. Ronald Speirs (Cambridge, MA, 1999), 33.

22. Friedrich Nietzsche, *On the Genealogy of Morality,* ed. Keith Ansell-Pearson, transl. Carol Diethe (Cambridge, MA, 1994), 118.

23. Robert Musil, *Der Mann ohne Eigenschaften* (Hamburg 1952).

24. Friedrich Nietzsche, "On the uses and disadvantages of history for life," in Daniel Breazeale, ed., *Untimely Meditations,* transl. R. J. Hollingdale (Cambridge, MA, 1997), 63.

25. Robert Musil, *The Man without Qualities,* transl. Sophie Wilkins and Burton Pike (London 1995), 10–11.

26. Ibid., 709.

27. Ibid., 21.

28. Ibid., 1509.

29. Ibid., Ibid., 484.

30. Author's translation. Alexander Kluge, *Schlachtbeschreibung: Der organisatorische Aufbau eines Unglücks* (Munich 1978), 140–141. Cf. Harro Müller, *Geschichte zwischen Kairos und Katastrophe* (Frankfurt/ M. 1988), 97ff.

31. Author's translation. Kluge, *Schlachtbeschreibung,* 85.

32. Author's translation. Jean Piaget, quoted in Alexander Kluge, *Der Angriff der Gegenwart auf die übrige Zeit: Das Drehbuch zum Film* (Frankfurt/ M. 1985), 83. On the relation between expansion and contraction of time in today's media age with its simulation games, cf. Götz Großklaus, *Medien-Zeit, Medien-Raum: Zum Wandel raumzeitlicher Wahrnehmung in der Moderne* (Frankfurt/ M. 1995), 11ff.

33. Author's translation. Kluge, *Schlachtbeschreibung,* 91.

34. Author's translation. Oskar Negt and Alexander Kluge, *Geschichte und Eigensinn* (Frankfurt/ M. 1985), 792.

35. Author's translation. Alexander Kluge, "Das Lesen des Textes wirklicher Verhältnisse" in Hans Dieter Müller, *Der Kopf in der Schlinge Entscheidungen im Vorkrieg. Mit einem Nachwort von Alexander Kluge* (Frankfurt/ M. 1985), 175–211, p. 195.

36. Negt and Kluge, *Geschichte und Eigensinn,* 1008.

37. Author's translation. Oskar Negt and Alexander Kluge, *Maßverhältnisse des Politischen: 15 Vorschläge zum Unterscheidungsvermögen* (Frankfurt/ M. 1992), 220f.

38. Author's translation. Christian Dietrich Grabbe, "Napoleon oder die hundert Tage," in Hans-Georg Werner, ed., *Werke in zwei Bänden,* (Berlin 1987), 204. See Harro Müller, "Subjekt und Geschichte: Reflexionen zu Grabbes Naploeon-Drama," in Müller, *Giftpfeile,* 143–158.

39. Author's translation. Niklas Luhmann, *Soziale Systeme: Grundriß einer allgemeinen Theorie* (Frankfurt/ M. 1984), 374.

40. Georg Wilhelm Friedrich Hegel, *Phenomenology of Spirit,* transl. A. V. Millar (Oxford 1977), 19.

41. *Danton's Death,* 3,7. Büchner, *Complete Plays,* 59.

42. Author's translation. Gottfried Benn, *Briefe: Briefe an F. W. Oelze 1932–1945,* ed. Harald Steinhagen and Jürgen Schröder (Frankfurt/ M. 1979), 165.

43. Friedrich Nietzsche, *Human, All Too Human,* transl. R. J. Hollingdale (Cambridge, MA, 1996), 324.

44. Robert Musil, *Diaries 1899–1941,* ed. Mark Mirsky, transl. Philip Payne (New York 1998), 63.

45. Author's translation. Kluge, *Schlachtbeschreibung,* 9, 150.

46. Hegel, *Reason in History,* 48.

Part II

HISTORY

CHAPTER 8

History, Culture and the Quest for Organism

AZIZ AL-AZMEH

Die schlechthin vorgeschichtliche Zeit ist ihrer Natur nach unteilbare, absolut identische Zeit, [die] daher, welche Dauer man ihr zuschreibe, doch nur als Moment zu betrachten [ist], d.h. als Zeit, in der das Ende wie der Anfang und der Anfang wie das Ende ist, eine Art von Ewigkeit, weil sie selbst nicht eine Folge von Zeiten, sondern nur eine Zeit ist, die nicht eine wirkliche Zeit, d.h. eine Folge von Zeiten ist, sondern nur relativ gegen die ihr folgende zur Zeit (nämlich zur Vergangenheit) wird.

—Schelling

Since the spread of modernity as a consequence of the French Revolution and the political forms and concepts, the ideologies, and the legal norms it exported, a succession of seekers have set out on a quest for organism, for notions of umbilical immediacy, that is thought to go beyond the arid snares and illusions of Reason, of Jacobinism, of Bonapartism. At the time of the French Revolution and in reaction to it, and after the revolutionary waves of the 1830s and of 1918–1920, no less than after the demise of Communism, voices have multiplied and achieved demotic hegemony, seeking to privilege sentiment over structure, organic continuity in history over change and progress, communitarian particularity over universalism, historism over evolutionist historicism, the value of the vital over that of coldly rational, Eckhart over Cartesius. In all cases, this option for the primal and prelapsarian over the cultivated and corrupted has been allied with a reclamation irrational.

In the mid nineteenth century, this broad trend, converging on the reclamation of historical and national particularity as against and prior to the uni-

Notes for this section begin on page 132.

versalizing trends of 1789 and its Napoleonic aftermath, was officiated politically as well as discursively most particularly in Germany, under the title of "culture."[1] This term has since become the standard retort to claims that history is universally human, which puts into effect changes in the cultures of human collectivities as meaningfully as it transforms the conditions of their material culture. In this anti-Enlightenment discourse, culture itself becomes nature, endowed with its warm fastness, its reliability, its regularity and predictability, and it becomes possible to construe history as a natural history politically resistant to political claims and movements deriving from the Enlightenment and the French Revolution: this is particularly well-reflected in the classic of Wilhelm Heinrich Riehl, *Die Naturgeschichte des Volkes als Grundlage einer deutschen Social-Politik*.[2]

At the close of the twentieth century, a similar quest for an irrationalist foundationalism is officiated under the title of a self-proclaimed antifoundationalist postmodernism that, in a curious concord between xenophiles and xenophobes, construes the nature of a historical mass and the consequent relations between historical masses as being overdetermined by "culture." The universality of the modern history of the world is dissolved either into cultural "incommensurability," or into a vulgar notion of "interest." According to this last, which is ubiquitous in current anti-Enlightenment polemics, and which authorizes itself by superficial references to Michel Foucault, propositions about history, aims of political and social movements, and cultural values, when they gather a universalist perspective and build up a universalist momentum, are regarded to be falsifiable by mere reference to a functionalist interpretation of the political, social, or other purposes of which they are part. Such has recently been a treatment accorded, among others, to the notion of universality, a treatment that reclaims a naive functionalism to construe the Enlightenment idea of universalism as a "myth," here used in a manifestly polemical sense.[3] This kind of treatment also nurtures itself with a not altogether illegitimate appropriation of the writings of Edward Said, Martin Bernal, and others, and sets up a self-styled "postcolonial" discourse: curiously, this results in the re-orientalization of Orientals in the name of an authenticity and singularity regained. Thus we find in operation a traffic in mirror-images between re-orientalized Orientals speaking for authenticity, and postmodernists speaking for difference: subaltern studies thus become a display of exoticism to outsiders, and a reclamation of archaism to insiders. Pushed to one extreme from the supposition that different "cultures" have different rationalities, the discourse derives the dissolution of reason by its dismissal as "hyperrationality," and the preposterous construal of the notion of myth as "so-called myth" by an eminent historical anthropologist.[4]

The natural corollary of this wholly illegitimate transposition of moral into cognitive relativism, is a culturalist determinism in the treatment of history, whose objective is seen to be the specification of particularities and their endowment with valences independent of the universality generally attributed to historical forces, such as economies, technologies, and cognate structural elements. One example of this is a recent textbook of the history of Muslim peoples that quite explicitly and deliberately excludes treatment of elements such as the economy, on the grounds that it is the nonmaterial elements that distinguish this history from other "societies" with similar ecological characteristics and modes of production,[5] and that constitute "the significant terms of historical individuation"—that is, from similar historical conditions. Similarly but far more reflexively, a firmer synthesis, along vaguely Weberian lines, is made of what are taken to be the institutions and conceptions constitutive of historical individuality in medieval Muslim and Western European histories with a view to construing impermeable and intransitive historical types. Needless to say, historical closure so conceived goes counter to historical reality.[6]

While the reality of particularity is incontestable, it is doubtful, as will become clearer below, that its seizure is an attainable epistemological utopia. Before broaching this subject and its correlates in the postmodern approach to history and society, some further remarks on culturalism are in order. I understand "culturalism" as the view that regards an entity termed "culture" to be the determining moment in the history and present condition of a historical mass. A historical mass, in this sense, is one whose imputed distinctiveness, individuality, and indeed whose very name—the West, Islam, etc.—is taken to subtend a culture so specific and irreducible as to be in itself constitutive of both the history and the present condition of this mass, or at the very least to be the element that overdetermines its other elements and totalizes them as mere manifestations of a prior substance. Culturalism is consequently the view that regards conceptual and imaginary representations to be the ultimate and irreducible constraints at work in the life of the historical mass in question, despite the fact that social and historical life demonstrates to us daily that not all social and historical constraints are conceptual, and that although cultures are in a certain sense genetically transmitted and do indeed use genetic traits—ethnographic detail, real or imagined, or propositions about the world, encoded in a genealogy and in a pseudo-history of uniqueness and continuity, of the West or of Islam for instance—as symbols and markers, yet this is not sufficient ground for asserting that societies, or historical masses, are perpetuated by cultures.[7] Still less is it legitimate to assert that what are imprecisely known as cultures have an absolutely determinant role in setting the constitution of this mass, or that culture is in itself a sufficient defining element in this mass, or that culture overdetermines a given history to such an extent that it is in itself not only the chief iconic marker of this mass, but is also substantively representative of an inner nature ascribed to it.

We have here of course a replacement of history by ethnology, in terms of a culturalist differentialism very much reminiscent of the natural-historical molds by which anti-Enlightenment trends construed history. This construal takes certain ethnological markers as icons of an abiding *Volksgeist,* and regards them as criteria for a historical and ethnological taxonomy in which the distinction between natural and human history is lost[8] and replaced by a notion of historical individuality so singular and irreducible that the multiplicity of historical masses is conceived in terms of impermeability. Thus, with Herder no less than with Fichte, Hegel, or Ranke, the consequential progress that occurs in human history is specific to nations that hand over progress to higher levels from the one to another, while preserving the essential spirit that might, after accomplishing world-historical duties, revert to historical abeyance, indeed to the ahistoricity of their essence. Thus Burke, for instance, spoke of the "method of nature," of a historical mass as a "permanent body composed of transitory parts,"[9] in order to indicate an instinctivist concept of the historical mass conceived as a nature. The far more systematic formulations of Herder fully deployed the organismic metaphor and conceived historical masses to be powered by *Kräfte,* vital genetic forces effected by but not determined by ecological and other factors. Further, and within the same medieval conception, Herder conceived the constitution of a historical mass—the nation—to consist of a permanent condition of perfection, an Aristotelian entelechy, whose maintenance by internal vital powers is the condition for historical stability, indeed for abiding historicity.[10] Nations or cultures are therefore utterly and irreducibly individual, according to a naturalist morphology of history described, with reference to the roughly congruent conception of Spengler, as "a deliberate and painstaking attempt to extrude from history everything that makes it historical."[11] History therefore becomes a vast space for the classification and tabulation of ethnological individualities in a manner that unites Romantic philosophies of history with nineteenth-century anthropology.[12]

This position yields not only a relativism, but a correlative exceptionalism, which regards others as being beyond the purview of the procedures and constraints of human histories, societies, polities, and cultures taken to be normative. This vitalist notion of history rests upon an ontology of substances and essences, which are taken to constitute cultures, identities, selves, and absolute historical subjects in the conceptual context of the vitalism and organism that form the grounds of this discourse on identity, subjectivity, and culture. The constituent notion of this substance is that of a historical subject that is at once self-sufficient and self-evident, that is self-identical over time, whose rhythm and tempo are prescribed by internal organismic mechanisms of system maintenance and essential continuity. The passage of time is therefore merely contingent, inessential; all change that is perceptible and that might appear consequential is not relegated to any proper notion of historicity, but is conjugated with the neo-Platonic metaphysical notion of materiality as privation. Time becomes the material element, whereas the essence—often confused with the textual or the historico-mythological, as in the discourse of religious and national fundamentalism—becomes the spiritual, whose diminution can never result from the inside, but is ever caused by heteronomous interference. Historical masses are thus conceived as suprahistorical masses that speak in the tones of a chronophagous discourse. Thus cultures and nations rise and fall, but do not change in any serious sense, and the wheel of fortune is animated, quite literally, by internal, intransitive, self-subsistent pneumatic impulses (Herder's *Kräfte*), which together can be described by the term *Volksgeist*. The term is wonderfully apposite, certainly, but it is grounded in a basis of ahistorical vitalism replete with associations with medieval natural-philosophical notions of somatic composites, sustained by the anima that is said to make possible a historical mass's realization of its very materiality. Furthermore, the term derives much of its rhetorical force from assertions called forth by its psychologistic metaphor of the individual, according to which the collective self is construed as individual subject.[13]

Fundamental to this conception of history as the history of a spirit or a culture is the implicit assumption of homogeneity. The other has a *Geist,* a soul, a genius, a totalizing notion whose appellation as "culture" was inaugurated by nineteenth-century anthropology, but which has also taken various other names, such as "pattern," "value," or "meaning,[14]" the latter being one of the least meaningful notions in the cultural sciences. With the transitions, under the signature of postmodernism, from notions of society to those of representation, from evolutionism to incommensurability, sight was lost of a number of fundamental facts concerning the social reality of history, and of the labor of time. Not the least of these facts is that knowledge and representation, even in small-scale societies, is distributed and controlled; that cultures are webs of mystification no less than of signification, and that local stratification and other forms of diversity render questionable the notion of "shared meaning," a notion that is also compromised by the surface pragmatism of daily life. Facts such as these invite the inversion of the Geertzean thesis that ideology is a cultural system, into the assertion that what are taken for cultural systems might be regarded as ideologies. The notion of culture as an unmodifiable system in which novelties are impurities and all disturbances lead to crises—of "identity"—has no justification in the historical reality of any historical mass. An epistemological correlate of this essentialist notion is an implicit causal hypothesis, according to which utterances and actions arise from and are held together by some uniform underlying intellectual project that has a propositional form and can be unraveled by a phenomenological description. Thus, the conservation of "traditions" is regarded in this kind of study as an observable property of events rather than "a hypothesis put forward (by the anthropologist—and by extension, the cultural scientist and historian—no less than by those who are taken for authentic speakers) in order to account for their actual repetition."[15]

There is clearly an epistemological complement to this essentialist notion of suprahistorical masses, as a means of access to it, leading to its narration. For correlative with the social instinctivism that arises out of this view of transhistorical abeyance is an irrationalist epistemology that I shall term *sympathetic differentialism*. The notion that knowledge is not absolute but is bounded by certain conditions of emergence—"cultural" conditions, according to the culturalist protocol, having replaced the psychologistic explanations of classical skepticism—and indeed, extending sensitivity to this notion to yield a skeptical epistemology approaching intuitionism, is combined with a Romantic notion of the immediacy of the act of knowledge if this act were to be adequate to its object. The result is that the proper procedure for attaining adequate knowledge of a life-world defined by its self-enclosure is a phenomenological description, an act of immediate apprehension whose primary vehicle is sympathy.

Thus, Hamann spoke of passionate understanding, Herder recommended *Einfühlung,* and Schleiermacher and, more pertinently, Dilthey, consecrated *Verstehen.*[16] This epistemology of relativistic life-worlds, really a model for the very fragmentary apprehension of these worlds rather than a full-fledged epistemology, rests implicitly on a historical and culturalist redaction of the innate ideas concept. It regards knowledge of cultures and their histories to be absolutely bounded by their essential conditions of emergence, and therefore postulates a certain correspondence between cultural-historical knowledge and its object, in such a way that being and knowing become, in principle, indistinct. In this system of *Lebensphilosophie,* life and history are connected in such a way that history is no more and no less than the realization of life, the *Volksgeist.* Reason becomes multiple, reflecting the multiplicity of life-worlds, and it becomes the voices of these life-worlds. In this perspective, knowledge is either autochtonous and thus spontaneously generated, or external, achieved by an act of inscription, by the knowing other, and through an act of differentialist sympathy, within the essence of his own other. From this location, the knower can register the spontaneous voice of the self become, albeit only virtually, object of knowledge, or rather of apprehension. In practical terms, this apprehended other can therefore be represented only in his or her own terms, that is to say, through a procedure of crude and unreflective empiricism. Such is the sense of privileging the "subject-position," in which the object-subject's self-representation, as the authentic voice, is taken as a register of essential reality that passes under the title of "meaning."

Of course, "meanings" are multiple, shifting, and situational, and no "meaning" describes a "culture," let alone prescribes its history; we have already referred to this above in connection with the notion of tradition. Yet the discourse, historical no less than pseudo-sociological in the cultural studies mold, on the radical otherness of cultures, times, and histories, by postulating that cross-historical and cross-cultural knowledge has conditions distinct from the conditions of knowledge in general, in fact performs primarily an act of classification. In its search for essences and inwardnesses, for meanings and authenticities, it in fact becomes a record of alterities. The act of sympathy is not only sympathetic, but is also differentialist, and differentialism, good intentions apart, is a taxonomic enterprise in which the language of the self is taken for a metalanguage of typology[17] whose elements are generically closed, being already pre-given as self-referential selves. The result of this can only be, in effect, an antithetical discourse, a system of generic classification whose structure is binary, composed formally of segmentary strings of *Ich* (I) and *Nicht-Ich* (Non-I). It constitutes, for instance, the analytics of classical orientalist discourse.[18]

It is not being suggested that this romantic postmodernism implies fully that self-representation necessarily corresponds to historical reality, although this suggestion is not excluded and is indeed operative in both the demotic and scholarly discourse on, for instance, the structure of Muslim histories. This is usually reduced to the history of Islam, which is itself is reduced to the canonical texts of Muslims regarded as embodying the essence of the life of Muslim peoples and the measure of their authenticity and invariance. What is being suggested, rather, is that the hyperempiricist approach operates at two possible levels: one that regards the event in the Rankean manner, and another that is increasingly coming into prominence, regarding the representation of the event as crucial: its meaning on the part of the other interlocutor (the source), or the Other writer, the historian or social or cultural scientist.

I do not wish to dwell on the narrativity theme in historical writing,[19] and will only indicate that the conclusion, based on the noncorrespondence between narrative and history, that no scientific history is possible, is premised on a notion of historical events as sheer flow, unseizable scientifically, and only representable narratively. The postulation that the properly correspondent account of historical facts is essentially mimetic rests on assuming event and writing to consist of different grades of the same substance, the one more abstract than the other in the Aristotelian sense. This cannot obtain, according to the strictly mimetic model, for experience has no beginning, middle, or end.[20] The fundamental fallacy, of course, over and above mimetism, is that structural studies need not be reducible to the play of rhetorical topoi. History is not historicity[21] and it can justifiably be argued not only that the construal of causalities in historical succession immobilizes duration, indeed eliminates the duration separating cause and effect, but also that the continuity of temporal phenomena is itself metaphorical, for in order for continuity to be in duration, its elements have to be in rhythm, i.e., to constitute a system of instants.[22] This should be taken to imply that here we have not a rhetorical play in the polemical sense, but rather a structure of apprehension and construal with determinate relations to its object, and also to the very construction of this object.

What is crucial for the present argument is that, in its quest for the resuscitation of the Rest, which the history of structures is alleged to have ignored, the neoromantic historiography now taking place under the signature of postmodernism is rejecting, wholesale, considerations of structure in favor of substance.[23] True, the Rest—the historical saliency of matters national, cultural, religious, and so forth—is "impervious to theoretical debunking."[24] But the revanchist revivalism correlative with the critique of the Enlightenment tends to ignore actually existing mutabilities, structural and other, that came as a result of the Enlightenment,[25] and results in its quest for organismic substance to extrude intelligibility from the process of knowledge and relegate it to the singular, and therefore relative, voice of the object of this knowledge.

The neo-romantic historiography under discussion, it is true, is less oriented toward large-scale units (unless they be others: other ethnoi, nations, and so forth), and more toward redemption of the local, the microhistorical, the hyperparticular, the quotidian, and other matters that merge with pure duration, with the eternity of substance. But this is conceptually equivalent to the preoccupation with grand historical categories: grand narratives of small-scale events, romances and epics of the singular. Both rest on notions of the individual organism. The postmodern version is congruent conceptually with the older, reactionary prototype and is largely derived from its traces, through demotic notions of *Gemeinschaftlichkeit* no less than the politics of populist demagogy, whose salience is most often underestimated. The connections sometimes claimed between it and notions of mentalités of the Annales School is fictitious, for this latter is understood in terms of different conceptual parameters with emphasis on structure,[26] and becomes serviceable only when severed from its conceptual discipline and endowed with the attributes of a substance: the longue durée is a structure, a mode of abeyance and of configuration, not a substance, just as France herself is not the substance of French history, and her history is not the history of a pre-given substance, but of evolving configurations that structure the connection between elements of the ecological, political, demographic, and other orders.[27]

Finally, historiographic neo-Romanticism of the individual, the partial, the local, and the marginal, while it does indeed restitute the sense of uncertainty, of nondetermination, has tended to do so by a denial of structure, not by its revision. This denial is implicit in its marginalization of structures, mutabilities, generalities, and permeabilities, in the name of restituting a prior and mightier, more durable, ontological weight. Instead of this being a change of scale of observation, with attendant conceptual and methodological consequences, it becomes a submission to the finality of the individual[28]—less the further refinement and therefore empowerment of the gaze, than its abandonment to the ostensible voice of nature.

This denial of putative determinism is premised on a subtext, the denial of Enlightenment teleology; the correlative finality endowed with the individual, be this a suprahistorical unit like the nation or an infrahistorical individual like a community, occurs invariably, therefore, in conjunction with a particular notion of time. Time becomes an element internal to the substance that constitutes the substrate both of history and of historicity; it carries no burden of synchronicity, anachronism, or asynchronicity with respect to other substances, other identities. Its time is bereft of chronometry, hence of universality, and is unconnected to diversity within that identity. It is, like Schelling's prehistorical time of myth, the time of its essence, and the chronometer is the sheer duration of its eternity and of its "inner meaning," of its "memory," to mention a topic much in vogue today.

Hence it becomes possible to speak of two categories of otherness with respect to a historical mass, other masses, and the past: as "temporal depth … and cultural distance."[29] This is a wonderfully rhetorical statement of the consequences adhering to the culturalist and other essentialist views of history: time within is profundity, resource, continuing origin, constant recommencement. Contemporaneity without is distance, difference, distinctiveness. History within is, in this perspective, not so much active time, but changeless essence—nation, culture, or whatever other historical mass may be in question—*sub specie temporalis.* History in fact becomes genealogy; and genealogy being formally the succession of one identity in separate moments, genealogical history becomes no more or less than typology, where past and present are sheer temporal repetitions of one another in distinction from other genealogical lines.

Thus, despite the reality within historical representation of an identitarian mnemotechnics that posits identity and continuity,[30] this, is a far cry from taking this representation as an adequate description, except in a perspective of phenomenological relativism like that indicated above, which regards the distinction between memory and myth as inoperative.[31]

Yet the collapse of history into variant historicities is, as I have suggested, rhetorical, and in more than one way: not only does it presume to convey affective conditions that are beyond the scope of linguistic expressions, these being the solipsistic tongues of inwardness, but the sheer need for rhetorical effect is in a backhanded way protective of historical discourse. That is to say, this whole discursive field of relativistic pronouncement upon history is most often absent from the actual writing of history, which has generally retained the large-scale classification of historical objects—Germany, the West, Christendom, for instance—yet at the same time has treated structure, mutability, change, and transformation to such an extent that the trope of immutability has not held, except in nominally or self-consciously nationalist redactions of history. On the other hand, this essentialist approach has very much held in discussions involving Others.

I do not need to treat at any length the standard textbook version of Arab history or of the history of what is known as "Islam." In brief, this is commonly seen to have arisen out of a small bibliocentric religion that exploded upon the vista of world history as a force of conquest, foisting a novel religion and a Levitical "way of life," with its sui generis system of belief, practice, legal organization, social structure, political form, and so forth, all of which correspond to one another, and all of which can ultimately be reduced to textual imperatives. That which is not so reducible is adjudged exogenous, heteronomous, the "influence" of previous times. Yet, the world being as it is, this psychodrama was soon defeated, and "Islam" started, very rapidly, on the road to decline, until today, when a "crisis of identity" is exemplified by what is portrayed as an essential incompatibility between authentic culture in reassertion, and forces of "westernization."

I am not here concerned with demonstrating that every one of these assertions is false. Rather, I am concerned with the matter of culturalism, with discerning the culturalist view of history, its essentialism, and hence its decidedly dubious value in providing a "cross-cultural" perspective on historical understanding. In order to do this, I shall look briefly and didactically at two modes of historical self-representation in Arab-Islamic tradition, one from the nineteenth century and the other from medieval historical writing.[32] I hope that this glance will produce a *Verfremdungseffekt,* and thereby suggest that matters are totally at variance with what might be expected of the culturalist view.

The nineteenth century witnessed the globalization of certain notions of society, polity, and history, which, albeit of European origin in their theoretical elaboration, became universal modules, everywhere produced and reproduced. Among these were ideas of representation, of the people, of sovereignty, of the nation, and the correlative notion of authenticity, truth to the essence of an abiding history. The Ottoman Empire, centrally and provincially (Greece, the Arab world, the metropolitan Armenian intelligentsia and commercial classes), was certainly no exception. Authenticity was a concept deployed in order to construe national political identity, and correlatively to identify others, in a process of historical formation. The names given to these units of historical identity in the process of formation were various, depending on political conjunctures, social and regional affiliation, and a variety of other matters. Suffice it to say here that the notion of an Islamic politico-cultural internationalism arose in the subaltern ranks of the Ottoman reformist state administration, similar in many ways to the German Bildungsbürgertum, in the middle part of the nineteenth century and was to be adopted later on, under Sultan Abdülhamid II, as state policy.

In common with other subaltern revivalist currents,[33] as with defensive, retrenching nationalisms and with populist ideologies, the notion of authenticity is widely used both in formal discourse on matters political and social and in the interstices of casual comment. The notion of authenticity is not so much a determinate concept as it is a node of associations and interpellations, a trope by means of which the historical world is reduced to a particular order, and a token that marks off social and political groups and forges and reconstitutes historical identities. In these senses the notion of authenticity has analogs elsewhere, doubtless officiated under different names, and a famous career in Germany starting from the end of the eighteenth century.

Asâla is the Arabic term for authenticity. Lexically, it indicates salutary moral qualities like loyalty, nobility, and a sense of commitment to a specific social group or a set of values. It also indicates a sense of exclusive individuality; and in association with the senses previously mentioned, asâla specifically refers to genealogical standing: noble or at least respectable descent for humans, and for horses the status of equine aristocrats. Combined together and transferred to an attribute of historical collectivities, Arab, Muslim or other, asâla becomes a central notion in a Romantic conception of history that evokes features commonly associated with such a conception. Of primary importance among these features is a vitalist concept of nationalism and of politics, replete with biological metaphor as discussed above and, occasionally, with sentimentalist populism.

Ultimately, therefore, the notion of authenticity is predicated on the notion of a historical subject that is at once self-sufficient and self-evident. Its discourse is consequently an essentialist discourse, much like the reverse it finds in orientalism, in discourses on the primitive, and in other discourses on cultural otherness. In common with these discourses, the discourse on authenticity postulates a historical subject that is self-identical, essentially in continuity over time, and positing itself primarily in distinction from other historical subjects, which are commonly construed in terms of alterity, as non-selves. For the historico-discursive viability of a historical subject such as this, it is essential that its integrity must be posited and asserted against a manifest backdrop of change of a very rapid and profound nature. It therefore follows that change should be conceived as contingent, impelled by inessential matters like external interference or internal subversion, the effects of which can only be confronted with a reassertion of the essence of historical subjectivity. History therefore becomes an alternance in a continuity of decadence and health, and historiographical practice comes to consist in the writing of history as a form of classification of events under the two categories of intrinsic and extrinsic, the authentic and the imputed, the essential and the accidental.

It is therefore not fortuitous or haphazard that the title under whose name this discourse (and its political implications) is officiated should be revivalism, *nahda,* in line with similar historical and ideological experiences among which the Risorgimento readily comes to mind. In fact, this entire ideological trope can be described as one of ontological irredentism, it being the attempt to retrieve an essence prior to superficial time, an essence that the vicissitudes of time and the designs of enemies, rather than change of any intrinsic nature, have caused to atrophy. The counterpart of this is that the degraded conditions of today are mere corruptions of the original cultural essence, the retrieval of which is only possible by a return to the pristine beginnings that reside in the early years of Islam, the teachings of the book of God, the Koran, and the example of the Prophet Muhammad. It must be added at the outset, however, that though revivalism was initially Islamist, and has tended to don the Islamist cloak in the very recent past, it received its most thorough grounding in the context of secular Arab nationalist ideology, which regarded Islam as but one moment in a glorious history that preceded it and continued after it. Indeed, just as modern European historical writing has assimilated ancient Greece to its history, without much real justification, modern Arab nationalist historiography has assimilated to its axis of continuity the histories of Babylon and Assur, and occasionally the essential moments in the universal history of monotheism.

Returning to Islamist historical conceptions, in historical terms this constellation of notions came into currency in the second half of the nineteenth century, first with the Young Ottomans in Istanbul and particularly Namik Kemal (1840–1888), and shortly thereafter in the writings of the remarkable Jamâl al-Dîn al-Afghânî (1839–1897). Afghânî was not a profound thinker, but a very potent speaker and charismatic conspirator. His careers in Istanbul, Tehran, Kabul, Hyderabad, Calcutta, Cairo, London, Paris, and St. Petersburg have left an important imprint on pan-Islamism in the Arab world, which in certain respects at this time can be regarded as a form of protonationalism. Afghânî left a body of miscellaneous writings, most notably his polemic against the pro-British Indian Muslim reformer Sir Syed Ahmad Khan (1817–1898),[34] with whose ideas, it must be stressed, he was not really at variance. He inspired the journal *Al-'Urwa al-wuthqâ,* a collaborative body of political, cultural, and reformist writing published in Paris in 1882–1883 with his then disciple, Muhammad 'Abduh (1849–1905), who was later to become the Arab world's foremost and most subtle Muslim reformist.[35] A section of 'Abduh's writings are in tune with the general theses of Afghânî but are far more finely tuned and retain none of Afghânî's crudeness of conception, and 'Abduh's disciples included some of the Arab world's foremost Muslim reformist and nationalist leaders in the early part of this century. This same constellation of notions was channeled into the mainstream of Arab political and social thought through the nationalism that was later to become Turkish nationalism, exemplified in Ziya Gökalp (1875–1924) and the Arab nationalism of his erstwhile associate, Sati' al-Husri (1880–1968),[36] although Husri was not a Romantic revivalist and populist like Gökalp, and Romantic revivalism was only to enter Arab nationalism between the wars in synchrony with European irrationalism, with strong opposition from Husri. Husri was a sober positivist and educationalist who believed nations, in the form of the nation-state, were the most advanced form of human association, and not the reassertion of a preexistent mystical entity.

The nation for Afghânî is akin to a body, although he changed his mind over what constituted a nation. In the final analysis he devalued ties of ethnicity, and to a lesser extent of language, to the advantage of the bond of religion. A nation consists of estates analogous to parts of a body, or of individuals whose organic unity is that of the parts of a vital organism. This organism is infused with a vital force like that which permeates its individual organs, and the power of this individual vitality is directly proportional to that of the whole organism.

This organismic, vitalist paradigm, like Herder's, has its major notions—if not its object, a sociopolitical order—in medieval natural philosophy. Equally important is that it naturally invites comparison with Herder's notion of *Kräfte* as inner sources of vitality and dynamic principles for the continued existence of nations; the question as to whether this Romanticism is medieval in its immediate conceptual inspiration is irrelevant to its modernity and to the vital part it played in nineteenth- and twentieth-century ideological tendencies. Though Afghânî's ideas initially were shaped in Iranian seminaries, they were received not in Qom but in Calcutta, Cairo, Istanbul, and Paris, where they were filtered through contemporary social and political categories. Also like Herder's, Afghânî's paradigm concretizes this vital principle for the unity and cohesion of bodies national in culturalist terms and, like Herder's emphasis on *Bildung,* finds in civic and moral education the key to the maintenance and resuscitation of national glory. The vital spirit in empirical terms is a yearning in the hearts of men for glory and a longing for the consummate realization of values. And this vital spirit is operative only when it impels bodies national with a desire for excellence and distinction in wealth as well as glory and might (*cízz*).

In situations of conflict brought about by pervasive Western interference in the Middle East, this perspective was not unnaturally invested with a social-Darwinist stance. It is well to bear in mind that the "conflict theory" of political sociology was emerging in Germany at about the same time—proponents of this theory, like Afghânî and the early 'Abduh, were keenly interested in Ibn Khaldûn's theory of the power of state, which they used in the construction of a nationalist Romanticism.[37] The struggle for existence, Afghânî tells us, pervades human history no less than it does the animal kingdom and inanimate nature. The reason for this is that "might is the visible aspect of life and of continued existence … and might is never triumphant and concrete except when it weakens and subjugates others." As illustration, Afghânî cites the powers of nations, and specifically the subjugation of the Ottoman Empire by the European powers.[38]

What, in this perspective, is history? And what does the passage of time yield? It can be noted that the subject of history is the body national. Each body national, as in Herder and many others, has a fixed nature that is, according to the characterization of Collingwood that was noted above, less the product of history than its presupposition. The historically significant unit is the national subject, and history therefore alternates between true historicity manifested in might, and historical desuetude manifested in subjugation. Might results from cohesiveness and unity, and if this unity were to be lost, the body national would lose its spirit or its general will, with the result that "the thrones of its might will fall, and it [the nation] will take its leave of existence just as existence has abandoned it."[39] It is indicative of Afghânî's style that he used the term *quwwa hâfiza,* which I have rendered as "spirit." The expression, literally "preservative power," is derived from the medieval Arabic natural philosophy in which Afghânî was deeply steeped, and whose concepts he often used, where it designates the subliminal quality that keeps together a somatic composite, as a variation on the Aristotelian notion of entelechy.

The cohesiveness and unity of this body national infused with a vital impulse that yearns for glory are maintained so long as the factors that originally constituted this *Volksgeist* are operative. But once corruption sets in, once the essence is diluted, the auguries of national calamity become manifest. Thus the glorious classical civilization of the Muslim Arabs was corroded from the inside by the snares of esotericist sects, which paved the way for conquest by Crusaders and Mongols. Similarly, the fabric of the Ottoman Empire was weakened by Ottoman westernizing reformists in the middle of the nineteenth century. As for the French, the glory of their royal past was corrupted by the seductions of Voltaire and Rousseau, which directly led to what Afghânî regarded as the calamities of the French Revolution, the Paris Commune, and defeat in the Franco-Prussian War. In the same class of universally destructive, disintegrative impulses are socialism, communism, and anarchism, which might cause the annihilation of humanity altogether, being for Afghani the ultimate forces of corruption and radical antinomianism, the antithesis of order and civilization.

There is no adequate response to weakness and destruction save that of revivalism: the retrieval and restoration of the original qualities that made for strength and historical relevance. No progress without the retrieval of pristine beginnings and the cleansing of the essence from the adulteration of history: such is the fundamental principle of revivalism. The Islam that results from the elision of history and the deprivation of time of any significant ontological weight will shortly be taken up; but before this is done it is necessary to take a closer look at the categories that subsist in the trope of authenticity, of absolute individuality and irreducible historical subjectivity.

The trope of authenticity, described above as less a determinate concept than a node of associations, is premised on a number of important notions and distinctions. Fundamental among these is a conception of history that posits a narcissistic continuing subject, mighty by virtue of its nature but enfeebled by subversion, inadvertence, and what Hegel termed "Oriental ease and repose." This same subject will regain its vital energy and continue the maintenance of its nature—its entelechy—by the revivication of its beginnings, which will subsist within it just as a nature, in the classical and medieval Arabic and European senses, inheres in a body.

But this subject is inconceivable in isolation from others, which exist alongside it, for essentially, the notion is formed in the context of political contestation, and conceptually, of antithesis. These others are construed to a very considerable degree to be absolute in their otherness, in that they are taken for antitheses of the subject. In order for them to be met, their subjectivity has to be constantly objectified, deprived of value except for that which, like forces of corruption, is inessential and contingent, hence transferable. Such was the attitude of Afghânî, and all those who adopted the hopes associated with his name, toward modern science and technology, of European provenance but not culture-specific and, moreover, necessary for the construction of national might. Throughout, the origin—the positive beginning—is adulterated, but still it flows as a subliminal impulse amid degradation and corruption, for the fall from previous heights is inessential, and the essence of this historical subject is in fact suprahistorical and still subsists in the innermost core of the cultural self. The revivalist project is simply one in which this core is again brought to the surface and to the forefront of historical existence, thereby restoring the historical subject to its true nature.

The truth of this nature is an ontological truth, one whose resistance to the vagaries of time is demonstrated by the revivalist belief in its capacity for resuscitation, and whose proposed durability is the measure of its truth. Indeed, this nature, the vital impulse of the body national, is the very reality of the subject in history; corruption is conceivable only as privation in this neo-Platonic cosmography of historical time. In the light of this, history consists of continuity over a time that knows no substantive causalities, hence no breaks. This continuity is in a constantly antithetical relation to all otherness: to all nations, which by virtue of the very nature of bodies naturally seek to subjugate the nation-subject, and to corruptions within, for these are privations of the essence that seek to subvert, and thus to nullify, the vital energy that uplifts and allows for glory.

Time is therefore cleft between origins and corruptions, between authenticity and the snares of enemies. Forces of privation, of foreign—that is, inessential—provenance, have no intrinsic extensions: they do not extend to the core of the historical self, for they have no avenues that lead to the fund of subjectivity, either in the past or in the present. They have bearings neither in the past nor in the ontological reality of the present. In contrast, extraneous influences disturb the homogeneity of the subject and confound the bearings of its historical course by repudiating the original inner indistinctness and homogeneity that constitute the stuff of authenticity.

Authenticity, for one contemporary philosopher who has been attempting a left-wing reclamation of Afghânî along the lines of a Muslim liberation theology, designates the self in contradistinction to the other, the essential as against the accidental, the natural as opposed to the artificial. Only thus can individuality and specificity properly be said to designate any genuine distinctiveness in opposition to "the loss of distinctiveness and dissolution in another specificity [of the West] which claims universality." Authenticity and its associated notions are, further, said to extend the cultural ego into history and endow it with "historical continuity and temporal homogeneity and the unity of the national personality."[40] This concordance with postmodernist discourse is noticeable but unremarkable, for they both partake of the same universal neo-Romanticist tropes.

Authenticity is therefore both past and future linked contingently by the ontological void of today. The past is the accomplished future and the future is the past reasserted; history is the past in the future anterior. History is an even continuum, on the surface of which fly chance eddies that counter the original energies of the continuum and work to suppress them, yet do not quite succeed in more than rippling the surface and disturbing its evenness. Only thus can teleology be assured: for a nature to consummate itself, for the future revival to close the circle of historical appearance and coalesce with the original condition, the end must be pre-given and inevitable in the sense that it is in accord with nature.

The body national is thus neither describable nor recognizable if measured against its contingent existence, or against the sheer temporality, identified with evanescence, and lack of perfection that characterize it today. Time is devoid of quality, corruption is purely vicarious, and the present is but a negative interregnum between a fully accomplished, entirely ample origin and its recommencement, which is also its consummation. History therefore takes place in "two modes of time, one of which has a decided ontological distinction,"[41] the one relevant to the essence and a measure of its duration, and the other dissolving into transience and contingency. The former is much like the time of myth as described by Schelling in his *Philosophie der Mythologie,*[42] quoted at the opening of this article, to which we shall come back later. The latter is the active time of history, whose activity is regarded as purely virtual.

The connection of these modes of time is the same as that of different bodies national: a connection of otherness that in a social-Darwinist world is one of subjugation and antinomy, essentially of negation, without the possibility of a mutual interiorization such as that inherent in, for example, the Hegelian dialectic of master and slave. Indeed, the polar structuration of the discourse on authenticity is what makes it possible not only to deny essential change in time, thus denying multiplicity over time, but also to deny what we might term spatial multiplicity of any essential consequence, this being the social, political, and ideological multiplicity at any one particular point in time, except in so far as such multiplicity is perceived as subversive of a homogeneous essence that requires evenness. Any unevenness, as has already been indicated, is perceived in terms of antithesis, privation, corruption, atrophy. This national subject is an essence that knows neither dysfunction nor transformation but only abeyance.

This subject is pure self-reference, a tautological circle, whose impenetrability to reason other than the reason of its own self-reference is very much in keeping with similar outlooks in the German *Lebensphilosophie,* where life is at once the subject and the object of the mind. The authentic self is immediately apprehended,[43] and knowledge of it by its own is a sort of pure and perfect *Verstehen,* an almost innate endowment in the mind of the components that make up this body national, whose self-enclosure is epistemological and not only ontological. Indeed, as we have seen above in connection with sympathetic epistemology, the epistemological and the ontological correspond perfectly, for knowledge of authenticity is but a moment in the life of this authenticity manifested as will. For what is such knowledge of a self-identical entity but a form of transcendental narcissism? Indeed, Afghânî specifically designates the *Bildung* of the renascent nation as one whose prime medium is oratory that exhorts and recalls the past.[44]

It should be clear from the foregoing that the subject that is being corrected by oratorical education, and that is Romantically conceived as both beyond history and underlying it, is indeterminate if its conception is left as presented. There are no indications of its determination except gestures toward historical particularities: events, names, dates. Beyond this there is reference to a name: Islam. There are analogs to this Romantic mode in virtually all cultures. In all these cases, in the absence of historical determination over and above the indication of a foundational first time of archetypes, the discourse of authenticity is socially open, in the sense that its essential emptiness, what Hegel might have termed the boredom of its concepts, renders it very versatile and protean. As this ontological self-identity is epistemologically reflected in solipsism, the result is that the construction of identities here is fundamentally an act of naming.

Naming is not an innocent activity but lies at the very heart of ideology, one of whose principal mechanisms is the operation of classificatory tokens that determine the memberships of sociopolitical groups. These operations also entail exclusions and inclusions by way of condensations, displacements, and associative interpellation of some complexity. The concrete images put forward as factually paradigmatic—the golden age, the glories of the Arabs, the Middle Ages in some European Romanticisms, the idyllic rusticity of Heidegger, of African nativist philosophers, or westernized Indian sages—serve as iconic controllers of identities and take on general values generated by a truncated and telescoped history; yet these are values that act as carriers of general attributes that no human collectivity can eternally possess and of paradigmatic value that is only imputed to them by the purveyors of the ideological messages. The versatility of the general name—such as Islam—lies therein; the abstract act of naming engenders as many distinct identities as there are constituted social and political groups that might claim the name as their own. The reality of the historical subject lies not in the head but in historical reality, and the key to this reality is not the conformity to some self-subsistent essence or some invariant historical Islam that does not exist, but the group that adopts the name by adapting it to its particular form and understanding of the historical paradigm evoked by the name, a paradigm that is metonymically suggested and not specifically indicated by the name itself. The connection between name and historical reality derives its validation and credibility from extrinsic criteria, from the capacity that the group adopting the name has to enforce and consolidate its interpretation and to perpetuate it within institutions both epistemic and social.

Evidently, the Romantic and the neo-Romantic culturalist notion of history as the history of individuality, described in connection with nineteenth-century European thought and with late twentieth-century postmodernism, is in all important respects congruent with the Romantic notions of history current among some circles in the late Ottoman Empire. That both find their organismic notions of history and of society in the demotic residues of medieval conceptions of nature (associations with entelechy, matter and form, being and privation) is incontrovertible. What is particularly striking is not the medievalism but the contemporaneity, a reflection of the universality, albeit uneven, of the rhythms of intellectual and political currents from the middle of the nineteenth century. Such medieval natural-scientific notions, before Herder, were quite unusual in the systematic treatment of history and society, one particularly notable exception being, of course, Ibn Khaldûn.[45]

Culturalism is therefore not relevant to gauging the manner in which another "culture" regards history: more germane to this is the analysis of historical discourse, with the full knowledge that this is one manner of conceptualizing history among others. In both the nineteenth and twentieth centuries the Arab World also produced other views of history, predominantly a generally positivistic, evolutionist, and nationalist history in the sense that it regards nations and nation-states to be the most advanced form of human social and political organization.

But this is a matter that might be treated elsewhere. For the purposes of this study, it is more appropriate to take our second glance at Arabic historiography, this time looking at notions of time contained in medieval works. This will reveal that Arab, or putatively Arab-Muslim "culture," cannot be regarded as one in continuity. Medieval and modern can in no way be said to be in continuity, a continuity ascribed to "culture." There are definite correlations in the typological construals of historical continuity between medieval and certain modern Arab conceptions of history, certainly. But this correlation is more general, extending to medieval and ancient typology as much as to nationalist historiography in Europe as well as to nineteenth- and twentieth-century Romanticism, whose contemporary form is certain strands of postmodernism. For the rest, residues and the radical imperfection and incompleteness of things notwithstanding, the predominant trends of modern Arab historical writing are academically grounded on conceptual principles that are today a cosmopolitan patrimony, and are distinct from medieval writing and not in conceptual continuity with it, although there are, here and there, some affective, symbolic, and mythological continuities.

The scale of medieval Arabic historical writing is colossal.[46] We shall here concentrate on the question of typology, in order to return to the start of this chapter, where the relevance of neo-Romantic culturalism was questioned, and with a view to showing that the quest for organism, in the notion of time implicit within it, is more in keeping with medieval typology than might be supposed and is, like medieval typology, amenable to study by means of the normal equipment of the humanities and social sciences. That the material used here is Arabic rather than Latin might appear somewhat fortuitous. Though I will leave it to scholars with a better knowledge of medieval European historical writing than mine to make comparisons between medieval Arabic and Latin (or Byzantine Greek) historiographies, I think that, in this as in many other fields, the formal and conceptual concordances are clear enough to dispel notions of the relevance of "cultural" difference.

Broadly speaking, there are two registers of historicity in medieval Arabic writing. One is technical, and structures annals, chronicles, dynastic histories, biographical dictionaries, and, generally speaking, universal histories. This technical repertoire of historical events depends for its narrative structure and indeed for its definition of historicity and of the event itself, on a chronometric notion of time. The other has a different notion of time: it is a typological concept of history that arranges events by category rather than by sheer flow of vulgar time, events that in some of their original sources may be dispersed among other, chronologically arranged events, but that can in other discursive locations be placed in relations of sequential repetition as prefigurations, figures, types, beginnings, recommencements, ends.

Medieval Arabic civilization shares an almost universal concern with origins and beginnings, of the world, of social order, of particular artifacts and customs. The cultures of antiquity cultivated this interest in origins[47] at least as much as does modern industrial civilization, which with its theories of social contract and its evolutionism is alone in claiming descent from savages or even from lowlier creatures, rather than from gods and heroes.[48] This pursuit of origins provided, among other perhaps more important things, explanations for the status quo of a historical type based upon a conception of generic continuity over time. Chinese literature knew a distinct literary genre of "techno-historical dictionaries" concerned with the first occurrences of things, systematized in the third century BCE.[49] Arabic literature has likewise left a rich patrimony on *awâ'il* (first occurrences), whose main purpose seems to have been to provide gems of recherché knowledge of an unusual and exotic character, a knowledge that formed part of courtly urbanity but that also figures predominantly in examples from the large body of Arabic *Fürstenspiegel.*

This register of "firsts" is a record of archetypes as well as exempla, salutary or nefarious, and ruinous acts of yore; these materials provide practical and moral education by means of example, very much like the artisanal learning process based on repetition. This demotic notion of education is enshrined in the exemplary register that was the *Iliad,* the "Homeric encyclopedia,"[50] in the body of exempla used in medieval European sermons,[51] or in the vast repertoire of Muhammadan actions and words that constitutes the *hadîth,* an important component of the Muslim canon.

This repertoire of archetypes can be both sacred and profane; in it, first occurrences inscribe the initial print of an act that is subsequently repeated, not renewed. They are acts of full inauguration. Thus, according to medieval Arab historians, Alexander was the first man to shake hands, having thus founded the custom. David was the first to have armor manufactured, incest goes back to a certain proclivity on the part of an ancient Chinese king, Zarathustra founded astrology. Each of these—and others—is a singular act that generically founds similar, subsequent acts, by a process in which the rhythm of time is internalized by each of these series, so that time is achronic, its rhythm being that of the sheer relation of precedent and consequent. The time of each series of events, like the time of myth, has neither continuous duration nor chronometric regularity but results in configurations of identical content, divided by boundaries akin in their mode of division to musical bars.[52] Succession in this sense yields identity; time only enriches this identity between precedent and consequent, between type and figure. The signaling of certain dynastic beginnings (among the Sasanians and others, of individual reigns) by the use of new calendars performs the complementary and opposite act of identifying difference, the beginning of a new act of inauguration.

Identity in succession is particularly enriched, and indeed is far more consequential in its effects outside this particular successive series, when the archetypal inauguration is of sacred and world-historical significance. These are inaugural events whose repeatability is imperative, and dictate the rhythms of salvation history and of cultic memory.

Inauguration and reenactment, the differentia of monotheism,[53] are structurally connected by a relation of repetition and, in terms of time, of recursivity, of closure, and ultimately of definitive closure in the great cycle of historical cycles that terminates with the apocalypse. Of these great cycles, according to Arabic traditions, the greatest and most durable is of course the history of prophecy, culminating in Muhammad, and definitively reenacted in apocalyptic time, when the inessential, the vicarious, and the merely temporal are ejected from the scheme of things. Indeed, as in Christian historiography, the Fall justifies the existence of historical time,[54] the passage of which ends in redemption.

Thus according to medieval Arab-Muslim historical writing, the first object created by God was the pen, which was commanded to write "everything" and thus described the entire future course of creation as a register of archetypes and reenactments. Adam was known by names that betoken his archetypal character. He was called Abû Turâb, the Father of Soil, out of which humanity is fashioned and to which it returns; he was also known as Abû Al-Bashar, Father of Humankind, and Abû Muhammad, the primeval Muhammadan appearance. God taught Adam the entire human lexicon from its most sublime to its most vulgar words, with its verbs and nouns, its superlatives and diminutives. Thus the linguistic order assuring human life was established, as were the skills of agriculture and metalwork, for perpetuity and in definitive generic form. In the same way, the punishments meted out to Eve (the pains of conception and menstruation, a lengthy pregnancy, inferiority to men, and so forth) marked the lot of womankind for all perpetuity, with a misogynist impulse that, however, almost pales into insignificance compared to medieval Christian writing on women. Not dissimilarly, it might be said that the far more benign creation of womanhood in Greek mythology—the fashioning of Pandora in the forge of Hephaestus, her endowment with life by Zeus, falsehood by Hermes, curiosity by Hera, along with beauty, grace, persuasiveness, skill, and other qualities by other gods—is also that of an invariant archetype, as were the contents of Pandora's box, and various other skills previously imparted to mankind by Prometheus.

Moreover, the sole creed available to Adam was *tawhîd,* an archetypal monotheism, and to him were revealed the fundamental interdictions on eating pork, blood, and carrion. Abraham was the primeval and timeless Muslim par excellence and received what are variously thought to be unnamed fundamental generic institutes of the faith or certain exemplary attributes of the faithful, particularly those concerning purity, such as circumcision and the removal of body hair. He also received and instituted cultic rituals, most specifically the pilgrimage rites at Mecca.

In these different articulations of the primeval religion—Islam—at different times and in different settings, the posterior event recapitulates the anterior. When modifications occur, they do not, in Muslim writing, imply invalidation of that which was subject to modification. Rather, they are characterized as consummation in generic continuity, an *Aufhebung.* The specific instances of ritual, as of prophecy, are calques of an invariant beginning completely in keeping with the structures of temporality addressed above. Such calques find completion and closure when profanity is wiped off the face of the earth in the history of the future: when this occurs a series of cataclysms will herald the coming of the Mahdî (the Messiah) and the restoration of the Adamic order in preparation for the day of judgement, a day that consigns profanity to the absurdity of Hell and recommences the Edenic order in Paradise. And indeed, this is truly a history of the future because in conception of fact and narrative it fully conforms to the conventions of historical writing in medieval Arabic letters.

Following the antinomian signs of the Hour and its subversions of Order—the rise of the sun from the west, the unleashing of the destructive force of the Gog and Magog upon the world, the reign of the *Dajjâl,* the Antichrist, and other events that occur in the history of the future—the decks are cleared for the recommencement of the Adamic order, much as they were with the Deluge. The Messiah is called "Muhammad Ibn 'Abd Allâh": he is the Prophet's namesake and clansman; in other religious traditions he might well have been considered his avatar. The armies ranged against the Dajjâl are commanded by Jesus, son of Mary, who consummates his primeval reality by overcoming the pre-Muhammadan historical specificity that rendered him inconsummate and shedding the vestiges of erstwhile imperfections following his initial appearances: he breaks the Cross and kills all pigs; he abrogates the toleration canonically extended to non-Muslim Peoples of the Book—Christians, Jews, Sabeans, and Zoroastrians—and accepts no further conversions to Islam, but rather kills all the unconverted. The cosmic counterpart of this recapitulation of the purity of the Adamic order is a similar recapitulation of the precreation order in preparation for the recommencement of all origins: at the coming of the Hour, God commands the angels to die, before they and the rest of creation are resurrected; until then, nothing remains but the Divine Face (Koran, 28:88 and 55:26–27).

Each instance of prophecy, therefore, is a realization of eschatology, a regeneration of the time of divinity, much as time was regenerated annually in the ancient Near East by kings with the Adonisian myth, or with the resurrection of Jesus. The detailed rhythms of these recapitulations in their turn recapitulate numerical and chronological accents of the archetype. The number of men who fought alongside Muhammad at the battle of Badr against the Meccans in 624 CE was the same as that of the Israelites who fought Goliath. The day of 'Ashûra, the tenth day of the Muslim month of Muharram on which the Passion of Husain is commemorated, is the equivalent of the Jewish day of atonement, and is believed by Shî'îte Muslims to be the day on which God forgave Adam and the day Noah's ark landed. In the month of Ramadân the Torah was revealed, as well as the initial verses of the Koran. And much as Christians invested a particular accent in Sunday, it being the Lord's day following the cosmic week and also the day of resurrection, for Muslims Friday is the day on which Adam was created and died and the day of the resurrection. Friday was so acutely accented that it is said of some persons of exceptional piety that they had a particular preference for sexual intercourse on Fridays. Al-Mas'ûdî in the tenth century quoted contemporary Arab Christians as believing the day on which Christ was crucified corresponds to the day Adam was ejected from heaven and the day he died.

Typological interpretation[55] serves as the hermeneutical complement and the mode of apprehension of this serial recursivity, according to which the history of the sacred is conceived as theodicy. The salience of typology is not sufficiently appreciated by historians, who may think it the preserve of theologians and students of literary genres: yet it is crucial for understanding the very notions of continuity and of influence in historical discourse and in discourses on the past in general. Just as the historical appearance of Jesus in early, and to some extent, in Patristic Christianity was a proleptic eschatology, so are all historical appearances of divine significance prefigurations of the end or recapitulations of the beginning. Moses could be and was read as *figura Christi,* Noah's ark as *praefiguratio ecclesiae,*[56] indeed medieval European kingship was read as Christomimesis, as the crowned *typus Christi,* just as medieval Muslim caliphs and sultans were figures of prophecy.

Type and figure are related in a time that stands outside chronometric time. That this time is also measurable is irrelevant, for along the chronometer the flow of typology—the succession of prophets, for instance—appears as accents of intensity that manifest an erratic immanence. Just as, for example, the Battle of Badr is removed from its proper register of local tribal raids and transposed to the perspective of epiphany when textually read as *Jihâd,* so are other elements in typological succession removed from history and transposed into the "perspective of eternity."[57] Contrary to the unqualified instants of chronometric time, qualitative continuity is maintained across spaces of this time by the succession rhythm internal to the purposes of theodicy. The time of Providence has its measure within itself. The time of the change is evanescent, unqualified, bereft of substance, ontologically dubious.

Not so the time of typology, whose instants are dense with essence, the essence transmitted by typological filiation, the register of genealogy. This time denies any specific gravity to the time of the clock, indeed to the time of history, for the latter lacks the generic determination and the generic closure, and is in itself fundamentally meaningless, the companion of events in themselves ungeneric with respect to the typological register, and therefore of an ontological status of irredeemable baseness.

This leads us back directly to the essentialist view of history to which reference has been made, in three conceptually congruent moments: a neo-Romantic, culturalist postmodernism; classical Romanticism; and nineteenth-century Islamist revivalism. This last brings us to mention the late-twentieth-century trope of Islamist revivalism, fully congruent with that of the nineteenth century, albeit with a very pronounced primitivist and nativist air.

If revivalism, religious or nationalist, and fundamentalism, Christian, Muslim, or Jewish, propound the view that a past can be resuscitated, this is possible only on the grounds of a conception of history that conforms to the essentialism that has been discussed throughout the present study. This is a notion of history that brings together an organismic view of history, one that came into circulation in European Romanticism and royalism in the nineteenth century, and a typological concept of time that is naturalistically grounded in the organismic conception of history. Both are, in their conceptual prototypes, ancient. But the translation of organism to the study of society, and its combination with typology, is a specifically modern achievement. In the case of contemporary Muslim fundamentalism, for instance, it can be shown that the integralism animating it on the basis of the presumption that a golden age can be integrally reinstated, is a phenomenon of the twentieth century, with no antecedents in actual Muslim traditions,[58] which relegated redemption to a moment in the eschatological future defining the end of time.

This achievement has been of great moment, constituting the bedrock of numerous ideologies: nationalist, paranationalist, and populist recividication; conservative reaction to the French Revolution; the revolutions of the 1830s and of 1848; and the Russian Revolution. That a similar Romantic quest for organism should again become popular in the context of a fin-de-siècle malaise after the end of the Cold War, and that this should manifest itself in the attitude to history, is itself amenable to study and understanding by means of historical reason. The quest for organism, in its culturalist form, is not the key to the understanding of itself, nor of other cultures.

Notes

1. See especially Woodruff D. Smith, *Politics and the Sciences of Culture in Germany, 1840–1920* (New York 1991).

2. Wilhelm Heinrich Riehl, *Die Naturgeschichte des Volkes als Grundlage einer deutschen Social-Politik,* 4 vols. (Stuttgart 1862–1869).

3. B. Giesen and K. Junge, "Der Mythos des Universalismus," in H. Berding, ed., *Mythos und Nation* (Studien zur Entwicklung des Kollektiven Bewusstseins in der Neuzeit, 3) (Frankfurt/M. 1996), 34–64.

4. Marshall Sahlins, *How 'Natives' Think, about Captain Cook, For Example* (Chicago 1995), 13–14, 180.

5. Ira Marvin Lapidus, *A History of Islamic Societies* (Cambridge 1988), xxiii. See in general, Aziz Al-Azmeh, *Reconstituting Islam* (Washington, DC., 1996).

6. Marshall G. S. Hodgson, *Rethinking World History* (Cambridge 1993), esp. chap. 8, and see B. S. Turner, *Weber and Islam* (London 1974).

7. Ernest Gellner, *Reason and Culture* (Oxford 1988), 14–15. It is characteristic of this author—and many others—to abandon this cautionary attitude, and to claim that "Muslim societies" are exceptional.

8. Cf. A. Al-'Arwî, *Mafhûm at-târîkh* (The Concept of History), vol. 2 (Casablanca 1992), 361, 367, 403.

9. Edmund Burke, *Reflections on the Revolution in France,* ed. C. Cruise O'Brien (Harmondsworth 1969), 120, and cf. David Simpson, *Romanticism, Nationalism, and the Revolt against Theory* (Chicago and London 1993), 57–58.

10. Johann Gottfried von Herder, *Reflections on the Philosophy of the History of Mankind,* abridged by F. E. Manuel (Chicago and London 1968), 96–97, and F. M. Barnard, ed. and transl., *J. G. Herder on Social and Political Culture* (Cambridge 1969), 272ff., 291ff. Cf. R. G. Collingwood, *The Idea of History* (Oxford 1946), 92.

11. Collingwood, *The Idea of History,* 182.

12. Johannes Fabian, *Time and the Other: How Anthropology Makes Its Object* (New York 1983), particularly 15–16, 19.

13. Marc Bloch long ago warned against this temptation in his review of Maurice Halbwach's *Les cadres sociaux de la mémoire* (Paris 1925): "Mémoire collective, tradition, et coutume: A propos d'un livre récent," *Revue de Synthèse Historique* 40 (N.S. 14) (1925): 73–83.

14. See the sober reflections of Paul Mercier, "Anthropologie sociale et culturelle," in *Ethnologie générale,* ed. Jean Poirier (Paris 1968) (*Encyclopédie de la Pléiade,* 24), 907–909, 915, 918–920.

15. See most particularly the sinuous and rigorous considerations of Pascal Boyer, *Tradition as Truth and Communication: A Cognitive Description of Traditional Discourse* (Cambridge 1990), 2–4, 10, 32–37, 79–86, 118, and cf. Marc Augé, *Le sens des autres* (Paris 1994), 28–29 and passim.

16. See in general the excellent account of Stepan Odouev, *Par les sentiers de Zarathoustra,* transl. C. Emery (Moscow 1980).

17. Juri M. Lotman, "On the Metalanguage of a Typological Description of Culture," *Semiotica* 14, no. 2 (1975): 97ff.

18. Abdallah Laroui, "The Arabs and Cultural Anthropology," in Laroui, *The Crisis of the Arab Intellectual* (Berkeley and Los Angeles 1976), 44ff.; Aziz Al-Azmeh, "Islamic Studies and the European Imagination," in Al-Azmeh, *Islams and Modernities* (London 1993), chap. 7; Al-Azmeh, "Ifsâh al-Istishrâq" in Al-Azmeh, *Al-Turâth bayn as-Sultân wa't-târîkh* (Beirut and Casablanca 1990), 61ff. For similar antitheses in other locations, see François Hartog, *Le miroir d'Hèrodote* (Paris 1980) and Aziz Al-Azmeh, "Barbarians in Arab Eyes," *Past and Present* 134 (1992): 3–18.

19. Suffice it here to refer to the discussion of David Carr, *Time, Narrative, and History* (Bloomington 1986), "Introduction."

20. For the structure of the event, see W. E. Bull, *Time, Tense, and the Verb* (Berkeley and Los Angeles 1968), 17.

21. Algirdas Julien Greimas, *Sémiotique et Sciences Sociales* (Paris 1976), 168.

22. Gaston Bachelard, *Dialectique de la durée* (Paris 1950), ix, 52–53.

23. For this distinction in history, see most saliently Krzyzstof Pomian, *L'ordre du temps* (Paris 1984), 211ff., and chap. 4, passim.

24. See the excellent account of Hegel and Marx's attitudes to romanticism, in Gopal Balakrishnan, "The National Imagination," *New Left Review* 211 (1995): 59–61.

25. Jörn Rüsen, *Zeit und Sinn. Strategien historischen Denkens* (Frankfurt/ M. 1990), 240–247.

26. See particularly Jacques Rancière, *The Names of History: On the Poetics of Knowledge*, transl. H. Melehy (Minneapolis 1994).

27. Fernand Braudel, *L'Identité de la France*, 2 vols. (Paris 1986).

28. Cf. the studies in Jacques Revel, ed., *Jeux d'échelle: La micro-analyse á l'experience* (Paris 1996).

29. Jan Assmann, *Das kulturelle Gedächtnis: Schrift, Erinnerung, und politische Identität in frühen Hochkulturen* (Munich 1992), 21.

30. Ibid., 89 and passim.

31. Ibid., 52.

32. In what follows, I have relied on points developed in more detail and with full documentation in the following works: Aziz Al-Azmeh, "The Discourse of Cultural Authenticity," in Al-Azmeh, *Islams and Modernities*, 39–59; idem, *Al-Kilâba al-târîkhîya wa'l-ma'rifa al-târîkhîya* (Historical Writing and Historical Knowledge), 2nd ed. (Beirut 1995); idem, "Chronophagous Discourse," in D. Tracy and B. Reynolds, eds., *Religion and Practical Reason* (Albany 1994), 163–211; "Histoire et narrative dans l'historiographie arabe," *Annales, Economies, Sociétés, Civilisations* 41 (1986): 411–431.

33. The following paragraphs are based on the author's *Islams and Modernities*, 42ff.

34. *An Islamic Response to Imperialism: Political and Religious Writings of Sayyid Jamal al-Din "Al-Afghânî,"* transl. Nikki Keddie (Berkeley and Los Angeles 1968). See also Homa Pakdaman, *Djamal-Ed-Din Assad Abadi dit Afghânî* (Paris 1969). On the great Indian Muslim reformist, see Christian W. Troll, *Sayyid Ahmad Khan: A Reinterpretation of Muslim Theology* (New Delhi 1978).

35. See especially Malcolm Kerr, *Islamic Reform: The Political and Legal Theories of Muhammad Ábduh and Rashid Rida* (Berkeley and Los Angeles 1966).

36. See Taha Parla, *The Social and Political Thought of Ziya Gökalp* (Leiden 1986); and William Cleveland, *The Making of an Arab Nationalist: Ottomanism and Arabism in the Thought of Sati al-Husri* (Princeton 1971).

37. For instance, Ludwig Gumplowicz, *Geschichte der Staatstheorien* (Innsbruck 1905), §59 and idem, "Un sociologiste arabe du XIVe siècle," in idem, *Aperçus Sociologiques* (New York 1963 [1898]), 201–226; Franz Oppenheimer, *System der Soziologie*, 2nd ed. (Stuttgart 1964) vol. 2, 173f. See A. Al-Azmeh, *Ibn Khaldûn in Modern Scholarship* (London 1981), 157ff.

38. J. Al-Afghânî, *Al-A'mâl al-kâmila* (Complete works), ed. M. 'Umâra (Cairo n.d.), 434.

39. Ibid., 153. One might instructively refer to Hegel's pronouncements about the exit from history of Muslim nations: Georg Wilhelm Friedrich Hegel, *Vorlesungen über die Philosophie der Geschichte.* Sämtliche Werke, ed. Eduard Gans and Karl Hegel, vol. 11 (Stuttgart 1971), 459.

40. Hasan Hanafî, *Dirâsât falsafiyya* (Philosophical Studies) (Cairo 1988), 52–57.

41. Waddâh Sharâra, *Hawl ba'd mushkilât ad-dawla fi'l-fikr wa'l-mujtama' al-'arabiyyayn* (On some problems concerning the state in modern Arab thought and society) (Beirut 1981), 71.

42. Friedrich Wilhelm Schelling, *Philosophie der Mythologie* (Berlin 1856), 182.

43. Abdallah Laroui, *L'Idéologie arabe contemporaine* (Paris 1967), 66.

44. *Al-A'mâl al-majhûla* (Unknown works), ed. 'A. Shalash (London 1987), 81.

45. On this point, see Aziz Al-Azmeh, *Ibn Khaldun: An Essay in Reinterpretation* (London 1982), passim.

46. An idea of the scope of medieval Arabic historical writing can be gained from Tarif Khalidi, *Arabic Historical Thought in the Classical Period* (Cambridge 1995).

47. See, for instance, Arthur Lovejoy and George Boas, eds., *Primitivism and Related Ideas in Antiquity* (Baltimore 1935).

48. Cf. Marshall Sahlins, *Culture and Practical Reason* (Chicago 1976), 52–53.

49. Joseph Needham, "Time and Eastern Man," in idem, *The Grand Titration* (London 1969), 267–268.

50. Eric Havelock, *Preface to Plato* (Oxford 1963).

51. Claude Brémond, Jacques Le Goff, and Jean-Claude Schmitt, *L' Exemplum* (Turnhout 1982).

52. Ernst Cassirer, *The Philosophy of Symbolic Forms,* transl. R. Manheim, vol. 2 (New Haven and London 1955), 108.

53. John Wansbrough, *The Sectarian Mileu: Content and Composition of Islamic Salvation History* (Oxford 1978), ix.

54. Karl Löwith, *Meaning in History* (Chicago 1955), 181.

55. For a particularly salient account, see Tzvetan Todorov, *Symbolisme et interpretation* (Paris 1978), 100ff.

56. For this matter, see most particularly Erich Auerbach, "Figura," in idem, *Scenes from the Drama of European Literature* (Manchester 1984).

57. Ibid., 42.

58. Aziz Al-Azmeh, "Utopia and Islamic Political Thought," in idem, *Islams and Modernities,* chap. 6.

CHAPTER 9

Competing Visions of History in Internal Islamic Discourse and Islamic-Western Dialogue

ABDULLAHI A. AN-NA'IM

This paper explores the prospects of a proactive approach to historical thinking in relation to the paradox of human difference and interdependence in a global context. The dual premise of my analysis is the reality and permanence of cultural (including religious) diversity of human societies, on the one hand, and the imperatives of peaceful and cooperative co-existence in an increasingly globalized environment, on the other. Competing visions of history, I suggest, have always been integral to conceptions of self-identity and relationship to the "other," in individual and communal interactions. But the history of any society would have been mixed, containing peaceful and cooperative as well as confrontational and hegemonic types of elements in human relations.

Different visions of history may emphasize one or another element of the ethical norms, social institutions, economic relations, or political organization and processes of a community, or present one view or another of its relations, with other communities. For example, different visions of history may present a positive or negative view of women and their status and role in society, or of relations with neighboring or distant communities, may emphasize a tradition of tolerance or intolerance of diversity of religious or political opinion and practice within society, and so forth. Whether consciously or subconsciously, such divergent visions of history influence, and have been manipulated in political discourse to influence, individual and communal behavior. The fundamental question I raise in this regard is whether it is possible to deliberately differentiate between various visions of history with a view to enhancing and promoting certain policy objectives as suggested in this essay.

Notes for this section begin on page 150.

I am also concerned with the role of historical thinking in cultural self-determination. Given the influence, and manipulation, of perceptions of history in the politics of communal self-identity, and intercommunal relations, how can communities articulate and realize the most relevant and constructive perceptions of self-identity in relation to other communities? To speak of cultural self-determination, it might be said, emphasizes difference and specificity, rather than similarity and universality, in human cultures. While appreciating the negative potential of the tension between the two, I do not believe that they are necessarily mutually exclusive. On the contrary, I suggest that similarity and universality should be premised on the realities of cultural difference, instead of pretending that such differences do not exist. As I see it, the question is not whether universality is possible, but rather on whose terms and how should it be sought.

I will begin with a discussion of some of the general conceptual and methodological questions raised by the proposed proactive and constructive approach to history, followed by a discussion of the interaction between competing visions of history of time, place, and community in internal Islamic discourse, and in Islamic-Western dialogue. With respect to inter-Islamic discourse, I am concerned with the role of competing visions of history in defining (and manipulating) Islamic identity in the dynamic of the relationship between a historically presumed center and peripheries of the world community of Muslims (Umma). I will illustrate my argument with reference to Sub-Saharan Africa in relation to the presumed "centrality" of the Middle East in Islamic discourse, but a similar analysis can be applied, I suggest, to Southern, Southeastern, and now Central Asia. A different approach, however, may be necessary in relation to Muslims in the diaspora of Europe, North America, and elsewhere. I am also concerned with the ways in which that internal Islamic dynamic shapes and informs Islamic-Western dialogue, while the latter also shapes and informs the former. In this respect, I seek to clarify the present circumstances of encounter in relation to the possibilities of peaceful co-existence and cooperation based on mutual respect for cultural identity and communal self-determination.

Cultural Difference, Internal and Cross-cultural Communication, and Historical Thinking

The basic question I wish to raise in this section is whether it is possible to utilize the possibilities of historical thought in promoting and enhancing mutual internal and cross-cultural understanding and recognition while respecting particularities of different cultural identities. Given the reality and permanence of cultural diversity, how can different human societies cooperate in pursuit of peace and mutual economic advantage, and so forth, without encroaching on each other's right to cultural and political self-determination? How can possibilities of mutual respect and cooperating be explored through an understanding of a community's historical thinking of its self-identity and relations with other communities?

This formulation of the issue assumes a dialectical relationship between historical consciousness and thought, on the one hand, and cultural identity, on the other. Perceptions of historical consciousness and thought are the basis of feelings of belonging to one self-identifying community in contrast to another, while cultural identity, in turn, determines perceptions of historical consciousness and thought about self and the other. People's perception of their history, and the way they think about it at any given point in time, are both integral to, and influenced by, their sense of collective self-identity and relations with other communities. However, perceptions of historical consciousness and thought themselves change in response to various factors and processes, including shifts in self-identity. Shifts in self-identity, in turn, contribute to transformations in perceptions of historical consciousness. Moreover, dominant perceptions of history and identity are constantly open to challenge and change at both the internal and external levels. The question is whether it is possible to understand and influence those shifts and transformations, and to what end?

To answer this question, a set of theoretical and practical issues need to be considered in relation to the dilemma raised by the ever growing and dense network of cross-cultural interactions and communications, on the one hand, and the body of scholarly disciplines and normative systems claiming universal validity, on the other. At the theoretical level, advocates of cultural autonomy argue that the integrity and particularity of cultures are in danger of being lost in a universalism of hegemonic cultures and their so-called rational methodologies, whereas universalists assert that an unreserved acknowledgment of differences of cultural identities threatens the possibility of cross-cultural understanding and moral judgments on the basis of universally accepted normative insights. An underlying practical issue, it may be added, relates to the nature and dynamics of power relations within and between cultures because this dilemma is often resolved through material and technical superiority of the proponents of one view or the other. If and to the extent that a people is capable and willing to defend the integrity and autonomy of its culture, it needs not worry about the universalist claims of others. Conversely, universalists can simply disregard the protests of others if they can impose their will on them.

In order to evaluate the reality, nature, and implications of the above-mentioned theoretical dilemma, it is necessary to question and seek to clarify both of its horns. Since there is no such a thing as abstract or neutral universalism of rational methodologies because both "rationality" itself and the methodologies it produces are culturally specific, what is at issue is the hegemonic "universalization" of a particular model of rationality and its own methodologies—whether within the same culture or of one culture over others—rather than an inherent contradiction between universality and acknowledgment of cultural diversity. In other words, the issue appears to be primarily the attitude and orientation of the proponents of competing cultural perspectives, as they seek to define and manipulate the terms and circumstances of internal and cross-cultural interaction. If apparent compliance with normative insights is achieved through coercive imposition, it is unlikely to last, and the human and material cost of maintaining it may be unacceptable. In contrast, the acceptance of insights through internal and cross-cultural understanding and accommodation is not only more likely to last, but would also be based on more humane and morally defensible ground. Do existing power relations and terms and circumstances of internal and cross-cultural interaction permit, or can they be modified to allow, the emergence of genuine collaborative universality of normative insights?

For universalists to speak of "unreserved" acknowledgment of cultural difference implies the imposition of limits in order to create and implement universality of normative insights. This proposition raises the question: who will determine the type and degree of "reservation," and how will it be done? If one perspective within a given culture, or one culture in relation to others, assumes the mantle of an arbiter of degree or type of limitation on acknowledgment of difference between cultural identities, then the understanding and insights thereby proclaimed will be, by definition, culturally specific and not universal. Internal and cross-cultural consensus on the universal validity of normative insights is possible, I believe, but only through an open-ended and mutually respectful process of negotiation, discourse, and dialogue within and between all perspectives.

In my view, cultural particularities, in and by themselves, do not preclude internal and cross-cultural understanding and consensus on mutually acceptable normative insights. Rather, serious barriers to understanding and normative consensus can arise from the tendency or capacity of some elites, at certain stages of the history of their societies and cultures, to adopt a hegemonic, superior, and imperialist attitude toward their opponents within the same culture or peoples of other cultures. It is certainly possible that hegemonic tendencies may become deeply entrenched in some cultures over time, but I would suggest that, since that would be the product of human agency, it can be changed through human agency by identifying and enhancing those points of view, principles, and rules of communication within each culture that can effectively challenge and combat a hegemonic, imperialist tendency or capacity.

The fundamental justification and guiding principle of this effort should be the universal moral and political concept of reciprocity—the Golden Rule of treating others as one wishes to be treated by them—as the basis of peaceful co-existence and cooperation between the cultures of the world. This principle is not only universally accepted by all human cultures as a "moral impertive," but is also supported by pragmatic common sense and historical experience. All human communities and societies are inextricably bound by their common environment and interdependent needs for survival and security in finite space and a world of increasingly diminishing resources. Hegemonic and imperialist relations and allocation of resources—used in the past at great human and material cost until they were defeated, as happened every time—have effectively been rendered redundant by the apocalyptic force of modern technology. Even so-called conventional warfare is no longer capable of achieving and sustaining hegemonic and imperialist objectives. Without peaceful co-existence through cooperation and mutual accommodation, humankind now risks total destruction, either immediately through nuclear war or gradually through serious environmental degradation.

To meet the imperatives of peace and cooperation in sustaining their increasingly fragile environment and managing its diminishing resources, human societies must acknowledge and respect each other's claims to human dignity and material welfare, that is, to treat other societies as they wish to be treated by them. Genuine and lasting reciprocity, however, must direct every conceptual and practical aspect of discourse and dialogue. Mere tokenism and superficial reciprocity will be seen as intellectually dishonest and politically patronizing, and therefore counter-productive. The search for common ground in a spirit of mutual respect and appreciation of the integrity and dignity of each culture must be, and be seen by all sides to be, the basis of all aspects of internal and cross-cultural discourse and interaction, in a genuine and meaningful sense.

For example, the achievements of the modern humanities, as organizing principles of human relations, should be seen as the product of a long history of continuing global human development, whereby civilizations and cultures build upon and integrate each other's experiences and achievements. In this way, each culture would see the modern humanities as a jointly constructed conceptual and methodological framework of mutual exchange and understanding, rather than the exclusive domain of European or some other culture. However, conceptions of the humanities for this purpose should avoid simplistic or sentimental proclamations that ignore or gloss over significant differences about the nature, and cultural and contextual framework, of notions such as "objectivity, rationality, critical analysis, empirical verification," and other conceptual assumptions and methodological tools of the modern humanities. In articulating and implementing these conceptions, participants in discourse within each culture, and dialogue among cultures, should also be open to understanding, incorporating, or adapting—as well as respectfully questioning and challenging when appropriate—epistemological and methodological systems of other cultures.

Similarly, physical sciences and technology must also be developed with a view to managing and protecting the global as well as local environments while maximizing the material well-being of all human societies. It should be emphasized in this regard that science and technology are always premised on a specific world-view along with the moral norms and economic, social, and political systems it generates and legitimizes. Furthermore, I would add, the underlying rationale, as well as the actual principles of science and their application in technology, are the product of human agency and choice. Human beings are responsible for the social, economic, political, and moral objectives they seek through the science they produce and technology they use. What world-view, moral norms, and systems should inform and guide science and technology today?

To summarize, the approach of this paper to the concept of competing visions of history in cultural communication and understanding is premised on the following seven propositions:

1. While significant cultural difference is a permanent and fundamental feature of human existence, internal and cross-cultural understanding is imperative for peaceful co-existence and cooperation. Power relations between relevant actors (within and between cultures) is very important, but to take that as definitive and exclusive of human choice and responsibility is to surrender to a destructive determinism of circumstances. Human willpower has repeatedly been cultivated throughout history to overcome and reverse negative power relations within and between cultures. Since confrontation and hostility are no longer tenable as defining principles of intercommunal and cross-cultural relations, human willpower must be utilized to achieve an equilibrium of power in the interest of peace and mutual accommodation. People can, and do, make a difference through the visions they have and positions they take regarding the options of understanding and cooperation, on the one hand, or confrontation and hostility, on the other. The question is therefore whether it is possible to achieve such a strong reorientation of human societies in relation to the use and abuse of history, and if yes, how that can be accomplished.
2. People's perception of their cultural (including ethnic and/or religious) identity is a product of continuing dialectical process of conflict and mediation between competing visions of their history. But these visions are, in turn, affected by shifts and transformations in the people's perception of identity. These shifts and transformations reflect the interaction of a wide variety of economic, political, social, psychological, and other factors within each culture, as well as the exchange of cross-cultural influences. But this process is far from deterministic, and is constantly shaped and directed by human choice and behavior.
3. Intercultural understanding and agreement on universally valid normative insights is possible, indeed essential, for peaceful co-existence and cooperation, but only when seen as a genuinely collaborative project, on the basis of reciprocity and mutual respect, and not as the hegemonic universalization of culturally specific (relativist) models, institutions, and processes.

4. Cultural difference, including differences in historical thinking, as such do not preclude intercultural understanding, but can be manipulated to that end by ambitious elites for their own political and economic advantage. Given the dynamic of competing visions of history, alternative voices and interpretations of cultural identity and historical experience do exist, and can be mobilized to counter the negative forces of confrontation and domination.
5. These alternative voices and interpretations should be identified and engaged in internal discourse within cultures, and cross-cultural dialogue, on the basis of mutual respect, and a desire to understand others on their own terms and in their own self-image, through jointly constructed conceptual frameworks and methodologies.
6. The processes of promoting internal and cross-cultural understanding and agreement on normative insights should include consensus on normative premises and methodologies of the humanities, social as well as physical sciences, and technology. All these and related fields of human knowledge and action should be seen as the product of, and deployed to serve, a universal community of humanity advancing peace and cooperation, and protecting its fragile environment.
7. The key to the whole process of intercultural understanding is a willingness to question one's own assumptions and motives, an ability to identify and address legitimate interests and concerns of individuals and groups—to come to the arena of understanding in good faith and candor, in search of peaceful and constructive co-existence—because the negative forces and experiences that need to be overcome are powerful and deep-rooted.

In the preceding discussion, I was referring to both the internal and cross-cultural dimensions of the process of understanding and consensus-building. In the next two sections, I will outline the role of historical thinking in an "internal" Islamic context, as well as with respect to Islamic-Western relations today.

Visions of History and Internal Islamic Discourse

As to what might be called internal Islamic discourse about cultural identity, I would argue that there has been a subordination of the history of so-called "peripheral" Islamic communities (of Sub-Saharan Africa in this paper) to the presumed "heartlands" of Islam in the Middle East. This subordination, it seems, was self-imposed by those communities themselves as well as externally created and encouraged in the dominant discourse of the Middle East. I am not suggesting that an Islamic identity was always the only one claimed by Muslim communities of either center or periphery, or that an Islamic discourse was the only one they practiced. Rather, my point is that, to the extent that some elites from those regions identified themselves and their communities as Islamic, and practiced an Islamic discourse to legitimize and internalize those perceptions, the relationship tended to be one of hegemony by the center over subordinated, unequal peripheries. I would also suggest that the general population of peripheral communities accepted subordination "voluntarily" whenever presented with it as a religious imperative, though they may not necessarily have perceived or articulated their own identities as exclusively Islamic at any given point in time. Visions of their own local and regional history were suppressed in favor of those from the more "significant" Islamic history of the Middle East.

From that perspective, "significant history" of time and place was represented by Islamic elites to have been that of the Middle East, especially of the seventh to the tenth centuries, as transmitted through Arabic oral traditions or recorded Arabic texts. With the Koran itself and records of the traditions of the Prophet and those of earliest Muslim communities rendered in classical Arabic, and the requirement that recitation of the Koran in prayer must be in the original Arabic, the cultural context of early Islamic time and place has come to acquire a sanctified religious authority. The authenticity and integrity of the religious experience of all Muslims came to be judged against the standards set by that "center" of significant time and place: the more closely a person or community is identified with the center, the "better Muslim" that person or community is deemed to be.

Thus, for example, most Muslim communities of the present-day Northern Sudan, despite their obvious African Nubian origins and complexion, claim direct decent from the tribes of Arabia, preferably that of the Prophet himself, and generally identify culturally and politically with North Africa and the Middle East, rather than with their own geographical region and historical origins in East and West Africa.[1] In my view, these features of Northern Sudanese consciousness are at the root of the chronic state of political instability and civil war, with the consequent economic weakness and underdevelopment that have afflicted Sudan since independence in 1956.

This phenomenon is also clearly illustrated by what are known as the Jihâd movements of the eighteenth and nineteenth centuries in West Africa and Sudan, where the concept of the state and society, and their underlying ideology, sought to reproduce, a thousand years later and in great detail, the model of the early Medina city-state of the Prophet in Eastern Arabia of the seventh century, and to imitate the rhetoric and discourse of classical Islamic theology of the Middle East of the eighth and ninth centuries.[2] Although those earlier forms of historical models were subjected to severe and sustained challenged by an alternative, "modern" European colonial and postcolonial hegemonic universalism, they have persisted in one form or another to the present time. These perceptions of identity and visions of history now appear to be poised to reclaim their earlier dominance in some parts of Sub-Saharan Africa, with drastic consequences for countries like Sudan and Nigeria today, and potentially serious implications for the whole of East, West, and Central Africa. The harsh and exclusive theology and politics of Wahhabism in West Africa are not only sustained by annual pilgrimage to Mecca and Saudi Arabian financial support; they are inspired by Islamic militancy in Algeria and also draw on a history of Jihâd radicalism in the region.[3]

By far, however, the most significant, pervasive, and enduring manifestation of Middle Eastern visions of history is the impact of that time and place on the conception and formulation of traditional Islamic Šaṛī'a law.[4] Because many of the Šaṛī'a concepts and principles were clearly derived from Middle Eastern customary institutions and practices of the seventh to the ninth centuries, the underlying customary social, economic, and political norms and structures of that time and place are now believed to have permanently acquired the sanctity of Islam itself. In particular, Šaṛī'a concepts of property, commerce, family, and status of women are clearly strongly influenced by customary norms and institutions of the Middle East of that time. Efforts to drastically reform and change those aspects of Šaṛī'a, or to replace them permanently by secular law, are seen as tantamount to apostasy (heresy) deserving the death penalty. If the law of Islam is believed to be as sacred as the normative dictates of the religion itself, why should the customary institutions and practices of the earliest communities acquire equal sanctity?

The underlying question, of course, is what is Islam, and how can it be distinguished from the community of believers, if at all? Any definition of Islam, from the believers' point of view, must begin with the Koran and Sunna (oral and practical traditions of the Prophet Muhammad), but the Koran was revealed to, interpreted by, and initially implemented by the Prophet, who was born, raised, and lived all his life in a specific community. To be understood and applied by that community, the Koran had to use the language and draw upon the institutions and experiences of seventh-century Mecca, Medina, and surrounding communities in Western Arabia. For subsequent generations of Muslims, the example set by the Prophet and his immediate community came to signify the ideal model of human understanding and practice of Islam. Nevertheless, as Šaṛī'a came to be developed by Muslim scholars living and interacting with their communities of Western Arabia, Iraq, Syria, and Egypt during the eighth and ninth centuries, the early Islamic model of the city state of Medina continued to enjoy great normative influence, although it was by no means definitive of Islam. The customary institutions and practices of the different "host" communities of the region, and indeed "alien" principles of Jewish, Persian, and Roman law, have all had their impact on the formation of Šaṛī'a. In other words, Šaṛī'a as it came to be known and accepted today by more than a billion Muslims, from Southeast Asia to Sub-Saharan Africa and beyond, is the product of the human experiences and understanding of the first and immediately following Muslim communities of the Middle East.

If that is true, what about the human experiences and understanding of subsequent Muslim communities of all time throughout the world? Is Islam, and any conception of its religiously sanctioned law, to remain forever defined by the customary practices and institutions of early Muslims? Assuming that those early Muslim communities, like all human societies, could not have been exclusively defined by Islam, is it possible to distinguish now between what was essentially Islamic and what was incidental to those founding communities but merely approved by Islam? Do these questions assume a separate existence of Islam, independent from the first and immediately subsequent communities of believers? Is that possible, given the fact that the fundamental sources, Koran and Sunna, were themselves expressed through a language and within the cultural context of a specific community, and that Šarī'a was elaborated through the customary institutions and practices of that and other early Muslim communities? If it is possible to make the distinction, what makes the outcome Islam in any valid sense, as opposed to another new and different religion?

These questions indicate to me the nature and extent of historic hegemonic discourse within the Islamic context. For Muslims throughout the world today to submit to this hegemony of the history of a specific time and place is to surrender the validity and relevance of their human existence and experience, the essence of their responsibility for their own individual and communal lives. Subsequent generations of Muslim communities, especially in the peripheries of the Muslim world, submitted to the hegemony of the history of the founding communities in the belief that it was necessary to do so in order to attain Islamic authenticity and continuity. Consequently, the challenge for those who wish to redress that hegemony is not only to disentangle Islam from the history of the founding generations of Muslims, but also to persuade their own respective present communities that the outcome is as Islamic as the practices and institutions of early Muslims.

This is a paradoxical and extremely complex task for someone who believes, as I do, that Islam, or any other religion for that matter, is what the believers accept it to be within the parameters of the religion, that is, the Koran and Sunna in the case of Islam. It is paradoxical in that if Islam is what the believers take it to be, then what the founding and immediately following generations of Muslims heard and understood the Prophet himself to say, or approve, is the most authentic conception of the religion. The complexity of the task to a believer lies in the normative message of Islam being totally embodied in the language and human experience of early communities. How to rediscover and redefine Islam out of the sources of its early Middle Eastern history?

In my view, the paradox can be resolved by holding that what early Muslims believed Islam to be is exactly what it was for them, without that conception of a particular, albeit special, time and place being necessarily and conclusively binding on subsequent generations of Muslims. The Prophet to them was what they perceived him to be and understood him to say and approve, while he remains Prophet to subsequent generations in the context of their respective time and place. Once that hermeneutical premise is accepted, it should become possible to disentangle the essential from the incidental in early history of Muslim communities by the direct interpretation of the Koran and Sunna in contemporary context, rather than exclusively through Middle Eastern historical sources. Although the Koran and Sunna are written in Arabic, a fresh and open-minded reading of them by contemporary Muslims will not lead to results identical to early interpretations because of the radical shift in the material and intellectual orientation of the modern reader and his or her local and global environment. Historical Middle Eastern sources should remain relevant and useful in this process of reinterpretation, without being definitive.

As to the question of Islamic authenticity of the new interpretation, the ultimate and only real safeguard will remain, as it has always been, acceptance by a living Muslim community of today. In all probability, I concede, the present living community of Sudanese, Nigerian, Senegalese, or other Muslims would wish to continue submitting to what I call historical hegemony, rather than asserting independent visions of their own history. In fact, contemporary Muslim communities will probably refuse to see the binding force of customary institutions and practices of early Muslims as historical hegemony at all, but rather as guidance and guarantee of Islamic authenticity. But I am not proposing that attitudes and perceptions would or should immediately change. Rather, I am calling for the possibility of articulating and advocating alternative visions of history, wherein the historical thinking of present Muslim communities includes their own independent history and other visions of regional and world history, instead of being totally determined by that of early Muslims. If that possibility is genuinely and freely open, I maintain, alternative visions of local, regional, and world history will gain acceptance and influence. Such alternative visions need not, and should not in my view, totally replace early Muslim history, but can simply be allowed to compete with it and modify its worldview and impact on some principles of Šaṛī'a relating to social, economic, and political matters, and perceptions of intercommunal and international (Islamic-non-Islamic) relations.

For example, I suggest, if alternative visions of history were taken seriously by Muslims today, their view of Muslim-non-Muslim relations would shift toward greater accommodation and acceptance of the non-Muslims on their own terms, instead of remaining locked in a framework of confrontation and hegemonic initiatives. Šaṛī'a law regarding the status and rights of women can be modified towards greater equality and justice on the basis of local experience and institutions rather than remain bound by customary norms and practices of the early Islamic Middle East.[5] Šaṛī'a law of property, commercial transactions, and international trade can be freed from the restraints of ancient and alien concepts of land tenure, enabling business associations and financial arrangements to become more responsive to local resources and needs, circumstances of global trade, and modern relationships of production and distribution of wealth.

I believe that the possibility of seriously considering competing visions of history is utterly indispensable if modern Muslims are to retain their confidence in the validity of Islamic precepts and their relevance to the practical lives of present and future Muslim communities who live in a globalized and interdependent world. In other words, introducing this possibility is imperative from an Islamic point of view. If early Islamic history is as vital and valid as conservative and fundamentalist Muslims claim it to be, it will withstand the competition and maintain its hold on Muslim minds. Otherwise, its influence should be modified and supplemented to the extent living Muslim communities of the present and future find necessary and appropriate.

Visions of History and Islamic-Western Dialogue

Islamic historical thinking is not the only type of hegemony seeking to dominate Muslim communities in Sub-Saharan Africa and elsewhere. More recent European hegemonic colonial and postcolonial relationships to the regions sought to replace the previous Islamic discourse with a new epistemological and philosophical paradigm of nation-states that are incorporated into, but peripheral to, a global economic, political, security, and ultimately cultural system dominated by Western powers, including the United States in recent decades.[6] Not wishing to openly risk unnecessary Islamic opposition, European colonial powers did not openly support active Christian competition with Islam in strongly Islamic regions, preferring to undermine Islamic influence through so-called secular education, the establishment of European political and economic institutions, and the promotion of European culture and lifestyles, especially among the educated elites. In non-Islamic parts of Africa, however, Christianity was an explicit part of the colonial hegemonic package at the expense of traditional African religions and beliefs. After independence, the Western hegemonic project continued in a variety of ways, including grossly unfair trade practices, political manipulation of local elites through military and security alliances, and so-called political conditionality of aid in favor of promoting and implementing universal notions of democracy and human rights.

Western secular education was particularly effective, with colonial state support and funding, in challenging and replacing traditional Islamic education, thereby introducing a radically different conception of the physical and social sciences. Far from attempting to understand and incorporate at least some elements of traditional systems of education, and their role in social, economic, and political institutions and processes of African societies, the colonial state sought to repudiate and totally marginalize preexisting education. In time, Western education had a profound impact on defining the person, and his or her value and contribution to the community, by controlling access to civil service, professional careers, salaried employment. Western education was also the foundation and controller of access to the mass media, the other powerful medium of social and intellectual transformation. Popular perceptions of the environment, economic and social relations, and political organization were all radically transformed.

Since the colonial period, and increasingly after independence, however, some "nationalist" elites advocated a resurgence and reassertion of Islamic identity in response and challenge to European colonial and postcolonial hegemony. That call had little impact in Sub-Saharan Africa as long as the Middle Eastern "center" did not present a viable alternative, and was itself subjected to the same European cultural and ideological hegemony, of either the liberal capitalist type or the Marxist, socialist, Arab nationalist variety. But with the failure of secular nationalist projects, collapse of Soviet Marxism, and general retreat of democratic socialism in Europe, among other factors, the dynamic of the whole situation appears to be changing. The success of the Iranian revolution and the strong emergence of political Islam throughout the Middle East, providing its own pan-Islamic funding and organizational networks, have now produced a purported Islamic alternative to the Western normative paradigm and its political, economic, and legal models and modes of international relations.

In light of this analysis, it is clear that competing visions of history were, and continue to be, of paramount importance in this ideological and economic Islamic-Western competition in the region, in Europe itself and elsewhere, with each side seeking to demonize the other and deny it any level or degree of legitimacy or credibility with the target constituencies. On the one hand, Islamic elites recall positive historical images of the Golden Age of Islam, which is really the history of the Middle East as appropriated by elites on behalf of their marginalized African communities. Islamic protagonists also invoke negative images of malicious and barbaric Christian crusaders of the past, and their modern European decedents whom they accuse of seeking to totally destroy Islam and Muslims forever. Their Western counterparts, on the other hand, relying on negative "orientalist" stereotypes of Islam and Muslims, seek to negate any value or relevance of Islamic visions of history and perceptions of cultural identity. Western-educated African elites who are believed to be unable or unwilling to adopt and support Islamic cultural identity and discourse are condemned by Islamists as renegades and traitors who have been corrupted and coopted by the West to promote its secular materialistic and exploitative agenda in the region. The struggle appears to be for total and exclusive victory, without allowing any room for compromise or seriously searching for common ground.

Ironically, the situation is complicated by the very success of Western powers in coopting the elites of colonized communities into a belief in universal modernity, which accords equality and justice to all. Significant numbers of African elites from formerly colonized communities sought fulfillment in Europe and North America of the promise of "the civilizing mission of the White Man," only to face rejection and disappointment through exclusionary immigration rules, social intolerance and racial discrimination, harsh labor conditions, and so forth. Westernized local elites, both abroad and at home in Sub-Saharan Africa, feel betrayed and abandoned by their former European supporters, without being adequately prepared to compete in an Islamic or indigenous discourse at home. As the confrontation and resentment deepens, options for reconciliation appear to be diminishing, and the rhetoric of "clash of civilizations" gains credence in official governmental and intergovernmental circles, and with the public at large.[7]

More recently, popular perceptions of Western bias and failure in the Bosnian conflict appear to reinforce claims of the validity of an Islamic hegemonic discourse in "self-defense." According to radical Islamists and nationalists, if the so-called international community, being effectively the United States, Europe, and their allies, is unwilling or unable to maintain the rule of international law in Bosnia, Muslims have to assert Islamic solidarity and counter-hegemony throughout the world because that is the only way they can survive and maintain their communal and global Islamic identity. Failures of the international community in Somalia, Rwanda, Liberia, and elsewhere in Africa and beyond are also quoted by Muslim protagonists in support of the need for self-reliance and confrontation.

Recalling earlier discussion, I suggest that more positive visions of Islamic, Western and global history must be articulated and deployed in pursuit of greater toleration, mutual respect, and cooperation. Colonial history in Africa certainly had its positive results in maintaining peaceful relations between ethnic and religious communities, improving economic conditions along with infrastructures and health conditions in general; Western colonial education also brought some benefits to the region. While colonialism itself most certainly had its own selfish motives, and was guilty of much oppression and exploitation of colonized peoples, it is untrue and unfair to dismiss the whole experience as totally and exclusively bad. A more historically valid and constructive approach would seek to identify and build on the positive contributions of colonialism, rather than unrealistically trying to repudiate all its achievements.

With regard to current international events, Muslims should also endeavor to fulfill their obligations as full members of the international community, and accept their share of responsibility for the severe crises in Muslim settings like Somalia and the Gulf, instead of continuing to blame the West for all the misfortunes of Muslims and others in the world. It is ironic that advocates of cultural and political self-determination should keep looking to the colonial and imperialist West for leadership and sacrifice, and yet expect that to be provided on their own terms and in the service of their specific objectives. If they wish an "international community" to emerge as a credible, if not totally impartial, arbiter and constructive force throughout the world, non-Western peoples must take the initiative and make the contributions that make such a vision possible.

Conclusion

The focus of this paper on internal Islamic discourse and Islamic-Western dialogue, with special reference to Sub-Saharan Africa, should be taken as illustrative of similar issues and concerns that can be raised with respect to other parts of the world. Whether it is Hindu-Islamic confrontation in India, Chinese-Islamic relations in Malaysia, the Korean conflict, Chinese, Japanese, and Western competition over Southeast Asia, or another situation, the competing visions of the history paradigm can be useful in understanding and addressing questions of peaceful coexistence and cooperation, economic and legal reform, protection of the environment, and other matters of general concern. In each case, the basic issue is to define the terms and circumstances of historical thinking about self-identity and perceptions of the significant other. Useful questions to raise in this regard include: Which visions of history about communal self-identity in relation to other cultural, religious, ethnic, or political groups are recalled and deployed in daily politics, the media, the educational system, and communal, intercommunal, and international relations? Do exclusivist and hegemonic images of history prevail, and if so, is it possible to present and promote alternative, more accommodating and mutually respectful visions of the history of the community in relation to the other? How are significant events, like the conflict in Bosnia, or the treatment of relevant ethnic or religious groups in other parts of the world, seen and interpreted in local discourse?

I would also emphasize the economic and security context of hegemonic discourse throughout the world. Militant and confrontational visions of history often exaggerate and thrive upon memories of perceived or real events and current realities of economic exploitation, military conquest, and political domination. These images and perceptions must be taken very seriously by all sides in internal discourse and cross-cultural dialogue in the interest of building mutual trust and confidence, and not dismissed as groundless or exaggerated. Due regard should be given to the collective psychology of denial and rationalization on both sides of each issue. While self-criticism, and even conceding more than what one may believe to be justified by the "facts" of history, is more likely to assist the other side in reciprocating, denial and rationalization will probably produce a similar counterreaction by the other.

In conclusion, I suggest that even as it is possible, indeed imperative, to reverse the dangerous prevalence of hegemonic and confrontational visions of history, that can only be achieved through a clear and deliberate strategy of promoting alternative visions of history, and perceptions of identity, that are more conducive to mutual understanding and respect as means for peaceful coexistence and cooperation. This should include, I maintain, redressing present local, regional, and global gross differentials in power relations, at least at the functional level, as well as struggling for the protection and promotion of human rights to create the political and social space for discourse and dialogue. Such strategies must also identify and challenge the forces and trends of hegemony or isolationism within all relevant cultures, whether Islamic, Western, Chinese, Malay, or any other. There is always the potential for good faith and peacemaking in every human society, but it is unlikely to materialize on its own, least of all in the face of strong perceptions of ambitions of hegemony. The ultimate and most practical guide for internal discourse and cross-cultural dialogue to promote peace and cooperation should be the Golden Rule, universal to all human cultures and religions and fully supported by common sense and pragmatic experience.

Notes

1. See, for example, Francis M. Deng, *War of Visions: Conflict of Identities in the Sudan* (Washington, D.C., 1995), chap. 2.

2. On the jihad movements of West Africa see Ibrahim Ado-Khrawa, *The Jihad in Kano* (Kano 1989); Robert Sydney Smith, *Warfare and Diplomacy in Pre-colonial West Africa* (London 1989), chap. 3; J. B. Webster, *The Revolutionary Years: West Africa Since 1800* (London 1980), chaps. 1 to 3; Elizabeth Isichei, *History of West Africa Since 1800* (New York 1977), chap. 2. On Islam in West Africa in general, see, H. Hiskett, *The Development of Islam in West Africa* (New York 1984); Peter B. Clarke, *West Africa and Islam* (London 1982).

3. The strong connection between the Wahhabi revival in late eighteenth-century Arabia and the Jihâd movements of West Africa was suggested, for example, by Murray Last, *The Sokoto Caliphate* (New York 1967), 56. See also Lansine Kaba, *The Wahhabiyya: Islamic Reform and Politics in French West Africa* (Evanston, Il., 1974), chap. 1.

4. See generally, Joseph Schacht, *The Origins of Muhammadan Jurisprudence* (Oxford 1950). Cf. Muhammad Mustafa Azami, *On Schachet's Origins of Muhammadan Jurisprudence* (New York 1985).

5. For an Islamic reform methodology that would achieve this objective see, generally, Abdullahi Ahmed An-Na'im, *Toward an Islamic Reformation: Civil Liberties, Human Rights and International Law* (Syracuse, NY, 1990).

6. See generally, Gayatri Chakravorty Spivak, *The Post-Colonial Critic: Interviews, Strategies, Dialogues* (New York 1990); and ibid., *In Other Worlds: Essays in Cultural Politics* (New York 1988).

7. The heated controversy around the recent article by Samuel P. Huntington, "The Clash of Civilizations?" *Foreign Affairs* 72, no. 3 (1993): 22–49, clearly illustrates the reality and serious consequences of these issues.

CHAPTER 10

Cultural Plurality Contending Memories and Concerns of Comparative History: Historiography and Pedagogy in Contemporary India

B. D. Chattopadhyaya

Setting the Agenda

Historical thinking has always been at the crossroads; it is never homogenous or unilineal. Thus, the dynamics and the heterogeneity inherent in the writing of history merit emphasis at the outset of this chapter, if only because it is sometimes assumed that history-writing has taken a common approach so far. The initial points I make are, first, that the domain of history has not really ever been rigidly defined. It is not only since Marc Bloch and Lucien Febvre effectively broke down some of its time-honored but artificial barriers that history has become open-ended; the two patriarchs of what is generally known as the mainstream Western history-writing tradition, namely Herodotus and Thucydides, represent heterogeneity within what is perceived as a broadly homogenous cultural tradition.[1] And second, we have to ask what qualities as the features of history are apply. The conviction with which historians set about their task—their sense of certainty as to what constitutes history, which defines sources ranging from archeological to archival—has been subjected to legitimate doubts, although it ought to be kept in mind that the craft of an innovative historian has always demanded deviation from the comforts of an established norm and format. The scenario today has been aptly summarized: "one thing is clear: the paradigms that have governed historical and literary study since the nineteenth century no longer hold unquestioned survey. The

Notes for this section begin on page 166.

confident, humanist belief that a rational, 'objective' investigation of the past permits us to recover 'authentic' meanings in historical texts has come under severe attack.... At stake in this debate are a number of concepts traditionally deployed by historians in their attempts to understand the past: causality, change, authorial intent, stability of meaning, human agency and social determination."[2] It is not that methods of source criticism did not exist earlier, but that the historians' sources of information, for example texts, are essentially suspect as sources for the representation of reality. This suspicion spurs persistent pleas for a shift from universality to fragments or, as Peter Burke has put it, from "the ideal of the Voice of History to that of heteroglossia."[3] It has been stressed that even if the result is a "discourse that is itself fragmentary, it is redundant to make apologies for it."[4]

The points following from the above relate to this volume's theme of time and history, and to this chapter in particular, in the following way. The expansion of the frontiers of history-writing—which is also linked to the question of which dimensions of the past are perceived as relevant for communication—has also long been seen as reflecting human history in general. This concern, I may point out, does not necessarily relate to a search for universal laws or strait-jacket patterns; rather, it can be seen as embedded in attempts at conceptualizating what is perceived as calling across histories of different segments of humanity. In comparatively recent times, the concepts of "comparative history"[5] and "social evolution"[6] can be cited as two examples of such attempts to take history beyond culture. However, when one comes across the emphasis on "fragments" as legitimate contestants in the arena of history, or across "cultural difference" in the context of historical thinking, one is propelled into rethinking comparative concerns and asking instead: Is history culture-specific, and if it is, how is culture-specific history likely to be constructed and transmitted, particularly for the purpose of pedagogy?

Cultural Spaces and Cultural Boundaries

If history is culture-specific, then it can be assumed that one must begin by looking for cultural boundaries to which historical thinking will relate and patterns of history will correspond. One is of course aware of the notions of cultural bipolarities, which are as much part of historical thinking as are stereotyped heuristic designs for construction and communication of history. In one such notion, that of "orientalism," the Orient is the subject of representation. As James Clifford puts it, "the Orient is textualized; its multiple, divergent stories and existential predicaments are coherently woven as a body of signs discernible to the skilled reader. This Orient, occult and fragile, is brought lovingly to light, salvaged in the work of the outside scholar. The effect of domination in such spatial temporal deployments ... is that they confer on the other discrete identity, while also providing the knowing observer with a standpoint from which to see without being seen, to read without interruption."[7]

The image of the Orient, in contrast to what constitutes the "unobserved" but categorized culture of the Occident, and representing a key bipolarity in relation to the constitution and communication of cultural images, is relevant to the present discussion for several reasons: (1) it is an essential ingredient in a hegemonic colonial construction of the history and culture of a colony like India, and although the colony no longer exists as a colony, the life of "the Orient" as a mode of historical thinking did not come to an end with the end of the colony. In other words, it has survived as a historical-cultural stereotype beyond the context of its historical origins and has been transmitted in some postcolonial, sophisticated versions; (2) the opposition to Western perceptions in the indigenous discourse in whatever form—perception of antediluvian national history, cultural nationalism, or even stress on subalternism—continues to use orientalism as a reference point; (3) use of the Orient-Occident bipolarity to legitimize the notion of bipolarity may be duplicated at different levels and down the scale. However, whether it is these broad bipolarities alone that accommodate the tremendous range of possible culture-spaces is a problem best saved for later in the discussion.

To return to the notion of bipolarity as it continues to relate to constructions of cultural differences, let me produce extracts from two comparatively recent works. While they may not do justice to the academic quality of the works from which they are taken, these extracts nevertheless represent an attitude embedded in thought about the "other"; both therefore are statements of what is perceived as genuine difference and therefore reinforce bipolarity.

I first refer to Wendy Doniger O'Flaherty's work titled *Dreams, Illusions and Other Realities;*[8] the study is culture-specific in its attempt to understand meanings of textual references in India, particularly one text's references to dreams and illusion. As a starting point, O'Flaherty refers to two kinds of common sense in the West, Platonic and empiricist, and points out that Westerners are constrained to choose one or the other. O'Flaherty goes on to write:

> In India, too, there are two modes of common sense: common sense A (*samsara*-linked, and materialistic) and common sense B (*moxsha*-linked, and idealistic). But Indians are often able to hold both kinds of common sense in their heads at once and to reconcile both kinds with perceptions of their lives. In the West, only a gaggle of professional metaphysicians or fanatics will have enough genuine familiarity with such ideas to be able to internalize them and act on them instinctively, *but Indian children learn about metaphysics as western children learn about sex, on the street.* (Italics added).

My other reference would be to Madelaine Biardeau's *Hinduism: The Anthropology of a Civilization,*[9] the title of the work implying that the civilization referred to, if it is Indian, has to be viewed essentially in terms of Hinduism as civilization's sole source; it can therefore be represented as the other. But in fact, in the "Introduction" Biardeau begins by confessing that an attempt

> to present a unified account of a culture and a history as vast as that of India, explicitly linking each part to a single totality, is no doubt a somewhat reckless undertaking. Would anyone dare to define a unit of the same kind within western civilization, still less presume to embrace the whole of western history in a single synthesis governed by a far simple norms that can be traced everywhere?

And yet, "India is … a particularity favorable case for an undertaking of this sort," the main reason being that India does not have a history.

> This amounts to saying that change, when it does appear, is only superficial and always refers back to a normative formation, the one source from which spring the most transient phenomena. It is then a question of adopting this language as far as one can, of entering into this world view, forgetting one's own categories so as to discover those of the others, accepting that its lines of cleavage are not ours, that its values are distributed in a way that for us is unexpected. Such a perspective in fact does no more than reconnect with the very modern and very western preoccupations of the historians of 'mentalities.'

I have produced these rather lengthy extracts because, in my understanding, they quite forcefully illustrate the continuity of the notion of an essential bipolarity between the Occident and the Orient, imbuing it with the strength of an axiom that remains a dominant feature of historical thinking about India's past. What is additionally and perhaps more radically significant is that this assumption, deriving from the West's perception of the East, is absorbed even in the notions that the indigenous response prepared as its defense against being perceived as the different "other," in the context of its colonial experience. Thus, when one encounters the notion of Indian "exeptionalism" there is perhaps no alternative to regarding it as another expression of the assumption of cultural bipolarity, which is a part of essentially orientalist assumptions.

Let me dilate on this a little more. It has been argued[10] that although India's struggle for freedom has been viewed practically only in political terms so far, the real resistance took the form a culturally constructed sovereign domain in which the colonial regime was not conceded any space to intervene. Whereas the colonial state wielded sovereign power over the material domain, the colony reserved for itself the domain that is spiritual and thus non-material, and cultural. "This domain of sovereignty, which nationalism thought of as the 'spiritual' or 'inner' aspects of culture, such as language, or religion or the elements of personal and family life, was of course premised upon a difference between the culture of the Colonizer and the Colonized. The more national engaged in its context with the colonial power in the outer domain of politics, the more it insisted on displaying the marks of 'essential' cultural difference so as to keep out the colonizer from that inner domain of national life and to proclaim its sovereignty over it."[11] However, it can be effectively argued that at least the nationalistic vision of the past, in the way it was being constructed in opposition to the material domination by the colonizers, was not really as schizophrenic as it is made out to be; the vision developed into that of an ideal, integrated world—both material and spiritual—which, however, was lost; the lost world that could match its strength with the colonial state. One must keep it in mind that Kautilya's *Arthasastra*[12] or the ancient Hindu "colonies" of Southeast Asia[13] represented an equally powerful dimension of this lost world. In other words, the lost world was a mirror image of what the colonized perceived the colonial sovereignty to have been.

Yet, from the historiography of our times, it appears that the nationalist vision of India's past could not have been as homogenous as it is sometimes made out to be, and the reason in one sense may be that there were parallel and conflicting (or what are perceived as parallel and conflicting) strands in the freedom struggle itself. Here we have to shift from intercultural perception of difference to what is essentially intracultural; if the anticolonial struggle was not homogenous, then the image of the lost world was not undifferentiated either. Let us see how this particular bipolarity within India's struggle for freedom is initially started: "The historiography of India nationalism has long been dominated by elitism—colonialist elitism and bourgeois-nationalist elitism. Both originated as the ideological product of British rule in India."[14] The suggested bipolarity of elite/subaltern sets us to rethinking our view not only of our immediate part in anticolonial struggle but of our precolonial past as well, because "far from being destroyed or rendered virtually ineffective, as was elite politics of the tradition type by the intrusion of colonialism, it continued to operate vigorously in spite of the latter, adjusting itself to the conditions prevailing under the Raj and in many respects developing entirely new strains in both form and content. As modern as indigenous elite politics, it was distinguished by its relatively greater depth in time as well as in structure."[15] I have been trying to argue that historical thinking and history construction, even in relation to what is perceived as a culturally bounded space like India, can neither be homogenously Occidental, nor homogenously indigenous; the "breakdown of the totalizing claims of a nationalist historiography" is thus inconsistent with the plea or the project for autonomous space for indigenous imagination.[16] To resolve this inconsistency, the project must answer the question: What is archetypically indigenous?

The Intercultural and Intracultural in the Construction and Communication of Contenting Memories

If, contrary to what is normally perceived, India does not constitute a homogenous cultural space, then what pluralities are articulated in Indian historical thinking? One must begin here by underlining that historical thinking in India did not originate in response to, and was not stimulated by, the colonial construction of its past. Neither can one talk of a sole mode of precolonial historical thinking and history construction. I am not really entitled to refer with any degree of competence to tribal and folk traditions of orally transmitted memories in India, but they are nevertheless important in their own spatial contexts. At the same time, they relate to general, transcultural patterns of transmitting what tribal or folk groups consider memories of their past and, specifically, to the way these memories contest certain assumptions underlying the classical Brahmanical mode or any other dominant modes of recording the past.[17]

Within the classical tradition too there is wide space for disagreement over how the Buddhists, the Jainas, and the Brahmins highlighted different, often contenting elements, in preserving what seemed significant to their different modes of thinking. But since this essay is not concerned with the details of this, I shall briefly refer to the Brahminic mode and its key elements as in the classical tradition itself, because the Brahminic mode of thought and method of recording history in some ways started gaining ascendancy. I would consider three interrelated elements as key elements in this mode: (1) the myth of creation and its extension into human history, (2) genealogy construction, and (3) preservation of social order.[18] The reciprocity of these three elements is perhaps best symbolized in the figure of Manu. The figure of Manu links creation with human history; he remains at the apex of all major efforts at genealogy construction. This genealogy is not sacred but temporal, yet genealogy construction extends to hagiography as well and is moreover seen as the essential precondition for any individual or any social group desiring a recognizable social status for passage to preserved memory. Manu's association with the perception of social order is that the relevant major *Dharmashastra,* codifying this order, derives from him, just as the codes themselves are perceived to have been ordained by him. I contend that the essence of the Brahminic mode of perceiving memory is virtually related to its device of legitimation, and it follows as a possibility that the device of legitimation, in various forms, has been resorted to even in the most recent phases of Indian history, notwithstanding the growth of other, different modes of historical writing over the centuries and, of course, what is seen as the modern phase of history-writing through exposure to the nonindigenous mode.

If one is looking for the existence of cultural plurality in historical thinking, however, it is not found within the Brahminic mode but instead relates to the way the nineteenth-century intelligentsia reacted to the assumptions of the politically sovereign colonial power. I have already briefly argued against the recent view that the reaction to the fact of being ruled was to construct a sovereign domain of culture. The bifurcation of what already was perceived as a nation is artificial; despite the often articulated accent on the spiritual culture of the East in contrast to the West's material culture, the historical vocabulary that was being used was consistently colonial; and if not by design, then at least in attempts to achieve approximation with the sovereign power, the major attributes of that power were being sought in the past. The search for the institution of democracy as well as a centralized, bureaucratic state in ancient India[19] and the "academic" rediscovery of the colonies that once constituted "Greater

India." This "Golden Age," now lost, was obviously located in the past, and the point that this view of the past did not radically depart from the colonial schema seems tenable when one considers the two interrelated issues—which will prove to be highly significant pedagogic issues as well—of the precolonial state and of periodization. It was these issues, and others deriving from them, that decided in large measure how the cultural plurality or diversity of India was to be treated in history and how the material would be incorporated into a system of pedagogic communication.

The periodization of precolonial India into Hindu and Muslim[20] found acceptance along with other notions corresponding to it, namely that the replacement of Hindu rule brought an end to one period of history and ushered in another. There is nothing exclusively cultural about this periodization; the perception of the loss of one's political sovereignty, and therefore of one's past, through the process of the imposition of another sovereignty was common to what came to characterize both Hindu and Muslim historical thinking, the only difference being that they related differently to the perceived distinction, between Hindu and Muslim, in this periodization schema.

The notion of the glorious lost past that needed to be retrieved constitutes an urgent plea for the writing of one's lost history. Here one finds that there are different spaces for highlighting and articulating one's own cultural affiliations; the plea for national history could therefore be quite contentions. When Bankimchandra Chattopadhyaya, considered a pioneer thinker in the context of India's growing cultural nationalism of the nineteenth century, made the emotive plea for the writing of Bengal's national history or the true history of the Bengali *jati,* he thought to strengthen his case by commenting: "Even the Oriyas have their history."[21]

The agenda for the writing of nationalist history could thus operate at two levels, not necessarily complementarily. "Region"—not academically defined, but in essence taken as representing a nation (*jati*) within a nation (*jati*) by identification with the dominant community resident in it—became a legitimate component of the enterprise of writing national history. This again is not a legacy of the precolonial mode of historical thinking or history-writing; it is derived from the way the response of the ruled under a colonial regime was being given shape by the English-educated intelligentsia that was familiar with the vocabulary of the history they encountered.

To recap before moving on: Despite often offering severe critiques of the colonial regime even as it often endorsed colonial rule, the historical thinking that began taking shape in the nineteenth century differentially constructed images of a past that was lost but retrievable, so history construction could also occur in relation to the vision of positive political action; the pioneers in historical reconstruction meanwhile related this lost past with what they perceived as their own cultural roots, most noticeably by imploring that histories be written of the regions or communities with which they culturally and politically identified themselves, even if this identification was an imposition of the present into the past. From the point of view of what I have called culture-specific history, this *identification-in-retrospect* is important, and I would like to cite here a statement about the sensitivity of the Muslim of the subcontinent, since the statement relates to a perceived homogenous Muslim history of the subcontinent, of which I was a part. I. H. Qureshi's *The Muslim Community of the Indo-Pak Subcontinent*[22], written after partition, articulates this homogeneity. Even though there were many key elements of difference within the community, as a preface to a political vision:

> The Muslim community of the Subcontinent has been treated as if it were only an irrelevant inconvenience, a mere obstruction in the even flow of events, a nuisance which raised its head at awkward moments. There has been little desire to understand the historical background of the emotions which have swayed the community in its great decisions.
>
> When nationalism began to stir the hearts of the inhabitants of the Subcontinent, it was taken for granted that since all of them lived in a common habitat, they would have the same desire of national unity; it was expected of the Muslims that they would dutifully reconcile themselves to the role of a religions minority and would otherwise identify themselves completely with the Indian nation. Such hopes were built upon the most deplorable ignorance of the psychology of the Muslim community, of the thinking of its leaders and masses, and of the history of its origins, development and integration. When history asserted itself and the Muslims claimed to be a nation, many were surprised and not a few even scandalized.

This "assertion of history" in the middle of the twentieth century is projected as deriving from an integrated community of the past that too had lost its political sovereignty and was now retrieving it through the "assertion of history."

Pedagogy and Communication of Images of the Past

Images of India's past are communicated variously, and this communication is not necessarily limited to agencies located in India or rooted in Indian culture. Even within India, the range of modes must be very wide, from the bewilderingly multiple images of India's past that are orally transmitted to an audience committed to a set of beliefs,[23] to the image by which India is seeks to become commercially attractive to prospective consumers. Both of these would differ from "academic" communication in classroom situations that may themselves be heterogeneous. In any case, here the concern is not varieties of communication but varieties of Indian history within classroom situations themselves. This is not to say that other channels of communication are less important—in fact, other channels are likely to assume greater importance in future because of what is communicated, in often tranquil narration of historical events, in classrooms. For the moment, however, prescribed syllabi still wield, so far as the "Voice of History" is concerned, a substantial measure of authority in the formal system of education,[24] and I shall now take up this aspect of communication alone, since, in my understanding, what is thus communicated does not seem to explicate cultural difference but rather congealed stereotypes with their roots going back to essentially one source: the colonial premises and their inversion. I will therefore not try to explain the infinity of the stereotypes only by pointing to the ever-increasing gap, a very rare absence of communication, between those who are involved in the writing of history and those who are its key communicators, but also take up the stereotypes themselves.

For in-depth engagement, it is necessary to go through a vast corpus of material prescribed as textbooks throughout the country, a task beyond the scope of any single enterprise. For a preliminary understanding, I have depended on two types of material: first, a review conducted by the National Council of Educational Research and Training in India (NCERT) and compiled in two parts, of school-level history textbooks prescribed and sometimes officially commissioned in seven different states of India[25]; and second, supplementary reading material on Indian culture, not always as a part of regular, prescribed syllabus but sometimes also determining the content of what is taught within the approved perspective of education.[26] Without going into details of the contents of the prescribed syllabi or the manner in which textbooks are prepared in the different states covered by the NCERT survey, I shall use the findings of the survey to understand certain key motifs of Indian history as it is taught at various levels and attempt to relate them to the other points raised in this paper.

It needs to be noted, as a preliminary point, that there is an implicit and sometimes expressed premise regarding what the teaching of history should mean; it has therefore an objective, which is sometimes stated to be to experience the pride of nationalism and patriotism through the study of history. One thus comes across such a statement as this, in a textbook commissioned and prepared by the government of Uttar Pradesh: "[T]he history of the past of a country enables it to defend its existence, its national unity and in making a proud political place for itself in the contemporary world; the teaching of history is thus to ensure that 'inspired by its proud past, any country can make its future a similarly proud one.'"[27]

Linking the writing and teaching of history with the objective of generating a feeling of national solidarity and pride derives, one can presume, from the earlier nationalist perception of history and history's correction with national and cultural identity. The nation as a given entity, representing an image of unity in diversity and the key reference point for history-writing, then has to continue giving rise to variations on certain motifs that pertain to the nation of the nation, and has to ensure their incorporation into textbooks. Two motifs that appear to be common, despite the difference in the ways they are projected, are the motif of invasions and invaders, and the motif of heroes, both national and regional, who represent unity in diversity. Invasions and invaders, implying fixed territorial borders of the nation projected into the past, relate to various patterns and phases of Indian history. When invaders attack, the culture of the country declines or its citizens put up heroic national resistance; there being a correlation between invasions and phases of decline, invaders must either be absorbed or remain outsiders, although contemporary sources may not have used the term "outsiders" in relation to them. Defeat at the hands of the invaders is national humiliation; one can thus see the device, of course in some extreme cases alone, of setting the order of things right and of constructing history by the negation of history through reversal. In this version of history, invaders like the Achaemedian emperor Darius or Muhammad Ghuri, are turned from victors into vanquished; history is rewritten through reversal of history.[28]

The perception of who is an outsider becomes a matter of contestation, following from the nation being the key reference point in history. The Aryans are seen as givers of civilization, so to regard the Aryans as invaders may be a sacrilegious proposition. There may have been different categories of critics of the British as a colonial power, but even in the eighteenth century there also were apologists for rule by the British; after all, they were invaders against invaders.

The second motif, that of heroes, can be seen in the projection of the roles of historical personalities who figure uniformly in all history textbooks. This, in a sense, goes beyond biography in history and may be seen as deriving from the icon-making process reflected in both works of creative literature and historical works from the nineteenth century onward. In fact, literature and history reinforced each other's efforts in creating icons, as indeed was so far other images of the past,[29] so their entry into textbooks was an inevitable process.

The motif of invasions perhaps created an initial space for the creation of icons, although one must note that simultaneously space was provided by the perception of the century's glorious cultural past for its cultural icons. Thus Poros, who lost to Alexander but retained his dignity, was a national icon, as was Chandragupta Maurya not simply because he established an empire through support of his minister Chanakya but also because he defeated Seleucus.

The later heroes, who all are heroes because they fought against invaders and foreigners, were Pushyamitra Sunga, Chandragupta Vikramaditya, Prithviraja Chauhan, Rana Pratap, Shivaji, and so on. Once the space for the making of icons was created, it could of course expand and also be contested. It could accommodate such figures as the Moghul emperor Akbar, but Akbar may also serve to show how sharply different perspectives may be brought to bear upon the historical evolution of the same personality. For example, in textbooks written for Muslim students of schools set up by Markazi Dargah Islami, Rampur Uttar Pradesh, and published by Markazi Maktaba Islami, Delhi, it is Akbar who is denounced and Aurangzeb, a villain in most history textbooks, who is projected as an ideal ruler. In textbooks specifically written for Muslim students, Aurangzeb is a natural subject for iconmaking because of his perceived religiosity; for the author of this history, "Muslims should study history with [the] true Muslim mentality."[30]

The space for icons is of course space for cult figures of regional history as well, with ranking and attribution of emphasis again contested in the context of both regional and national history; any alert historian working in India today knows that most icons transcend any review since they are invariably seen to represent national identity and aspirations.[31]

The textbooks on Indian history (not uniformly, though in isolated cases even pointing to future trends) suggest the possibility of the creation of national symbols and tracing them to a single cultural root. The device of appropriation inherent in this is already evident in the incorporation, in a prescribed book, of the fantastic theory that the important monument Qutb-Minar was built not by Qutbuddin Aibak but by Samudragupta.[32] History is constructed here, as in the case of the supposed defeat and humiliation of Darius, through the negation of history.

Taking my evidence essentially from the findings of the textbook review, I have been trying to argue that the reference point for the pedagogic strategy of communicating history, as for historical thinking, has been the nation, but it is a nation which can be and is perceived differently. These perceptions, which do not necessarily follow the officially expressed "cultural perspective"[33] calling for an "understanding of the diverse cultural and social systems of the people living in different parts of the country," create spaces for contending memories. I have perhaps chosen some extremely deviant samples from a variety of lessons through which history is transmitted. But the possibilities of the varied uses of this space are nevertheless very real, and I have, in the appendix to this chapter, tried to show how the space can even be used to chose a "national" symbol first and then construct history for it.

Summing Up

To sum up, the Indian experience can be viewed through an attempt to understand pedagogic strategies of history transmission in relation to the varieties of positions that historians take. I am aware that my coverage of textbook materials is highly selective, and that my reference to historiography is mainly incidental, so it is necessary to explain that my reference to historiography has only been in relation to a critical query: How do historical thinking and historical writing tend to perceive, and respond to what is perceived as, cultural difference, and transmit images of cultural difference?

My impression is that the problem cannot be simply situated at the level of pedagogy and formulated in terms of objective and non-objective or secular versus communal brands of history.[34] Instead, one must probe into the roots as well as the rationale of persistent aberrations and deviations from what are still regarded as standard forms of history-writing.

It seems to me that up to a point, we have to go along with the post-orientalist critique of the orientalist discourse and its derivative, the nationalist discourse. Many of the premises that mark the transmission of history, in a form which has large spaces for deviations and for relapsing into myths, may be traced to these sources. The post-orientalist critique, however, does not seem to take us much further because breaking down "the totalizing claims of a nationalist historiography"[35] is no guarantee, by itself, that things that seem to have gone wrong in the past will correct themselves.

Also producing apprehension is the position that the construction and transmission of images of the past, done in cultural terms in the expectation that the "concept of culture," in its wide, anthropological sense, may serve as a possible basis for the "reintegration" of different approaches to history,[36] may be a viable alternative to the unrealistic goal of historical objectivity. The big question, in the context of history-writing, would still be: Whose culture? I have been trying to show that cultural difference may be accentuated as effectively as cultural unity, even within one perceived boundary of culture; one appropriate example of this would be the classic work on the history of Hindi literature *Hindi Sahitya Ka Itihas,* by R. C. Shukla, cited in an article by Krishna Kumar.[37] Seen as representing a "pedagogic strategy," but really in keeping with a line of thinking identifying nationalism with Hindu nationalism, in the work first published in 1929 Shukla took a "strong ideological position indicating the irrelevance of the Urdu-Persian tradition for the development of modern Hindi and *performed a decisive symbolic act in shaping the cultural identity* of college-educated men and women for generations" (italics added).

This is just a reminder of why historians perhaps ought always to remain apprehensive, despite the assurance of "heteroglossia," about the cultural meanings and symbols that will be projected in historical interpretations of cultural difference. One can understand the growing sensitivity in Western scholarship toward understanding and communicating heterogeneity in cultural patterns, particularly those of "others"; one may put it in Clifort Geertz's words as the "systematic unpackings of the conceptual world" of other countries.[38] It must then again be asked: If culture, like language, is a system of differences, then where do cultures converge in history? The interpretation of cultures can be an "act of interpretative violence,"[39] from both within and without. That this apprehension, in the context of the scene of national culture, is shared will be evident from the following quote from the introduction to a symposium on "Dimensions of Ethnic and Cultural Nationalism in Asia":[40] "[E]thnic or culturally based nationalism calls upon those elements of identity that we have limited ability to deny or alter, such as language, religion, place of origin or residence, skin color, and gender. It also asserts the claims of certain cultural characteristics and traditions, almost always containing a strong primordialist element."

Meanwhile, the folly of offering a remedy in the form of the counter—the violence of a straitjacketed brand of history—can no longer merit credence; the "heteroglossia" is here to stay. Perhaps it is time again to remember what Marc Bloch foretold not so long ago, that "even the future of our discipline" may come to depend on what he, following an earlier tradition, called "comparative History." If we attempt to make history culture-specific or nation-specific, there will always remain the possibility of history crashing through the limits of its own possibilities, which the concerns of comparative history, if not the universal voice of history, set before the path of historians.

Appendix

Questions and Answers on the Place of the Sri Rama's Birth:
A Text and Comments on It

The Text

Question 1: Who was the first to build a temple on the birth place of Sri Rama in Ayodhya?[41]
Answer: The great king Kusa, son of Sri Rama.

Question 2: Who was the foreign invader who first destroyed the temple of Sri Rama?
Answer: Menander, the *Yavana* (Greek), 140 BCE.

Question 3: Who built the present temple of Rama?
Answer: The great king Chandrargupta Vikramaditya, 380–417 CE.

Question 4: Which Muslim plunderer led an attack on the temple of Ayodhya in 1033 CE?
Answer: Salar Masud, the nephew of Mahmud of Gazni.

Question 5: What was the consequence of Salar Masud's attack?
Answer: The devotees of Rama massacred his army.

Question 6: Which Mughal invader destroyed the temple of Rama in CE 1528?
Answer: Barbar.

Question 7: What is the structure built by Babar on the Rama temple known as?
Answer: Babri Masjid.

Question 8: Why is Babri Masjid not a Masjid?
Answer: Because to date no Muslim has offered *namaz* prayers at the structure.

Question 9: How many devotees of Rama sacrificed themselves between 1528 and 1914 to free the Rama temple?
Answer: Three hundred and fifty thousand.

Question 10: When did the army of Sri Guru Govind Singh leave for Ayodhya to make the "Rama temple free?"
Answer: 1680 CE.

Question 11: Followers of which protectors of religion (*dharma*) defeated the army of the Mughal king Arangzel and made the Rama temple free?
Answer: Sri Guru Govind Singh and Barba Vaisnavdasji.

Question 12: When did the Muslims happily decide to hand over the birthplace of Sri Rama to the Hindus?
Answer: 1847 CE.

Question 13: How did the English aggravate the Temple-Mosque quarrel?
Answer: The English put up a wall between those who called [the place] *Rama mandir* (temple of Rama) and those who called it a *Masjid*.

Question 14: When did the district judge order that the lock on the Rama temple had to be opened?
Answer: 1 February 1986.

Question 15: When did the activity of the worship of stones (*sila pujam*) at different places begin for the purpose of the construction of the Rama temple?
Answer: 30 September 1989.

Question 16: When did the *Karserak* (volunteer) devotees of Rama unfurl the *Bhagwa Jhanda* (the saffron flag) at Sri Rama's place?
Answer: 30 October 1990.

Question 17: Give the names of those teenagers who sacrificed their own lives when unfurling the Bhagwa flag at the temple of Sri Rama.
Answer: Ramkumar Kothari, Sarad Kothari.

Comments

As an example of "pedagogic strategy," the text, given in a literal English translation, appears remarkable. In essence, what is communicated is also present in other parts of this series on the cultural Heritage of India in which too the temple of Rama occupies the central position;[42] this particular lesson is remarkable because of the way it can function as a strategy for communication.

At the center of the text is the temple of Rama. It is projected as a symbol of Indian national culture through the fact of its repeated destruction in invasions; in other words, the motif of invasions,[43] which are seen as bringing destruction to Indian society time and again, relates equally validly to the temple of Rama: Like the regeneration of India, the temple too has to be regenerated: the two regenerations are implicitly parallel.

The temple is given a genealogy. The original construction of the temple is traced to Kusa, the son of Rama; this certainly attributes to it a proximity to Rama himself and thus a very desirable chronological status. What is similarly significant is that the rebuilding of the temple, or making it free in different periods, was done by outstanding figures of Indian history like Chandragupta Vikramaditya and Guru Govind Singh.

The destroyers of the temple and its restorers are identified, so one can make out which category of Indians relates to the essence of Indian culture and which attempts to destroy it. At the same time, however; this distinction collapses because of the overarching significance of the temple as a symbol for the whole nation. After all, it was restored by mighty kings in the past, thousands sacrificed their lives to make the site free, a Sikh *Guru* too led an army to set it free, and, in 1847, the Muslims happily handed the site over to the Hindus. In one sweep, the site of the temple, became the locus of national consensus, and if the reality was conflict, the conflict was an aberration caused by the divisive act of the British and not a real issue, because after all, the Muslims had never offered *namaz* prayers at the shrine, which was being irrelevantly called a *masjid*.

If the remarkable achievement of the text, for its invented audience, is to establish the supremacy of a temple as a national symbol, in a country in which almost every village has a temple, there are a few additional points regarding the "strategy" that may be noted.

1. The format of the text is that of questions and answers, neither the questions nor the answers being in the slightest degree ambivalent; they constitute therefore a most affirmative assertion of reality of the country's cultural history, leaving no space for any further query. Through simple memorization, one absorbs a definitive version of a historical-cultural reality.

2. Juxtaposition of characters, dates, and events whose the historicity can be verified with those whose historicity cannot, is used effectively to collapse the barrier between myth and history; both have become open-ended, myth having gained the temporal legitimacy of history, and history the timeless quality and sanctity of a myth.

3. The location of the lesson in what purports to be a book of knowledge concerning the cultural heritage of India and the great personalities of India adds to its legitimacy.

4. The motif of invasions and seizures to which the temple was subjected brings to the fore the value of sacrifice. It is the sacrifice of national heroes that brought about the regeneration of Indian society in the past; it is the sacrifice of the devotees of Rama that will restore to this symbol of national culture its past. It should be noted that the text does not refer to any positive political action in present times. It associates both sacrifice and political action in the context of the past, and does not make any explicit connection between them in the context of the present.

Notes

1. For the vision of the founders of the Annales school see S. Clark, "The Annales Historians," in Q. Skinner, ed., *The Return of Grand Theory in Human Sciences* (Cambridge, MA, 1986). For Herodotus and Thucydides see Walter Blanco and Jennifer Tolbert Roberts, eds., *Herodotus—The Histories* (New York 1992); M. I. Finley, "Thucydides the Moralist," in Finley, *Aspects of Antiquity: Discoveries and Controversies* (Harmondsworth 1972), 48–59. Also Francois Hartog, *The*

Mirror of Herdodotus: The Representation of the Other in the Writing of History, transl. Janet Lloyd (Berkeley, Los Angeles, London 1988).

2. Gabrielle M. Spiegel, "History, Historicism, and the Social Logic of the Text in the Middle Ages," *Speculum* 65 (1990): 59.

3. Peter Burke, "Overture: The New History, Its Past and Its Future," in Peter Burke, ed., *New Perspectives on Historical writing* (Cambridge 1992), 1–23.

4. Partha Chatterjee, *The Nation and Its Fragments: Colonial and Post-Colonial Histories* (Delhi 1994).

5. Marc Bloch, "A Contribution towards a Comparative History of European Societies," in *Land and Work in Medieval Europe: Selected Papers,* transl. J. E. Anderson (London 1966).

6. Note the notion of comparative history present in the writings of V. G. Childe; see his *Social Evolution* (London 1963); *What Happened in History* (Harmondsworth 1957); *Man Makes Himself* (New York 1957).

7. James Clifford and George E. Marcus, eds., *Writing Culture: The Poetics and Politics of Ethnography* (Delhi 1990).

8. Wendy Doniger O'Flaherty, *Dreams, Illusion and Other Realties* (Chicago and London 1984), 12.

9. Madelaine Biardeau, *Hinduism: The Anthropology of a Civilization,* transl. Richard Nice (Delhi 1989), Introduction.

10. Chatterjee, *The Nation and Its Fragments.*

11. Ibid.

12. Nationalist historiography, and its extension into other brands of writing too, centering around the text of *Arthasastra,* attributed to Kautilya, is voluminous; for a sample, one may refer to comparison instituted between the social and economic policies of Kautilya and the social legislation in Bismarck. See H. C. Ray, "Was State-Socialism Known in Ancient India? A Study in Kautiliya's *Arthasastra,*" in *Sir Asutosh Mookerjee Silver Jubilee Volumes,* vol. 3, part 1 (Calcutta 1922), 429–446; idem, "Economic Policy and Functions of the Kautilyan State," *Journal of the Departments of Letters* 13 (1926): 1–30.

13. Note, for example, the titles of such works as: R. C. Majumdar, *Ancient Indian Colonies in the Far East,* vol. 1: *Champa* (Lahore 1927); vol. 2: *Suvarnadvipa,* parts 1–2 (Dacca 1937–38); idem, *Hindu Colonies in the Far East* (Calcutta 1944).

14. Ranjit Guha, "On some Aspects of the Historiography of Colonial India," in idem, ed., *Subaltern Studies I: Writings on South Asian History and Society* (New Delhi 1982), 1.

15. Ibid., 4.

16. Note Partha Chatterjee's plea: "The project then is to claim for us, the once colonized, our imagination." *The Nation and Its Fragments,* 13.

17. See, as a sample of historical memory of groups that remain largely at a distance from Brahmanic hegemony: M .L. K. Murty and Gunter D. Sontheimer, "Prehistoric Background to Pastoralism in the Southern Deccan in the Light of Oral Traditions and Cults of Some Pastoral Communities," *Anthropos* 75 (1980):163–184; Gunter D. Sontheimer, *Pastoral Deities in Western India,* transl. Anne Feldhaus (New York 1989), passim.

18. For relevant references, Romila Thapar, "Society and Historical Consciousness: The Itihasa-Purana Tradition," in S. Bhattacharyya and Romila Thapar, eds., *Situating Indian History for Sarvepalli Gopal* (Delhi 1986), 353–383.

19. Note that the society that was founded to study early India's relations with Southeast Asian countries was named the Greater India Society and its journal the *Journal of the Greater India Society.* See H. B. Sarkar, *Cultural Relations between India and Southeast Asian Countries* (Delhi 1985), chap. 2.

20. See Romila Thapar, "Interoretations of Ancient Indian History," in idem, *Ancient Indian Social History: Some Interpretations* (Hyderabad 1978), 1–25.

21. Cited in Chatterjee, *The Nation and Its Fragments.*

22. I. H. Qureshi, *The Muslim Community of the Indo-Pak Subcontinent (610–1947): A Brief Historical Analysis* (Delhi 1985), viii, 88.

23. That oral transmission, backed by visual images communicated through technologically advanced media, can have tremendous impact on the formation of images of the past is suggested by the recent eminently successful TV serials on Indian epics.

24. This is very strongly conveyed by efforts to officially control or intervene in the area of history textbooks prescribed in schools, and by occasional attempts to suppress textbooks prepared by the National Council of Educational Research and Training. See M. H. Siddiqi, "History-Writing in India," *History Workshop Journal* (1980): 184–191.

25. *National Steering Committee on Textbook Evaluation: Recommendations and Report,* parts 1 and 2 (typescript), (New Delhi n.d.).

26. References may be made, in this context, to the series *Akhila Bharatiya Samskriti-Jnana-Pariksha* (All-India Test on Cultural Knowledge), (Kurukshetra n.d.)

27. *National Steering Committee on Textbook Evaluation: Recommendations and Report,* part 1, Appendix 1.

28. Ibid., Appendix 2.

29. The genre of "historical" novels and poems and plays either dedicated to or woven around history's heroes and heroines had a rapid development in India from the nineteenth century onward. The literary writings have had profound impact on historical consciousness, and the "icons"of history are identical with the "icons" of literature.

30. *National Steering Committee on Textbook Evaluation: Recommendations and Report,* part 1, Appendix 3.

31. The trend, which is new, is to make the achievements of a political bigwig an integral part of a textbook on Social Studies at school level.

32. *National Steering Committee on Textbook Evaluation: Recommendations and Report,* parts 1 and 2.

33. Ministry of Human Resource Development, *National Policy on Education—1986* (*with modifications undertaken in 1992*) (typescript), (New Delhi n.d.), 6–7.

34. For the first coherent statement of how communal assumptions have permeated writings on Indian history, see the pamphlet by Romila Thapar, Harbans Mukhias, and Bipan Chandra, *Communalism and the Writing of Indian History,* 2nd ed. (New Delhi 1977).

35. Chatterjee, *The Nation and Its Fragments.*

36. Burke, "Overture: The New History, Its Past and Its Future," 1–23.

37. Krishna Kumar, "Hindu Revivalism and Education in North-Central India," *Social Scientist* 18, no. 10 (1990): 4–26.

38. See the discussion in Spiegel, "History, Historicism, and the Social Logic of the Text in the Middle Ages," 59, and Vincent Crapanzano, "The Hermes Dilemma: The Making of Subversion in Ethnographic Description," in Clifford and Marcus, eds., *Writing Culture,* 12.

39. Ibid.

40. Refer to David D. Buck, ed., "Dimensions of Ethnic and Cultural Nationalism in Asia—A Symposium," *Journal of Asian Studies* 53, no. 1 (1994): 6.

41. This is lesson 19, translated by me into English, of *Akhila Bharatiya Samskriti-Jnana-Pariksha,* part 7.

Similar lessons are given in the form of capsules in other parts of the series. According to an estimate given in a publication of 1990, the number of school, at which the text given here is taught would number around 3,000. However, the estimate need not be taken to relate to the distribution of such texts. See Kumar, "Hindu Revivalism and Education in North-Central-India." The number seems to have gone up after 1990; it was being claimed during 1993-94 that the total number of students taking the cultural knowledge tests, was 355,282, and that the total number of schools run by the organization was 6,000 with 1,200,000 students and 40,000 teachers; *National Steering Committee on Textbook Evaluation: Recommendations and Report,* part 2 (typescript), 90.

42. The same lesson figures in slightly different forms in parts 8 and 10 of the series.

43. According to one version of the text, the temple suffered seventy-seven invasions in the past. *National Steering Committee on Textbook Evaluation: Recommendations and Report,* part 2, 93.

CHAPTER 11

Politics of Historical Sense Generation

D. L. SHETH

I

It is a truism to say that sources of historical sense generation in a society are not confined to history.[1] The recognition that history's own established procedures of making sense of the past cannot remain insulated from the influence of ideas and action moving the wider society is growing within the discipline of history itself. This has made the discipline pliable to modes of understanding the past developed in other disciplines such as arts, aesthetics, literary theory and criticism, ethnology, sociology, philosophy, linguistics, and so on.[2]

This, however, does not mean that the discipline has lost, or is in the process of losing, its sense of boundary or its self-image as a rational, cognitive pursuit of objective truths about the past. Nor has history's claim to dominance over other modes of making sense of the past, if not to universality, become negotiable. What has changed is that the discipline of history now accords recognition to "nonhistorical" modes and shows a certain readiness to sift data and to process concepts of other "nonhistorical" disciplines through the historical mode.[3] This is true more at the level of the theory than at the level of actual procedures of writing history. But in this process, often the procedures are expanded, even modified, to accommodate the insights and approaches of other disciplines. This has given rise to important controversies within the discipline, which, in my view, have implications for changing the discipline's very orientation to viewing the past.[4]

This new catholicity of history, however, lives very uneasily beside in one vast arena of modern life, in truths about the past are contested made to settle

Notes for this section begin on page 177.

social, cultural, and economic equations in the present. This is the arena of politics. History's ambivalence toward politics arises from the fact that if historians do not participate in contemporary political discourses about the various conflicting reconstructions of the past, the established dominance of the historical mode may diminish. But if they participate in these controversies, or even try to mediate, as they often do, they cannot avoid taking sides in the ongoing battle for owning history in the society; in the process they become as partisans, compromising their claim to objectivity. The reason for this vulnerability of history to politics is rarely sought in the structure of the discipline itself; more often, it is attributed to manipulative politicians abusing history for their power needs.[5]

It is easier for history to open up to other fields of knowledge, as has already happened, than to politics. It is particularly difficult for history to explicitly recognize and accept the role of politics in shaping historical consciousness. Although history's actual relationship to politics is very old, its selfimage as a discipline devoted to sifting objective truths about the past and formulating, on that basis, principles of collective human behavior for the present and the future is relatively new. Historians fear that by moving closer to politics, modern history's assiduously cultivated image and claim to objectivity and scientificity would be severely undermined; the discipline would be thrown back in the lap of politics, from which it has theoretically differentiated itself in modern times. Historically, in premodern times, accounts of the past were by and large constructed by historians working directly under the political patronage of rulers or on behalf of the social and political groups to which they belonged. One outcome of this mode of doing history was that various accounts of the past—though today they would be considered "nonhistorical"—became integral to history. In fact, the distinction between historical and nonhistorical modes of viewing the past was not just tenuous, but almost nonexistent.[6]

It is, therefore, understandable that history's growth as a modern, scientific discipline with cognitively oriented procedures led to its self-consciously putting itself at a great distance from politics. This theoretically cultivated distance, it seems, has resulted in the very denial of the relationship between the two spheres, imparting a degree of opaqueness on the part of the discipline of history to political issues. Put another way, the distance has created a sort of cognitive blindness in the discipline to certain political formulations and data, making it possible for politics to enter the field of history directly, and history to move into the political arena in ways not sanctioned by tenets of the historian's discipline.

My intention in this paper is not to search for reasons for history's vulnerability to politics in the history of history, though this may be an interesting line for a historian to pursue. I am a political sociologist. I feel more comfortable examining this relationship from the other end, through the lens of politics. In doing this, I am also interested in searching out terms of recognition, and if possible retrieval and redefinition, of the cognitively denied relationship between history and politics. In my view, such a redefined relationship can keep the legitimacy of politics by and large confined to political arenas, while at the same time opening up history further to some political concepts and data in a manner that prevents politicization of the discipline itself. This process may lead to a more productive interdisciplinary debate and, hopefully, to a more constructive mode of intercultural communication in the global society.

Insofar as competing contentions exist in a society about the legitimacy of ideas, institutions, and representations (symbolic, ritual, or as arrangements for powersharing), claims on "history" and to "historical truths" are constantly made in politics by the contending social and political forces of the time. And, insofar as the contending parties making claims on history either challenge the dominance of the historical mode or themselves refer their disputes to historians in search of historical legitimacy for what are essentially political-cultural formulations about the past, politics is drawn into the domain of history. Thus the periodically changing historical revaluation of ideas, events, and institutions, the dimming of a certain image of the past and the illumination of another, the shrinking of some strands of consciousness and the expansion of others, have to do not only with new historical data coming to light, but also with history's constant involvement in politics.

Put briefly, both history and politics are involved in the ongoing process of contention and settlement of legitimacy claims in the society. In modern times, such contention has produced a confrontation between the two: politics seeking to redefine historical consciousness in more explicit terms of contemporary political discourses and history wanting to interpret, even ideologize, contemporary political consciousness in "pure" historical terms. Of course, in practice the contending forces are not so neatly divided. In fact, the confrontation creates new divisions both in politics and in the discipline of history; often the entire discourse is hijacked by politicians masquerading as historians, and politically oriented historians purveying politics as history.

If, however, the relationship between history and politics is properly recognized and if possible codified, it may, in my view, lead to creative controversies in the respective disciplines. It may contribute to raising new questions, generating new data, and stimulating thinking on finding new modes of interpretation, often questioning the discipline's established procedures and methods. This is already happening to history and to other disciplines interacting with it, but its relationship to politics and the discipline of political studies remains by and large unexplored. My intention here is to highlight some elements of this relationship in the Indian context. In doing this, I shall confine myself to the politics of the independence movement in India, and show how the historical mode of viewing the past became integral to the political consciousness of modern Indians, i.e., to viewing India as a nation-state. This change in consciousness, from the cultural to the historicol-political, had far-reaching implications for India, one of the world's most complex multiethnic societies governed by the nation-state.

II

The politics of competing legitimacy claims on "history" and to "historical truths" surfaced during the movement for India's independence. In fact, it can be said that this politics was born with the rise of colonial, broadly Western, historiography, which had begun to challenge and replace the then prevalent modes of making sense of India's past.[7]

In the initial phase of colonial rule, before the independence movement acquired the pronounced political character of a "national" movement for self-rule by Indians, new constructions of the past were made by what was then a very small, English-educated class of Indian elites, who had been exposed to colonial historical writings. In the absence of any mobilizing political movement, these constructions were not embodied in politics. They, however, did contribute to the growth of a new cultural consciousness, loaded with political images of India as a land and people subjected to foreign rule of the British then, and earlier to the power's of the alien culture of Islam. These images of subjugation and resistance were articulated more in terms of regional-linguistic cultures than of an India representing one territorial cultural unity. The preferred mode was mainly narrativist or fictional. It relied more on the power of images than on factual accounts of the past they sought to recreate. The narratives wove rich cultural imageries. Often they came close to depicting a parade of pictorial images in writing, more in the epic-Purāṇic tradition than in the new historical mode. By and large they thus relied on the old mode for new constructions of the past, despite the fact that these new constructions were inspired in the first place by the historical mode introduced by colonial historians.[8]

Most of these narratives and fictional accounts were focused on concrete historical figures, ranging from King Puru fighting back Alexander's armies to a Rana Pratap or a Shivaji resisting the empire-building campaigns of the Mughals. It was thus around the historical figure of a hero that events, and ambiences, were created that permitted the depiction of a hero's great deeds, exemplary courage, and sacrifices in resisting the assaults by the "foreign" invaders or empire builders, on the audience's native land and culture. These narratives, in form much like sagas or what modern historiography describes as hagiographic accounts, drew more on the memories of these heroes preserved in the minds of people, either through an oral tradition of generationally transmitting memories of the past or via other "nonhistorical" sources such as bardic lore or the accounts of genealogists. The effort was to recreate a poetic imagination about the hero's exemplary life and preachings (i.e., the hero's *kavan*), rather than to construct biographic accounts of the actual life he or she lived (i.e., the hero's *jivan*). These constructions also succeeded in conveying a new image of India as a land and a people beleaguered by "foreign" invaders and rulers, all of whom were valiantly resisted by native heroes.[9]

In short, these accounts of assaults and transgressions by what were now perceived as foreign invaders or rulers, and of the resistance offered to them by "India's" own heroes representing concrete historical figures of the past, were depicted more in socio-cultural and religious terms than in political national-territorial terms. The "historical consciousness" invoked in these narratives was culturally pronounced but politically subdued. Indeed, the idea of "peopleness" of the "native" Indians and a sense of common cultural, often even religious, bonds having been inherited from the past was very much a part of this consciousness. But rarely, if ever, was this consciousness invested with any sense of bounded territory or idea of a political center for and around which the people fought back the assaults and resisted transgressions. Of course, in quite a few of these narratives the lack of a political center or a sense of territory is indicated as the prime reason for the defeat of individual heroes.

Sometime in the mid nineteenth century this cultural consciousness began to acquire political trappings, leading eventually, among other things, to the founding of a national political forum, in the form of the Indian National Congress. Starting as a forum to articulate grievances and aspirations of "native Indians" under colonial rule, by the turn of the century, the Indian National Congress had become a political center leading the movement for India's independence from British rule. It was during this period that the consciousness of India as a territorially constituted and historically endowed cultural unity, i.e., a nation, subjugated by alien rule entered popular consciousness. These new claims to historically produced cultural unity and territorial continuity made by the leadership of the independence movement were not just cultural; they were pronouncedly political, and obviously unacceptable to the colonial rulers. Aided by British historians, the colonial authorities pointed to the lack of historical reasoning in the making of such claims and doubted the historicity of the cultural constructions of the past on which the claims by and large still depended. Colonial politics, aided by modern historiography, thus posed a new challenge to the politics of the independence movement, i.e., to historically establish territorial and political continuity as the basis of the cultural unity of India as a nation. In the colonial view, the cultural constructions of the past privileged the dominant religious community of the Hindus, which if seen historically was a heterogenous mass of population divided sharply within itself, both culturally and politically. Thus, the idea of India representing a nation in the image of the European nation-states was found historically unsustainable and therefore politically illegitimate. Its grounding in the politicol-cultural consciousness of the "natives" was insufficient basis for India's claim to nationhood. Paradoxically, a new group of nationalist historians and historically inclined political leaders of the movement also found this to be a valid argument. Like the colonial historians, they too believed that if the historicity of India's cultural unity was not established in terms of territorial and political continuity supported by the "historical consciousness" of its people, India could not become a nation and therefore couldnot politically justify its claim to independence.

A new project and a new discourse were born: historicizing India's nationhood. Through this process politics entered the domain of history and history came to the center stage of politics. It led to endless controversies between colonialist and nationalist historians about sustainability of the historical claims to India's nationhood. Both depended on the same basic procedures of writing history, but for different political purposes. Now, over four decades since independence, the debate has resurfaced, in a radically different political context created by India's democratic system. In this post-independence debate, whether India is a nation or not is not at issue. The question is now framed: In what terms and to what purpose should India's nationhood be articulated in politics? This point is not just a locus of debate but of intense political conflicts.[10]

To return to the politics of the independence movement, the new project of historicizing India's past gave rise, among other things, to intense competition and conflicts in the society among various religious, cultural, and linguistic communities. The competition and conflict among them became increasingly centered around claims to "historical truths." These groups' political claims to representation now needed support from history. In making these claims the communities began to conceive of themselves as historically differentiated "nationalities," rather than as culturally overlapping and interlocked socio-religious identities interested in preserving some of their distinctive symbols, rituals, and practices. In this process, history became the arbiter not only of all contemporary claims to political legitimacy by various groups in the society, but of the whole society's future.

When Gandhi returned to India from Africa and led the independence movement, he rejected the new historical mode of constructing India's past. He refused to conduct his politics of the independence movement by joining the prevailing discourse about the past, whose terms had been set by the colonial rulers. He found it unnecessary, even undignified, for a people with a long civilizational past to justify their claims to independence on such vacuous terms.[11] He even saw historicization of the past as a self-defeating project for Indian nationalism. He sought to revive the old cultural mode but in radically different terms, by emphasizing traditions of co-living among different communities within a political arrangement they consented to. Ghandi brought many political and cultural issues that had been pushed into history's lap back to the political arena. Rather than weakening political mobilization for the independence movement, Gandhi's abandonment of the debate over history, his privileging cultural-political consciousness over historical consciousness, expanded and intensified political mobilization on a truly national scale.

Paradoxically, while this strategy won the political battle for India's independence, the cultural vision it represented was defeated. At the moment of independence, the historical mode emerged as triumphant. This does not mean that the attempts by nationalist and modernist historians to historicize India's nationhood had succeeded. What succeeded was their effort to make the idea of the nation historically integral to the state structure. The nation-state logic that was implicit in the historicization project, probably unintentionally privileged and politically legitimized the historically presented claims of various cultural and religious entities to political autonomy. In effect, historicization legitimized the ethno-religious principle of nationalism and eventually led to the partition of India based on that principle.

It would be mistaken to think that modern nationalist historians either envisaged or endorsed the principle of ethno-religious nationalism. In fact, they, sought to delegitimize such claims to nationalism by showing, for example, how historically the pre-British state of the Mughals had acquired a secular character in the process of building an empire in India and how British colonial rule, in a large part, had inherited the Mughal state structure. In their view, the cultural consciousness of India as a nation, which inevitably emphasized the ethnic diversity of its population, by itself could not provide a basis for India's political unity. They held that it was necessary to impart a sense of political and territorial continuity of the state to the politically unformed cultural consciousness. Crucial to this thinking was the idea that a cultural entity claiming politically to be a nation, in this case the whole of India, must have its own state.

The modernist leadership of the movement saw in this formulation a great potential for political mobilization for India's independence. Firstly, by making the idea of India's nationhood integral to the state structure, or representative of one political center, it could mediate and possibly superordinate the various competing ethnic claims of political representation that had surfaced in the politics of the movement. This was necessary to sustain the claim for independence on behalf of the nation rather than for any single linguistic or religious community. Secondly, the linking of nationalist consciousness to state power was conducive to the modernist leadership's aspiration to inherit power after independence, along with the continued legitimacy of an already established modern state. They believed this was necessary for governing the multiethnic society that was India, if and when the British left India. Thirdly, and more importantly, they found that the nation-state logic enabled them to practice their politics of confronting and containing the emerging claims of ethno-religious nationalism and thus to lay the foundations for civic and secular nationalism in India.[12]

Unfortunately for the modernist leadership and for India, the nation-state logic did not work in this direction. In the view of the British colonial rulers, historically more powerful claims to "nationhood," and therefore to political power, had been made by the ethno-religious communities. Of course, the point that a nation must have its own state did find validation in Western historiography and was eventually conceded by the British. But the modernist leadership of the independence movement argued that the Indian nation must have a basis in a secular state, and that India could find such a basis through the independence movement. This, they believed, would make it possible for the Indian nation to inherit the Indian state, which was presently under the alien rule of the British.

This argument did not cut any ice with the British concept of "historical consciousness." In British colonial eyes, it was a wrong, wishful reading of history by the modernist Indian leadership. In their part of the world, i.e., Europe, the nation represented a historically evolved political entity that tended to be culturally homogeneous in terms of race, religion, and language. Only such a nation could shape the state for itself. This could be a secular state conferring citizenship rights on all, including minorities. But it could take other forms as well. The British colonial authorities, therefore, did not see anything wrong with, and in fact encouraged, claims to nationhood by different ethno-religious and linguistic communities in India. Eventually they recognized such claims not only to nationhood but to a separate statehood for the major ethno-religious community in India, namely the territorially based Muslims (i.e., for the provinces in which Muslims were in the majority).

Thus, independence was won through the creation of two nation-states on the principle of ethno-religious nationalism: India and Pakistan. The British authorities left it open for other linguistic and regional-cultural identities to sort out their claims to nationality and statehood in the newly created nation-states of India and Pakistan. This logic has since been at work, in both India and Pakistan almost inexorably pushing many an ethnic entity to claim nationhood and often beyond that to demand separate statehood. Along with this, the use of history in the politics of nationhood and statehood and as well as that of international relations between the already separate nation-states of India and Pakistan has become intense, threatening a breakdown of the old, established inter-cultural communications among various communities within India and Pakistan as well as between them.

It was easy to sociologize or politicize the cultural consciousness of Indianness or of Indian nationhood for the politics of the independence movement. The idea of "India" as a continuous civilizational entity binding Indians through a sense of a common past was credibly available to such efforts. This consciousness was not just about the past that was dead and gone; rather, it could find validation in the everyday life of its people: through memories of the past symbolized in the present (in places and heroes they worshipped and in festivals they celebrated), and in rituals and practices, through which they sought sanctification of various economic, social, and even political activities. Despite the different cultural representations in different communities' rituals and practices the meaning system of one cultural representation was seen and felt to be commensurable with other representations. But historicization of such cultural consciousness in a multicultural society, in linear terms of political and territorial continuity, inevitably gave rise to different and opposing claims on history, with various regional and religious identities trying to find their own separate historical and cultural legacies. This led to the legitimacy claims for representation in politics soon getting converted into claims and counterclaims aimed at establishing cultural and political hegemony, especially by the large territorial or ethno-religious communities. After this had already happened it became irrelevant whether these claims were, in fact, sustainable historically or not. In this battle for hegemony between communities, both the historical and the cultural modes of interpreting the past were subverted by politics. History was now used to establish the claims to nationhood of the two major religious communities of the Hindus and the Muslims, claiming to represent separate pasts and even culturally incommensurable meaning systems. This is what I mean by the triumph of the "historical" mode over both the politics and the culture of the independence movement. The partition of India was a consequence of history's political project. Its consequences for the discipline of history itself tell another story.[13]

Notes

1. The author is grateful to his colleagues Giri Deshingkar and Ravi Vasudevan for their suggestions and criticisms.

2. Works of Ranajit Guha and Bernard Cohn provide important instances of cross-disciplinary influences in Indian historiography. Guha is influenced by structural linguistics and narrative theory. For example, see Ranajit Guha, "The Prose of Counter-Insurgency," in Ranajit Guha, ed., *Subaltern Studies,* vol. 1 (Delhi 1982); and Guha, *Elementary Aspects of Peasant Insurgency in Colonial India* (Delhi 1983). For a dialogue between anthropology and history, see Bernard Cohn, *An Anthropologist Amongst the Historians and Other Essays* (Delhi 1987).

3. For creative use of social-anthropological concepts in writing history of ancient India see Romila Thapar, *From Lineage to State: Social Formations in the Mid-First Millennium B.C. in the Ganga Valley* (Delhi 1990); for a critique of historical writings using sociological concepts see, A.M. Shah, "Social Anthropology and the Study of Historical Societies," *Economic and Political Weekly,* Special Number (July 1959); and Shah "Towards a Sociological Understanding of Ancient India," *Contributions to Indian Sociology* 20, no. 1 (1986): 117–133.

4. This change is reflected in the later subaltern studies. In earlier volumes, this group of historians aimed to recover histories from below in the mode of the radical social history of E. P. Thompson, Eric Hobsbawm, and Georges Rude. While this constituted a critique of "high" or official history in that different voices were brought to bear in their accounts, subsequent studies were premised on a critique of history itself as governed by a hegemonic Enlightenment rationality that serves the drives of nation-states, their developmental paradigms and culturally homogenizing impulses. Examples would include Partha Chatterjee, *The Nation and Its Fragments* (Delhi 1994); Dipesh Chakrabarty, "Post-Coloniality and the Artifice of History: Who Speaks for 'Indian' Pasts?" *Representations* 37 (1992); and Chakrabarty, "The Difference-Deferral of Colonial Modernity," in David Arnold and David Hardiman, eds., *Subaltern Studies,* vol. 8 (Delhi 1994). Also, see Gyan Prakash, "Writing Post-Orientalist Histories of the Third World: Indian Historiography Is Good to Think," in Nicholas B. Dirks, ed., *Colonialism and Culture* (Ann Arbor 1992), 353–388.

5. See, for instance, Sarvepalli Gopal, et. al., *The Political Abuse of History* (New Delhi n.d.), pamphlet; also Romila Thapar, Harbans Mukhia, and Bipan Chandra, *Communalism and the Writing of Indian History* (New Delhi 1969), pamphlet.

6. For premodern modes of writing on the past see V.S. Pathak, *Ancient Historians of India* (Bombay 1966); U. N. Ghoshal, *Beginnings of Indian Historiography and Other Essays* (Calcutta 1944); F. E. Pargiter, *Ancient Indian Historical Tradition* (Delhi 1972); Romila Thapar, *Interpreting Early India* (Delhi 1992).

7. See Partha Chatterjee, *Nationalist Thought and the Colonial World: A Derivative Discourse,* especially chap. 3, "The Moment of Departure" (London 1986).

8. For examples see Chatterjee, *The Nation and Its Fragments,* chap. 4; and Ranajit Guha, *An Indian Historiography of India: A Nineteenth Century Agenda and Its Implications* (Calcutta 1988), 27–36.

9. For an account of earlier traditions of this writing, see Pathak, *Ancient Historians,* chap. 1; for the impact of hero-centered narratives of the past on the novel see Meenakshi Mukherjee, *Realism and Reality: The Novel and Society in India* (Delhi 1985), 38–68; and Sudhir Chandra, *The Oppressive Present: Literature and Social Consciousness in Colonial India* (Delhi 1993), chap. 3, especially 126–130; and on representations in folk theater see Kathryn Hansen, *Grounds for Play: The Nautanki Theatre of North India* (Delhi 1992).

10. For an illuminating discussion on this issue see Sudipta Kaviraj, "Crisis of the Nation-State in India," *Political Studies* 42 (1994): 115–129.

11. As early as in 1924, Gandhi had declared himself opposed to using the modern-colonial mode of history for reconstructing India's past: "I believe in the saying that a nation is happy that has no history. It is my pet theory that our Hindu ancestors solved the question for us by ignoring history as it is understood today and by building on slight events their philosophical structure. Such is the *Mahabharata*. And I look upon Gibbon and Motley as inferior editions of the *Mahabharata*." In Raghavan Iyer, ed., *The Moral and Political Writings of Mahatma Gandhi,* vol. 1 (Oxford 1986), 187.

12. The most influential among those modernist leaders of India's independence movement whose writings (and politics) shaped this vision of Indian nationalism was Jawaharlal Nehru. His *The Discovery of India* (New York 1946), which he wrote during periods of his imprisonment by the British, comprehensively articulated this vision. For an excellent discussion of Nehru's nationalist thought, see Chatterjee, *Nationalist Thought and the Colonial World,* chap. 5.

13. For a critique of the historical mode of reconstructing the past and its implications for the discipline of history itself, see Ashis Nandy, "History's Forgotten Doubles," *History and Theory* 34, no.2 (1995); for a discussion on alternatives to nationalism see Ashis Nandy, *The Illegitimacy of Nationalism: Rabindranath Tagore and the politics of self* (Delhi 1994).

CHAPTER 12

Communalism, Nationalism, Secularism: Historical Thinking in India and the Problem of Cultural Diversity

Michael Gottlob

The Political Context

"Communalism" in the political language of modern India denotes first and foremost the strivings of religious, ethnic, social, or linguistic communities to assert their particularity against the generalizing elements in the concept of the nation.[1] Thus during the course of India's struggle for independence the communalism of the Muslims in particular distanced itself from the more broadly defined all-Indian nationalism. But other minorities, such as the Sikhs, the Tamils, the lower castes, etc., have also exhibited communalist tendencies out of fear that the postcolonial Indian state would be dominated by the Hindus, the North Indians, the upper castes, leaving them with no option but to insist on extensive autonomy or even an independent state.[2]

The dynamism of the rising national movement was such that with the revival or "invention" of indigenous traditions and their defense against colonialism and westernization, even internal divisions were reinforced. Thus, toward the end of the nineteenth century, a political Hinduism began to take shape,[3] its exclusivist claim to Indian tradition giving rise to the fears of minorities: The increasingly articulate forms of the latter's particularism, in turn, reinforced the conviction of the Hindus that theirs was the only authentic Indian culture.[4] Communal tendencies on both sides, organized above all in the All-India Muslim League (established in 1906) and the Hindu Mahasabha (Grand Assembly of the Hindus, established in 1915) intensified. The idea of a separate Muslim state in South Asia, proposed by Muhammad Iqbal (1877–1938)[5]

Notes for this section begin on page 193.

in 1930, was taken up by Mohammad Ali Jinnah (1875–1948) and the Muslim League in 1940. The Muslims (and other minorities) were confronted with Hindu nationalists like Vinayak Damodar Savarkar (1883–1966) and Madhav Sadashiv Golwalkar (1906–1973); according to the ideology of Hindutva, India was the holy land of the Hindus and those who would not assimilate or accept being subject to the majority had to leave the country.[6] The identification of India as a Hindu nation implied the "exteriorization" of the non-Hindus.[7]

Against the backdrop of the emerging two-nation theory, the secularism of the Indian National Congress (INC, established in 1885) took shape. The INC sought to segregate the political sphere from the religious and cultural and to subordinate the communal differences to a joint resistance against colonial rule. Only a secular form of nationalism seemed likely to guarantee the ushering in of India's autonomy against a colonial regime that, with its policy of "divide and rule," knew how to exploit the internal conflicts to secure its own power and, at the same time, to present itself as a peacemaker.[8]

The advocates of this policy of secularism, however, found themselves caught in a continuous tension between the claims of indigenous traditions facing threats from outside and their mutual mediation inside. As secularists, they relativized the traditions that as nationalists they sought to consolidate. Moreover, the minorities perceived the tensions differently from the majority. The smaller communities, worrying about the survival of their traditions (which were not necessarily opposed to the Hindu majority) were often suspected of being communal, despite their overwhelming orientation toward a secular state. The Hindu communalists (who were as a rule anti-Muslim), on the other hand, were able to hide behind the democratic majority principle. Secularism itself thus could become a criterion of communalist demarcations.

Even after the attainment of independence in 1947 this confrontation did not lose any of its explosive potential. With the overcoming of colonial rule and the simultaneous partition of the country, sovereignty asserted itself in a communal and in a secular variant. In contrast to the Islamic Republic of Pakistan (this name was assumed in 1956), the Indian Union expressly enshrined its secular and multicultural character in the constitution (with an amendment in 1976); and the Indian government declared its political aim to be national integration as per the slogan "unity in diversity." To the Hindu nationalists, on the contrary, it seemed only logical that as the two-nation concept was now sealed with approval, the congruence of religion and nation should be established for India as well. Free now from the compulsions of considering concerted action against the colonial power, they could address the task of Hinduizing India all the more determinedly.

Communalism now is no longer a question of asserting particular traditions over an all-Indian nationalism. The "majority communalism" usurps nationalism, which itself becomes increasingly communal, both merging to a certain extent and thus confirming that "communalism is only a form of nationalism."[9]

It is secularism that now becomes the crucial counter-term, whose advocates present themselves more than ever as guarantors of internal peace and forward-looking development.[10] However, scrutiny of one issue—buried during the struggle for national independence—becomes imperative in postcolonial India: whether the indigenous tradition is present at all under a secularism that stands above all religious affiliations. Struggling to establish a perspective that does not imply the "conversion and the loss of one's culture,"[11] the Hinduization of India is presented as a continuation of the anticolonial struggle for self-assertion. It is seen as defending the particularity of India in the face of the persistent global process of its marginalization and its subordination to Western hegemony. Even as the secularists fight Hindu communalism as a reactionary or fascist force, the Hindu nationalists view the secular "progressive humanists" as accomplices of foreign powers, from which the Hindus' autonomy and equality is yet to be gained.

The communalists oppose secular universalism in the name of indigenous tradition; nevertheless, they complain also about the governements' alleged pseudo-secularism in dealing with religious diversity. They demand a "positive" secularism that would put an end to the special rights of religious minorities.[12] This dual direction in the arguments of today's communalists matches in a way the "double standards" of some avowed secularists who, while imputing fundamentalism to political Hinduism, exploit religious loyalties and occasionally practice a form of "soft Hindutva."[13]

These multiple confrontations testify in the first instance to the difficult balance of self-assertion between particularity and universality or, for that matter, tradition and modernity. But the dominant factor in the reality of communalism is the exclusion of the other rather than the assertion of the self. The recurrent excesses of violence testify to a hatred aimed at subjugation. A peak was reached in December 1992, when the Babri Mosque in Ayodhya, which was supposed to stand exactly at the birthplace of Lord Rama (Ram Janmabhumi), was stormed and torn down. The calculated consequence was a wave of communal clashes in many Indian cities that killed 1,800 people (according to official figures), predominantly among the Muslim population. What becomes evident in the dynamics of communalism, nationalism, and secularism is above all the loss of tolerance. The defense of particular traditions vis-à-vis impending alienation has led to the suppression of minorities, which are denied equal rights in their own state. What should have served as an assertion of cultural diversity has sometimes taken up forms of totalitarianism.[14]

The Communalization of History

Most critics of communalism in Indian politics are in agreement that it is closely linked to the interpretation and representation of the past. For many it is the communal view of Indian history that forms the core of the phenomenon: "without it little would be left of communal ideology."[15] The beginnings of the process of communalization have been related to the manner in which Indian history was treated by Western researchers and administrators at the time of establishing colonial rule: the discovery of early Indian history as Hindu history by William Jones, the contrasting of Hindu and Muslim civilizations and the periodizing of Indian history based on it by James Mill, the portrayal of the Muslims as invaders and despots by Henry M. Elliot, etc.[16] The gradual shift in Western interest from the more easily accessible Mughal period to the early Hindu culture, identified with Indian history as such, encouraged an understanding of tradition among Indians that highlighted the antagonism between Hindus and Muslims and increased the rift between them.[17]

In popular historical writing after 1850, Hindus often were viewed as the hereditary inhabitants of the country who constantly had to ward off foreign aggressors. The historical novels of Bankimchandra Chatterji (1838–1894),[18] the poems and dramas of Harishchandra of Benares (1850–1885),[19] and other works recalled primarily episodes from the period of Mughal rule, portraying the Muslims as violent conquerors and cruel tyrants and the Hindus as fighters for religion and the motherland. This portrayal was reversed by Muslim writers in Urdu literature, in which the Hindus figured as the villains and as enemies of true religion. Social, ethnic, and linguistic differences were played down so that the heroism of the Rajputs and Marathas could be adopted by the Bengalis too as a part of the national tradition, whereas religious differences became more distinct. In basing their sense of identity mainly on religion, the Indians were encouraged also, directly or indirectly, by Western tendencies to contrast, for nostalgic (Jones), emancipatory (Mill), or even self-critical motives (Max Mueller), Indian spirituality with European rationalism or materialism. The Indian struggle for self-assertion had started as religious reform movements; Bankimchandra too conceived the Hindu tradition as a "national religion" (*anushilan*) whose defense became almost a fight for salvation.

The incidental and diffuse hint of communalism in the representations of the past toward the end of the nineteenth century became more explicit one generation later. For Savarkar it was on the day "when Mohammad of Gazni [sic] crossed the Indus" that the "conflict of life and death" began, which from then on determined the destiny of India. The Indian past now was seen exclusively as the history of the Hindus in their fight against intruders from outside, which in turn meant that the nation was identified with the Hindu community: "all suffered as Hindus and triumphed as Hindus."[20] History was also rendered as a continuation of the legendary deeds of Rama; Indian territory became the "holy land" (*punyabhu*) and the writing of history the "holy work of the historians."[21] Among the Muslims, the idea of a persistant antagonism was confirmed. Jinnah referred to the experience of twelve centuries of history that "witnessed, during the ages, India always divided into Hindu India and Muslim India."[22] Moreover, the Muslim separatists could also count on the motivating power of an Islamic tradition of historiography.

To counter the pronounced communalist interpretation of history, which only confirmed British opinions on Indian disunity and oriental despotism, the advocates of secular nationalism brought up the experience of encounter and coexistence of religious communities in India's history. The policies of religious harmony practiced by the Buddhist ruler Ashoka (295–232 BCE) and the Mughal Emperor Akbar (1542–1605 CE) were cited as paradigms of a tradition of tolerance and syncretism.[23] According to the secularist reading, it was the very acceptance and integration of the diverse influences that went to make up the uniqueness of India.

This was supposed not only to mediate the inner cultural diversity with the assertion of Indian particularity but also to present the politics of modernization as a continuation of the national heritage. Cultural openness thus was offered by the secularists as a proof of the indigenous strength and future potential of India. The communalists, on the other hand, seized on it as precisely the reason for India's present weaknesses. The Hindu nationalists held the Muslim influence responsible for the decline of society and its backwardness as compared to the West. The Muslims, for their part, lamented the weakening of Islam due to the contaminating association with the idol-worshipping Hindus. Once the scene for primordial communities as subjects of historical agency was set, even hints of change and advancement became marks of demarcation and forms of disparagement. Thus the Muslims contrasted the superiority of their empires on Indian soil with the premodern level of Hindu political organization.[24] And in the language of Hindu nationalists, the very stereotype of the "tolerant Hindu" could have communal implications when contrasted with that of the "aggressive, fanatical Muslim." The communalization of the past also served as a strategy for sections of the traditional elites who in the name of indigenous culture and religion claimed the expected increase of agency in view of national independence.

Not very much changed in this confrontation of secular and communal uses of history, even after the end of colonial rule. In his "Presidential Address" at the first session of the Indian History Congress after attaining independence, Mohammad Habib held the false interpretations of the past partly responsible for the atrocities of partition. He demanded "that our historical vision will and must undergo a complete change with reference to all our past."[25] The Nehru administration recommended that the instances of interreligious harmony in the accounts of Indian history be emphasized, for this was required now in the interest of national integration and socioeconomic modernization. Tara Chand, a confirmed secularist, was entrusted with a government-sponsored account of the independence struggle;[26] he was preferred over R. C. Majumdar, who regarded the Hindu-Muslim conflict as a constant factor of Indian history and as an insurmountable antagonism.[27] The Hindu communalists, after a separate state of Muslims had become a reality, sought all the more to render the entire Indian past exclusively from a Hindu point of view.[28] Some went so far with the Hinduization of Indian history as to subject it to a sort of ethno-religious cleansing, highlighting the misdeeds of the Muslims and disputing their accomplishments or even, as in the case of the Taj Mahal, attributing them to the Hindus.[29]

In the continuing public debate over the representation of India's past, members of the academic community objected to communalist interpretations, stating that they contradicted the findings of research and even perpetuated certain patterns of colonial historiography.[30] The communalists, on the other hand, not only claimed indigenousness and authenticity for their accounts[31] but, capitalizing on the scientific prestige of researchers like Majumdar, who took great pains in his work to confirm the communal view of the past at the empirical level, they went on to accuse the secularists themselves of being biased by ideological prejudice.[32]

Even the most obvious forms of mythical renderings of the past are accompanied by the pointed insistence on empirical evidence. This can be observed also in the dispute over the Rama temple in Ayodhya. Some of the recent histories of Ayodhya, written in Hindi and narrated in the style of precolonial local chronicles, reproduce the Rama myth and the struggle for Ram Janmabhumi as national history.[33] According to these, in the first century BCE King Vikramaditya had built a temple at the birthplace of Lord Rama, which had been revealed to him in a miraculous manner. After the destruction of the temple by Muslim invaders the site became a major scene of Hindu resistance against Muslim rule, a struggle that is seen both as revenge against historical injustice and as a continuation of Rama's battle for the restoration of the eternal moral order (*dharma*). On the one hand the story is presented as a myth above all scientific doubts; on the other hand empirical evidence is used in presenting Rama as a national hero.[34] The contentions regarding the existence of the temple constructed by Vikramaditya and its destruction by the Muslims are based on a series of investigations and archaeological findings that contradict the allegedly distorted interpretation in secularist historiography.[35]

According to secular critics, it is the mixing of myth and history is a crucial strategy of communalism. Secularists plead instead for rationality in dealing with the past as a factor of tolerance: "When beliefs claim the legitimacy of history, then the historian has to attempt a demarcation between the limits of belief and historical evidence. When communal forces make claims to historical evidence for the purposes of communal politics, then the historian has to intervene."[36]

The Historicization of Community

In the eyes of secularist critics, communalism is based on an account of history that is distorted by religious fundamentalism and an irrational attitude and is to be corrected by means of historical research and reflection. Over the past decades, however, more and more skeptics have linked communalism with the concept of history itself, not only with its distortion and misuse. According to them, it is the historical mode of reasoning as such that has caused or at least encouraged intolerance and violence between the religious communities. The "historicization" of the forms of collective identity among Indians, as a reaction to colonial rule and by adoption of the Western approach to the past, is supposed to have changed the community fundamentally and turned communal awareness into communalism.[37]

Traditions pertaining to religious communities existed long before communalism came into being. Collective remembrance had primarily local allusions, and even if the Hindus as a whole perceived the past differently from the Muslims, this did not accentuate existing forms of conflict. But beginning with the arrangement of the first historical archives in the context of Western cognitive interests and administrative practices, the contrasts grew sharper.[38] The urge to reveal and record in the written word arcane and oral traditions, and the establishing of the past as an object of research, drove a further wedge between the communities.[39]

The past came to be perceived by the Indians in a manner that was hitherto unknown to them, and factual details acquired a new importance for practical issues. As the "orientalist empiricism"[40] of the Europeans gave rise to a nationalist empiricism on the Indian side, knowledge about the past became an element of collective self-understanding. When researching the reasons for the subjugation of Indians, Bankimchandra attributed it in particular to their lack of historical consciousness or its distortion through the writings of foreigners. The systematic recollecting of former greatness was supposed to transform the "sad memory" of the Hindus into the much-required "glorious memory."[41] The remembrance, especially, of their resistance to conquerors and invaders during the course of history came to be seen as a means of reattaining agency and autonomy.

The historical reconstruction or imagination of the struggle against the invaders had both integrating and disintegrating effects. Like other forms of communal agitation, they encouraged "a sense of 'oneness' among Hindus," and at the same time, "as a necessary corollary, tended to divide 'Hindus' from all 'non-Hindus.'"[42] It was no longer an issue of local communities. An "all-India Muslim community" was confronted with an "all-India Hindu Community." The multitude of coexisting and overlapping religious, cultural, and social structures or networks gradually turned into nations that rivaled and threatened each other.

That it was particularly religion that came to demarcate the self from the other was conditioned not just by Western classifications, for administrative or culture-comparative motives, but also by traditional forms of cohesion. Communalism was formed from these in particular through the communities' claim to historicity. The contrasting of the spiritual, world-renouncing, and unhistorical East with the materialistic, worldly, and purposeful West worked in a manner that linked religious and political aspects. Both Muslims and Hindus adopted the Western characterization of the East as spiritual,[43] but most of them refused to accept the lack of history that accompanied it. The preservation of tradition now became an issue of public opinion and an aim of concerted action.[44] It was not confined to the realm of literature and historical research; the public campaigns for cow protection and temple building, the controversy over Hindu or Urdu as official and national languages, etc. were all expressions of an increasing awareness of history. It always was related to concrete issues of power and economic interest.

The new historical consciousness of the Hindus was bound to contend with the historical self-understanding of the Muslims, who were committed to the glory of the Mughal Empire and the tradition of Islam. By losing power to the British, India, for the Muslims, had already become a foreign, non-Islamic country. And it could hardly become their own again through a process of decolonization that would result in the elevation of the Hindus as predominant community.[45] The moderate reformer Sayyid Ahmad Khan (1817–1898) therefore had believed that Islamic identity and tradition were most likely asserted under the protective rule of the British. For the new generation of Muslims raised under the impression of the national movement, a separate state became also imaginable.

Early nationalists like Bankimchandra, Harishchandra, and M. M. Malaviya (1861–1946) were painfully aware of the dilemma that the search for homogeneity of religion, language, and territory[46] reinforced the contrast with the nonhomogeneous. They not only included the Sikhs, Jainas, and Buddhists in the Hindu fold but also appealed to the Indian identity and the anticolonialism of the Muslims.[47] The demarcation of the self from foreign traditions varied considerably in this exploratory phase of nationalism. The recollection of the struggle against the Mughals could also be intended to serve the mobilization of the colonized population against the British.[48] Conversely, however, the actual experience of humiliation at the hands of the British Raj also renewed the grief over the earlier Muslim rule.

What at first seemed like a regrettable side effect of the new Hindu consciousness became its main characteristic, the more the religious community was redefined as a subject of history. This was the case with Savarkar who, in the face of the apparent variety of Hindu castes and sects, conceived *Hindutva* from the prerequisites of agency. According to him, because it was the experience of struggle for existence and power that constituted the political community of the Hindus, it was in the long and furious conflict with the Muslims that "our people became intensely conscious of ourselves as Hindus and were welded into a nation."[49]

In postulating historicity Savarkar went beyond the sphere of religion more manifestly than Bankimchandra. The history of the Hindus was "not only the spiritual or religious history of our people as at times it is mistaken to be ... but a history in full."[50] The appeal to religion here served rather to give an ultimate intensity to the feeling of oneness of the people.[51] When Savarkar laid claim to the religious quality of the Indian freedom struggle—"in the East all revolutions take a religious form"[52]—it was to create a spiritual aura for the actions of a community constituted by political will. Adherents of the Pakistan idea displayed a similar attitude; as with the Hindus, the genuine religious characteristics remained in the background. Jinnah was no more a confessor of religion than Savarkar.[53] And in both cases, while politics became equipped with religious symbols, the religious communitiy took on a secular militancy. "Hinduize all politics and militarize Hindudom"—this slogan of the 1940s expresses the dialectics of sacralization of politics and politicization of religion that made up communalism.[54]

The communalist interpretation of Indian history, which was pagan rather than religious, was countered by a secularism that confirmed this move to paganization. The experience of communal harmony is attributed by the modernist critics not so much to the religious, as for instance the Buddhist or Islamic, traditions; Ashoka and Akbar serve rather as proto-secularists whose politics of tolerance and reconciliation should be considered a paradigm of modern statehood.[55] Tolerance conceived in this way stands in every respect above the religions. The secularists envisioned not only an all-Indian nationalism against the particularism of religious communities, but also modernity as a guarantee of tolerance against the tradition invoked by the communalists.

It was also the priority of modern statehood over regional and local forms of community that determined the model of the Muslim or the Hindu nation.[56] Just as the sacralized history of the communalists neglected the religious tradition that they laid claim to,[57] the "new religion of secular nationalism,"[58] in the eyes of its critics, narrowed the space for diversity instead of expanding it. The postulated hegemony of the nation over the community and of secular tolerance over religious forms of lived coexistence strengthened communalism instead of subduing it. While the defense of particularity on the part of the religious minorities now appeared often as antinational and reactionary, the imposition of the majority culture could convey the impression of being modern and in the national interest. The communities that rooted out their nonmodern elements emerged as all the more irreconcilable, thus paving the way for partition.[59]

To many of those who conceived the overcoming of colonialism as an assertion of Indian historicity, the simultaneous experience of partition represented an ultimate challenge to it. And in any case it intensified the already existing tensions over how to define the identity of postcolonial India.[60] While secularists aspired to social emancipation in the framework of a multicultural state, Hindu nationalists all the more demanded a consequent Hinduization, thus completing the struggle against foreign rule. Nevertheless, as becomes evident in the polemics about the protection of religious minorities as a sign of "pseudo secularism," communalists even challenge the secularists' own claim to modernity. And notwithstanding their general distrust of academic history, they also present themselves occasionally as advocates of scientific rigor.[61]

The pervasive claim to historicity is also discernible in the conflict over the temple and mosque in Ayodhya. By an extreme reduction of the entirety of Indian history not only to Hinduism but to the city of Rama and to the fate of Ram Janmabhumi,[62] the freeing of the temple symbolizes the regained historical agency of the Hindus. If this is given the religious meaning of a struggle for the rule of Rama (*Ram Rajya*), Rama is not so much a mythical figure here as predominantly a national hero.[63] Certainty and efficiency, as demonstrated by the presentation of archaeological evidence and the exact enumeration of the seventy seven wars over Ayodhya and, as expressed also in the strategic planning of the warlike attack on the mosque, characterize the new coldbloodedness and ruthlessness of modern policy managers who exploit religious zeal for their purposes.[64]

To sum up: for those who look critically at the effects of the historicization of Indian society, it is the nation rather than the community, the secular and not the genuinely religious elements, that appear to be the problem. It is the "force and consistency of the nationalist-modernist project,"[65] that account for intolerance and violence. The communalists' call for "real history"[66] marks a breaking away from tradition, rather than its recovery. Modernization and secularization by themselves are causing an uprooting that in turn triggers the search for roots, the call for nationality, ethnicity, and other "pseudo-solidarities" as well as also the call for the revenge of history.[67]

Historicity and the Recognition of Cultural Diversity

The findings on the relationship between the communalization of history and the historicization of community fit into a more general suspicion about the link between communal violence and modernization. It was only with the disenchantment of the traditional sphere of life, which had not characterized them as religious entities that the hard antagonism between the communities was created. It was the precise enumeration, for instance, and the fixing of previously "fuzzy boundaries" that created the exact distinction between majority and minorities.[68] And it was these very same demarcations that the Hindu communalists used to confront the alleged "minority appeasement" of secularists with the majority's right to define the character of the state as Hindu. Even the government's attempt of the to prevent possible conflicts through guarantees for the minorities—quotas, reservations, etc.—can prove counter-productive in its practical administration.

This experience questions the secularists' monopolistic claim to tolerance and calls for a revision of the respective positions in the dispute over the rights of minorities. N. C. Saxena objects to the "secular" model of explaining communal collisions, stating that it does not adequately grasp the problem of legitimate group loyalties. According to him, an impartial stock-taking shows "that communalism *per se* does not threaten the stability of culturally plural societies, and that it need not be interpreted in a pejorative sense."[69] Ashis Nandy, for whom the intolerant Hindu nationalism itself is "an illegitimate child of modern India, not of Hindu traditions,"[70] pleads persistently for the rehabilitation of the nonmodern communities and their "internal principles of tolerance," often overlooked or undervalued by secularists.[71] Instead of projecting modern constructs of a state-assured tolerance back into the past, one should remember the forms of Hindus and Muslims' mutual acknowledgment at the level of "folk theologies" and the common grammars the communities had produced in the continuous interpretative efforts to live together.[72] These paradigms tried and tested in the past, along with the creative potential for mutual understanding and appreciation that existed in Indian society, are in danger of being sacrificed to the supposed superiority of a concept that meanwhile has empirically proved to be itself inadequate.

Indian society, so goes the criticism, in the process of its modernization has lost both authenticity and traditional pluralism, without having accomplished the internal peace pretended by the secular state.[73] The modern concept of history is regarded as a key element in this process: "[H]istory is today, not implicitly, but in the most explicit way possible, the pretext for violent political conflict in India." The demands that were raised in the Ayodhya conflict "become possible only within the modern forms of historiography."[74] During the last decades a grave suspicion of "historian's history" has grown—even among historians themselves. According to Gyanendra Pandey, the long standing lack of interest in research on atrocities during the time of partition too, indeed the relegating of the history of partition to that of independence, follows from the historiographical agenda itself.[75]

Such a general suspicion regarding the historical as an inauthentic and inappropriate presentation of the Indian past, however, may be countered with a different suspicion and a query regarding the underlying claim to authenticity: "Who speaks for 'Indian' pasts?" The analysis of the involvement of historical thinking in the manifestation of intolerance and violence could also being with the consideration that "we will need 'history as critique' in order to develop 'critiques of history.'"[76] Moreover, one could also ask in which sense the representation of the past in question is actually historical. Perhaps, to strengthen tolerance, the prevailing concept of history has to undergo even further historicization. Nandy himself, in his criticism of the historical discourse, undertakes a greater differentiation pointing in this direction when he links the "older traditions of social healing" in the "lived world of Hinduism" with a "principled forgetfulness (as an antidote to the ravages of modern historical consciousness)" and poses it against "another kind of 'principled' forgetfulness" of those who claim a monopoly in the matter of ethnic and religious tolerance and therefore neglect all traditional forms of resistance against religious violence.[77] Forms of forgetting correspond to forms of remembrance. Instead of pitting "historical" against "nonhistorical," which in a way inverts the claim to monopoly, it would be worth the effort to make the dialectic of forgetfulness and remembrance itself an object of historical reflection, and to confront the respective dispositions to tolerance or intolerance with principles of historical thinking.[78]

The claim to Hindu agency manifests itself especially in the form of a nativism that is not historical in the modern sense, i.e., an interpretation of change, but rather denounces as alienation any deviation from what is believed to be the unchangeable essence of traditional culture. The spreading ideas of the Hindus being the original inhabitants of India and of the discovery of modern statehood in the early Hindu empires express a striving for detemporalization. Even the historical sequence of seventy seven wars over Ayodhya[79] is inserted into the timeless setting of the Rama epic.[80] And the destruction of the Babri mosque, intended to give expression to historical agency, is portrayed equally as a reestablishment of the eternal order of *dharma*. Portrayed as the unchangeable and the original, the self takes on a sacredness that intensifies the struggle for self-assertion in a holy war. However, while human actions are deprived of their orientation toward temporal change, at the same time the divine order is drawn into the historical perspective and religion made a factor in the political power equation. The secular and the religious, the historical and the eternal are in a relationship of mutual displacement.

Such a relationship of mutual displacement between the temporal and the essential is typical of the "invented traditions" of nationalism. European national histories too are often fixated on origins, looking to find the "invariant parts" in a world of "constant change and innovation" and emphasize the "eternal and unchanging character" of the nations.[81] The inevitable change should be mediated with self-assertion by interpreting it as a renewal of mythical beginnings or a Golden Age. This inner conflict is especially pronounced in the context of colonialism. It is particularly the colonized, who, in a situation of forced change, seek to comprehend themselves as a modern nation from time immemorial in order to reconcile historicity and identity.[82]

The historical ideology of the colonizer contributes to the dilemma, inasmuch as it negates the historicity of the colonized and regards his connection to history as a gift of the West. In fact, what characterizes the colonizer's position toward India's past is not so much the "appropriation, but the denial of a valid history to the colonized."[83] The "denial of coevalness" and the "otherization of actions" have been identified as Western attitudes or strategies that aim at the systematic dehistoricization of the other[84] in order to legitimize colonial rule as the historicization of the nonhistorical, as a mission of civilization. The advancement of the West is given an exclusiveness that makes the relationship to the East's backwardness appear as one of essentially different types of society, graded in the hierarchy. This clearly contradicts the more historicizing tendencies in the Western self-conception that temporalized cultural differences and characteristics by relating them to stages of development.

The nativism of the colonized seems to reiterate this unhistorical typology of cultures insofar as it does counter the Western verdict with the assertion of indigenous historicity but remains within the purview of the same essentialist "thematic."[85] What is conspicuous about nationalism and what is especially discernible under the conditions of colonial rule—the interplay of historical and essentialist elements—is also characteristic of communalism. Hindus and Muslims are opposed to each other as both historical actors and eternal essences, modern nations and religious traditions. The modernist redefinition of the Hindu and Muslim communities as modern nations thus claiming equal status with the West lowers their ability for mutual understanding.

Secularists, instead of countering the colonizer's strategy of dehistoricization within the same typological (i.e., unhistorical) framework, attempt to conceive historical change as opening up new choices—directed equally against internal as well as external resistances. And they outline a rationalization of all spheres of life that, according to them, secures tolerance as much as social progress. For critics of secularism like Nandy, it is precisely this, the hegemony of the modern over the traditional, of science over myth, that causes intolerance—and not only in its colonial deformations. According to Nandy, modern historical consciousness as such is characterized by a determinism that suppresses alterity and reduces diversity. Even in the benevolent development theories of the Enlightenment, the European project of modernity, which aims at emancipation from conditions of violence and coercion, at equal rights and tolerance, is subverted by making that which is different disappear as a phenomenon of an earlier stage of evolution.[86] Skepticism is advisable with regard to a concept of tolerance, which arises not from the religions' mutual respect for each other, but exclusively from their absorption in the secular consciousness.[87]

By showing the links between the Western concept of history and the establishment of colonialism as well as the rise of communalism, Nandy exposes the ideology of a modernist power elite that almost completely neglects the traditional religious forms of dealing with alterity. The secularism it practices aims more at safeguarding the political sphere from supposedly retarding values, cherished only for tactical purposes, than at ensuring tolerance. Nevertheless, instead of countering the hostility to myth in the secular consciousness with the hostility to history of the nonmodern, the critique of historical determinism could be based also on the concept of history itself. History always includes an interest in alterity: after all, it is one's own experience of having been different that opens up the prospects for self-induced processes of change and thus makes agency historical.[88] In a manner akin to the alterity of one's own past, the presence of the other could also be understood as a source of the experience of historicity. The posteriorization of the other as preliminary stages of the self, however, not to mention its classification as unhistorical, does not take it into account as an alternative history at all. The exclusive reference to the development of the West, disregarding any other possible course of events, deprives the other of the openness of the future that makes change historical. The Indian secularists and development strategists who swear by progress and modernization agree implicitly to reducing their future to the present of the West; they reproduce within Indian boundaries the constricting determinism of colonial history, instead of countering it with their own historicity.[89]

The critique of both colonial historiography and the modernization paradigm of the secularists (but also of some of the "easy constructions" of the "national histories"[90]) could, therefore, be proposed as a quest for further historicization, leading to a concept in which the nonmodern is no longer reduced to an anachronism, but appears complete with its unrealized future. Instead of presupposing a self-denying assimilation into the "advanced" culture for emancipating the supposedly nonhistorical and making it historical, its historicity must be based on the openness of its own perspective.[91]

A sustainable historicization of this type, which does not repudiate alternative futures, serves the decolonization of the colonizer as much as the self-assertion of the colonized. It is in the historical interest of the former to counter the exclusion from history of the alternative and the repressed, as discernible in elite and mainstream accounts, in the unilinear evolution theories and especially in colonialism, with systematic efforts to regain "marginal voices and memories."[92] Oppressed, forgotten, or unknown dimensions of historical experience and orientation are thus made accessible: "plural pasts" can open up "plural futures."[93]

It is in the interest of the colonized and his or her "recovery of self" to counter the systematic dehistoricization from the side of the colonizer with a historically substantiated claim to the openness of his future. In view of the persistant global tendencies of colonization, the experience of the victims of historicization is to be situated within the very discourse of those who imagine their deeds to be in accordance with the universal sense of history. The oppressive presence of a deterministic view of history, of course, lies not only in tying the colonized down to Western patterns of modernization. It is also at work where the future is shaped exclusively as the renewal of a supposed origin or *status quo ante.* From the point of view of the subaltern, any form of representing the past that does not allow for change is suspect.[94]

As for the recognition of diversity in the Indian discourse on communalism, nationalism, and secularism, we can conclude that instead of postulating unqualified secularization as a necessary precondition for tolerance, in the context of a society distinguished by its religious traditions it must be possible rather to acknowledge the nonsecular paradigms of understanding and to bring them to bear, through historical reconstruction, as realistic political options in the horizon of expectations.[95] Historical thinking, which evidences thus a critical awareness of its colonizing elements and an interest in its own openness, precludes the need for the colonized to retreat defensively to the nativist bastions that are conducive to communalism, and it stimulates an articulation of oppressed and alternative pasts with a view to the future. This would (re-)establish history as a means of producing and safeguarding diversity. Even a tolerance rooted in tradition, then, need not be based exclusively on the hoped-for recovery of a nonmodern pluralism but can also realize its potentialities within the framework of historical thinking.

Translated by Nalini Adinarayanan

Notes

1. Besides the term "communalism" and its derivations, terms such as "sectarian conflict," "fundamentalism," "ethnic nationalism," "religious nationalism," "Hindu nationalism," "casteism," etc. are also used, thus indicating differences in the demarcation of the phenomenon, the context of its thematization, and the strategy of interpretation. It was only after completion of this study that I came across the instructive overview of the historical and social-scientific approaches to communalism by Peter Heehs, "Indian Communalism: A Survey of Historical and Social-Scientific Approaches," in idem, *Nationalism, Terrorism, Communalism: Essays in Modern Indian History* (Delhi 1998), 124–142. With good reasons, Heehs calls for a better incorporation of the findings of sociology, anthropology, and psychology into the historical explanation of communalism. The following reflections, however, focus on how the specific mode of historical thinking itself is involved in the conflict. This too has to be included in an interdisciplinary theoretical framework.

2. For an overview of nonreligious "subnational" movements, see Subrata K. Mitra and R. Alison Lewis, eds., *Subnational Movements in South Asia* (Oxford 1996). For the articulation of the demands for regional autonomy against the backdrop of the Hindu-Muslim conflict, see Dietrich Reetz, "In Search of the Collective Self: How Ethnic Group Concepts Were Cast through Conflict in Colonial India," *Modern Asian Studies* 31 (1997): 285–315.

3. Cf. Hans-Joachim Klimkeit, *Der politische Hinduismus: Indische Denker zwischen religiöser Reform und politischem Erwachen* (Wiesbaden 1981).

4. Gyanendra Pandey, "The Appeal of Hindu History," in Vasudha Dalmia and Heinrich von Stietencron, eds., *Representing Hinduism: The Construction of Religious Traditions and National Identity* (New Delhi 1995), 369–388, here 369.

5. Cf. Iqbal's "Presidential Address" (1930) before the Muslim League in Allahabad (William Theodore de Bary and S. Hay, eds., *Sources of Indian Tradition,* vol. 2, 2nd ed. [New Delhi 1991], 218–222); and Jinnah's speech before the Muslim League in Lahore, which led to its "Pakistan Resolution" (in de Bary and Hay, *Sources,* 228–231).

6. Cf. the programmatical texts of Hindu nationalism: V. D. Savarkar, *Hindutva: Who Is a Hindu?* 6th ed. (Delhi 1989 [1923]); M. S. Golwalkar, *We or Our Nationhood Defined* (Nagpur 1939). Savarkar was president of the Hindu Mahasabha from 1937 to 1942; Golwalkar from 1940 on was a leader of the militant voluntary organization Rashtriya Swayamsevak Sangh (RSS). Christophe Jaffrelot, *The Hindu Nationalist Movement and Indian Politics 1925 to the 1990s: Strategies of Identity-Building, Implantation and Mobilisation* (New Delhi 1996), gives a general description and analysis of the development of Hindu nationalism, showing its simultaneous stigmatization and emulation of the threatening other.

7. For the concept of exteriorization see Joachim Matthes, "The Operation called 'Vergleichen'," in idem, ed., *Zwischen den Kulturen? Die Sozialwissenschaften vor dem Problem des Kulturvergleichs* (Soziale Welt, supplement vol. 8) (Göttingen 1992), 75–99.

8. This came about, for instance, through the Communal Award (1932)—the establishment of separate constituencies for some communities of minorities, which was also included in the Government of India Act (1935).

9. Peter van der Veer, *Religious Nationalism: Hindus and Muslims in India* (Delhi 1996), 22.

10. For developments after 1947 see Gita Dharampal-Frick, "Zwischen 'Säkularismus' und 'Kommunalismus': Identitätsprobleme in der indischen Politik und Gesellschaft seit der Unabhängigkeit," in Wolfgang Reinhard, ed. *Die fundamentalistische Revolution: Partikularistische Bewegungen der Gegenwart und ihr Umgang mit der Geschichte* (Freiburg 1995), 133–149.

11. T. N. Madan, "Secularism in Its Place," *Journal of Asian Studies* 46 (1987): 747–759, here 754.

12. It is not just the Hindu nationalists who regard it as a betrayal of secular principles in favor of obscurantism and fundamentalism that in the realm of "personal law," which is differentiated according to religious communities, practices such as polygamy among Muslims or the unequal treatment of men and women in the law of inheritance are protected.

13. See Malini Parthasarathy, "Soft Hindutva and Nationalism," *Seminar* 411 (November 1993): 19–22.

14. See Pandey, "The Appeal of Hindu History," 370.

15. Bipan Chandra, *Communalism in Modern India* (New Delhi 1984), 209; see also idem, "Nationalist Historians' Interpretations of the Indian National Movement," in Sabyasachi Bhattacharya and Romila Thapar, eds., *Situating Indian History: For Sarvepalli Gopal* (New Delhi 1986), 194–238. As per K. N. Panikkar the "construction of communal history … [is] used as a powerful ideology for creating communal solidarity." K. N. Panikkar, ed., *Communalism in India: History, Politics and Culture* (New Delhi 1991), 2. Romila Thapar also observes the "interlinking of communal ideologies and the interpretation of history." Thapar, "Communalism and the Historical Legacy: Some Facets," in Panikkar, ed., *Communalism,* 17, 31. Asghar A. Engineer holds "vested interests" primarily responsible for the communalization of medieval history and sees herein a "political misuse of history": Asghar A. Engineer, *Communalism in India: A Historical and Empirical Study* (New Delhi 1995), 1.

16. William Jones, "On the Hindus" [orig. pub. 1786], in Peter J. Marshall, ed., *The British Discovery of Hinduism in the Eighteenth Century* (Cambridge 1970), 246–259; James Mill, *The History of British India* (London 1817); Henry M. Elliot, "Original Preface" to *The History of India As Told By Its Own Historians* (Calcutta 1867). Mill's dividing of Indian history into the periods of the Hindus, Muslims, and British is considered to be a prototype of communalism in Indian historiography.

17. For the role of Muslims as the "quintessential other," not only for the West but also for the Hindus, cf. Peter van der Veer, "The Foreign Hand," in Carol Breckenridge and Peter van der Veer, eds., *Orientalism and the Postcolonial Predicament* (Philadelphia 1993), 33–34. For the earlier forms of communal conflict see Christopher A. Bayly, "The Pre-History of 'Communalism'? Religious Conflict in India, 1700–1860," *Modern Asian Studies* 19 (1985): 177–203.

18. See, for example, the novel *Anandamath* of 1882. An English version translated by Nares Chandra Sen-Gupta appeared under the title *The Abbey of Bliss* (Calcutta 1906). For the significance of Bankimchandra see Sudipta Kaviraj, *The Unhappy Consciousness: Bankimchandra Chattopadhyay and the Formation of Nationalist Discourse in India* (Delhi 1995).

19. See, for instance, the play *Bharat Durdasha* (Poverty of India) of 1880. For Harishchandra's conception of national identity see Vasudha Dalmia, *The Nationalization of Hindu Traditions: Bharatendu Harishchandra and Nineteenth-Century Banaras* (Delhi 1997).

20. Savarkar, *Hindutva,* 42, 45. This identification was confirmed by the insinuated exterior view: "the enemies hated us as Hindus." Ibid., 45.

21. Ibid., 111 passim; Vinayak Damodar Savarkar, *The Indian War of Independence 1857* (Bombay 1947 [1909]), 1.

22. Quoted in Chandra, *Communalism in Modern India,* 215.

23. The secular reading of history becomes especially clear in the report of the inquiry committee set up by the INC over the communal riots in Kanpur in 1931. Here, the "historical misconception" of an enmity existing between the Hindus and the Muslims from time immemorial

is made out to be itself an important factor of the unrest. See *Congress Kanpur Riots Enquiry* (Allahabad 1933), XXVI–XXVII, 75–6; quoted in Gyanendra Pandey, *The Construction of Communalism in Colonial North India* (Delhi 1990) 251–252. Other examples of this secular picture of history are: Tara Chand, *Influence of Islam on Indian Culture* (Allahabad 1963), written in the 1920s; Beni Prasad, *The Hindu-Muslim Question* (Allahabad 1941); Humayun Kabir, *The Indian Heritage* (Bombay 1946); Jawaharlal Nehru, *Glimpses of World History* (New Delhi 1982 [1934–1935]) and *Discovery of India* (Calcutta and New York 1946).

24. In the political self-awareness of the Muslims the Hindus appear occasionally as merely a "subject race." Cf. Chandra, *Communalism in Modern India,* 229, 219–220. The British similarly legitimized themselves not only as founders of political unity and stability in India, but also as the ones who abolished the backwardness of the Hindus, for which the Muslims were to be blamed.

25. Mohammad Habib, "Presidential Address," *Proceedings of the 10th Session of the Indian History Congress* (Bombay 1947), 9–21, here 18.

26. Tara Chand, *History of the Freedom Movement in India* (New Delhi 1961).

27. According to Majumdar, alien rule had not begun with Clive and the battle of Plassey (1757), but with the conquest of India by the Muslims many centuries earlier. Cf. Romesh Chandra Majumdar, *History of the Freedom Movement in India,* 3 vols. (Calcutta 1961–1962). In the foreword and in an appendix to the first volume Majumdar, who felt that his draft had been overlooked for political reasons, gives a detailed report on the handling of the issue by representatives of the government.

28. See Pandey, "The Appeal of Hindu History," 369ff., who uses the term "Hindu history" for those accounts that are intended to be "Indian history" but not at the same time "Muslim history."

29. See P. N. Oak, head of an "Institute for Rewriting Indian History," in his book *Some Blunders of Indian Historical Research* (New Delhi 1966).

30. See especially the inquiry by Romila Thapar, Harbans Mukhia, and Bipan Chandra, *Communalism and the Writing of Indian History* (Delhi, Ahmedabad, and Bombay 1969).

31. The acceptance of a monolithic "Hinduism," however, meant the application of an alien—read Christian or Semitic—model of religion on to a completely different Indian religiosity that was characterized by the diversity of sects. Cf. Romila Thapar, *Interpreting Early India* (Delhi 1992), 71, and "Syndicated Moksha," *Seminar* 313 (September 1985): 14–22.

32. As a consequence, in the textbook controversy during the Janata Government (1977—1980), critical and Marxist historians became victims of government interference. Secularist textbooks on Indian history, and also the inquiry by Thapar, Mukhia, and Chandra referred to above, were blamed for producing distorted versions of the past. Cf. Vijay Chandra Prasad Chaudhary, *Secularism Versus Communalism: An Anatomy of the National Debate on Five Controversial History Books* (Patna 1977); L. I. Rudolph and S. H. Rudolph, "Rethinking Secularism: Genesis and Implications of the Textbook Controversy, 1977–79," *Pacific Affairs* 56 (1983): 15–37.

33. For the Ayodhya conflict as a problem of historiography see the special issues of *Seminar* ("Mythifying History," October 1989) and *Internationales Asienforum* (November 1994); with specific reference to the intertwinement of myth and history: Peter Heehs, "Myth, History, and Theory," *History and Theory* 33 (1994):1–19; Gyanendra Pandey, "Modes of History Writing: New Hindu History of Ayodhya," *Economic and Political Weekly* 29 (18 June 1994): 1523–1528. A general overview is offered by Sarvepalli Gopal, ed., *Anatomy of a Confrontation: The Babri Masjid—Ramjanmabhumi Issue* (New Delhi 1991).

34. "Myth has acquired the secular and a time-bound legitimacy of history and history the timeless quality and sanctity of a myth." B. D. Chattopadhyaya, in this volume. For the significance of "historicity" in the struggle for Ayodhya see Romila Thapar, "Historical Realities," *Seminar* 329 (1987): 42–45.

35. Pandey, "The Appeal of Hindu History," 370–371.

36. Sarvepalli Gopal et al., "The Political Abuse of History," in Vinay Chandra Mishra and Parmanand Singh, eds., *Ram Janmabhoomi Babri Masjid: Historical Documents, Legal Opinions and Judgements* (Delhi n.d. [1991]), 177–184, here 177.

37. While is mostly academic historians who figure among the first group, the position of the last is expressed primarily outside the discipline, partly in a critical turn against it in general. See Ashis Nandy, Shikha Trivedy, Shail Mayaram and Achyut Yagnik, *Creating a Nationality: The Ramjanmabhumi Movement and Fear of the Self* (Delhi 1995).

38. Bernard S. Cohn, "Notes on the History of the Study of Indian Society and Culture," in idem, *An Anthropologist among the Historians and Other Essays* (Delhi 1987), 136–171; Nicholas B. Dirks, "Colonial Histories and Native Informants: Biography of an Archive," in Carol Breckenridge and Peter van der Veer, eds., *Orientalism and the Postcolonial Predicament* (Philadelphia 1993), 279–313.

39. Dipesh Chakrabarty, "History as Critique and Critique(s) of History," *Economic and Political Weekly* (14 September 1991): 2162–2166, here 2162.

40. David Ludden, "Orientalist Empiricism: Transformation of Colonial Knowledge," in Breckenridge and van der Veer, eds., *Orientalism and the Postcolonial Predicament,* 250–278.

41. For the "sad memory" of the Hindus see Chandra, *Communalism in Modern India,* 220.

42. Pandey, *The Construction of Communalism,* 163. In the interests of the political integration of the Hindus into a "majority religion," the differences among the sects were ignored as much as the traditional exclusion of the outcasts, causing continuing tensions between the reformers and the orthodox. Singling out the Aryans as "immigrants" who had brought culture to India in contrast to the inimical "invaders" was as problematic and arbitrary as the inclusion of all the religions that originated in Indian territory as Hindu. This could also lead to communal reactions, as the example of the Sikhs shows.

43. Iqbal asserted the spirituality of the East as much as did Bankimchandra and Vivekananda. See the "Presidential Address," in de Bary and Hay, eds., *Sources of Indian Tradition,* 218–222.

44. "'History' became a means for asserting the identity of a community and identifying its rights and status." Pandey, *The Construction of Communalism,* 151. For the "Community as History" see ibid., 109ff.

45. For Muslim melancholy see Chandra, *Communalism in Modern India,* 234f.

46. As to the quest for homogeneity, Pratapnarayan Mishra proposed the Mantra "Hindi, Hindu, Hindustan" to all those "who don't know their own identity." Quoted in Dalmia, *The Nationalization of Hindu Traditions,* 27.

47. Harishchandra occasionally included even the Muslims when he addressed the Hindus. Cf. his famous "Ballia Speech," published in Pandey, *The Construction of Communalism,* 272–278, here 277. Also in Michael Gottlob, ed., *Historical Thinking in South Asia: A Handbook of Sources from Colonial Times to the Present* (New Delhi 2003), 173–175. To Malaviya, cofounder of the Hindu Mahasabha and the Banaras Hindu University, it was obvious that India would remain a country for Muslims even in the future. Cf. Pandey, "The Appeal of Hindu History," 369.

48. Bipan Chandra, in this connection, speaks about the cautious "backdoor" or "vicarious nationalism" of intellectuals, many of whom (like Bankimchandra) were in the service of the colonial masters: *Communalism and the Writing of Indian History,* 48. The regret about this dynamic becomes apparent in some formulations about the history of nationalism: "The history of Indian national movement, unfortunately, is also a history of the communalisation of Indian society." Panikkar, *Communalism in India,* 8.

49. Savarkar, *Hindutva,* 44.

50. Quoted in Gyanendra Pandey, "Which of Us Are Hindus?" in idem, ed., *Hindus and Others: The Question of Identity in India Today* (New Delhi 1993) 238–273, here 248.

51. "The ideal conditions, therefore, under which a nation can attain perfect solidarity and cohesion would, other things being equal, be found in the case of those people who inhabit the land they adore, the land of whose forefathers is also the land of their Gods and Angels, of Seers and Prophets; the scenes of whose history are also the scenes of their mythology." Savarkar, *Hindutva,* 136.

52. Savarkar, *The Indian War of Independence 1857,* 10.

53. Cf. Nandy's typology of "believers" and "nonbelievers": Ashis Nandy, "An anti-secularist manifesto," *Seminar* 314 (Oct. 1985): 14–24, here 15.

54. See for this tendency—denoted as kshatriyaization or masculinization of Hinduism—Nandy et al., *Creating a Nationality,* 68. This could even imply something beyond the borders of India and lead to the idea that India "dictate terms to the whole world." Savarkar, *Hindutva,* 141.

55. Akbar, who, according to Nehru, could be considered the "father of Indian nationalism," "placed the ideal of a common Indian nationhood above the claims of separatist religion." Nehru, *Glimpses of World History,* 306.

56. The ambivalent impact of the concept is also manifested in contemporary historical research, which was guided by the modern idea of the nation and sought to prove the ever existing "fundamental unity" of the Hindus or Aryans, describing the early institutions of "Hindu-Polity" in a clear analogy to the modern Western state. See Radha Kumud Mookerji, *The Fundamental Unity of India* (London 1914); Kashi Prasad Jayaswal, *Hindu Polity: A Constitutional History of India in Hindu Times* (Patna 1924).

57. Orthodox Muslims like Syed Abu'l-Ala-Maududi (1903–1979) saw in the idea of nationalism itself an element of alienation.

58. Cf. Pandey, *The Construction of Communalism,* 255.

59. Nandy, "An anti-secularist manifesto," 20. That the modern demand for the whole bequeaths only a schism in the place of the traditional, conflict-ridden juxtaposition of syncretic practices and orthodox transcendence is shown by Jan C. Heesterman, "Ein geteiltes Haus: Die hinduistisch-muslimische Beziehung," in Shmuel N. Eisenstadt, ed., *Kulturen der Achsenzeit,* vol. 2 (Frankfurt/ M. 1992), 275–293.

60. Regarding the historiography of post-independence India, Pandey states that "our senses of community and of history are reconstituted quite remarkably in and through Partition." Gyanendra Pandey, "Partition and the Politics of History," in Madhusree Dutta, Flavia Agnes, and Neera Adarkar, eds., *The Nation, the State: Indian Identity* (Calcutta 1996), 1–26, here 4. According to Pandey the partition of India was experienced as "the handiwork of some other people; *we* were not really involved." Ibid., 2.

61. Majumdar counters the "perspective of recent politics" with his own "strictly historical point of view" and "official history" with "real history." Cf. *History of the Freedom Movement in India,* vol. 1, xii–xiii. According to Majumdar the tradition of "Hindu-Muslim fraternity" desired by the government did not correspond to reality. An interpretation in this sense would therefore be "not only a great historical error, but also a political blunder." Ibid., xix.

62. Cf. Gyanendra Pandey, "Nationalism, Communalism, and the Struggle over History," in Mehdi Arslan and Janaki Rajan, eds., *Communalism in India: Challenge and Response* (New Delhi 1994), 50–60, here 56.

63. For the changing image of Rama see Romila Thapar, "The Ramayana Syndrome," *Seminar* 353 (Jan. 1989): 71–75. Savarkar himself was very explicit about this: "Some of us worship Rama as an incarnation, some admire him as a hero and a warrior, and all love him as the most illustrious representative monarch of our race." *Hindutva,* 93f.

64. For another case study, focusing on the 1990 excesses in Hyderabad, see Sudhir Kakar, *The Colors of Violence: Cultural Identities, Religion, and Conflict* (Chicago 1996).

65. Partha Chatterjee, "Secularism and Toleration," *Economic and Political Weekly* 29 (9 July 1994): 1768–1777, here 1771.

66. Pandey, "The Appeal of Hindu History," 370f.

67. Ashis Nandy, "History's Forgotten Doubles," *History and Theory* 34, no.2 (1995): 44–66, here 56.

68. Cf. Kaviraj, *The Unhappy Consciousness,* 113. The proto-Gandhian nationalism knew of neither an irreconcilable antagonism to the West nor a differentiation of the mainstream from the periphery. In Gandhi's thinking the enemy was always included. Cf. Nandy et al., *Creating a Nationality,* 80.

69. N. C. Saxena, "Historiography of Communalism in India," in Mushirul Hasan, ed., *Communal and Pan-Islamic Trends in Colonial India* (Delhi 1981), 302–325, here 303.

70. Nandy et al., *Creating a Nationality,* 78.

71. Ashis Nandy, "The Politics of Secularism," in Veena Das, ed., *Mirrors of Violence: Communities, Riots, and Survivors in South Asia* (Delhi 1990), 84.

72. "These grammars survive, in spite of the efforts of learned scholars to read them as folk theologies—as inferior, peripheral versions of Hinduism and Islam. They survive as a mode of mutuality and a major source of Indian creativity." Nandy, "An anti-secularist manifesto," 20. Cf. also Kaviraj, who differentiates between the "thin religion" of communalism and the "thick religion" of rituals and sect formations. Sudipta Kaviraj, "Kolonialismus, Moderne und politische Kultur: Die Krise Indiens," in Joachim Matthes, ed., *Zwischen den Kulturen? Die Sozialwissenschaften vor dem Problem des Kulturvergleichs* (Göttingen 1992), 219–238, here 236.

73. Ashis Nandy, *The Illegitimacy of Nationalism: Rabindranath Tagore and the Politics of Self* (Delhi 1994), 3.

74. Partha Chatterjee, "Claims on the Past: The Genealogy of Modern Historiography in Bengal," *Subaltern Studies* 8 (1994): 1–49, here 1, and "History and the Nationalization of Hinduism," *Social Research* 59, no. 1 (1992): 111–149.

75. Gyanendra Pandey, "The Prose of Otherness," *Subaltern Studies* 8 (1994): 188–221, here 204–205. Similarly D. L. Sheth, in this volume.

76. Dipesh Chakrabarty, "History as Critique and Critique(s) of History," 2166; idem, "Postcoloniality and the Artifice of History: Who Speaks for 'Indian' Pasts?" *Representations* 37 (1992): 1–26.

77. Because to remember them would deny the importance of their own categories. Cf. Nandy et al., *Creating a Nationality,* vii–viii.

78. Nandy himself admits that the process of secularization has brought about tolerance and respect for others that can be emulated. See Nandy, "The Politics of Secularism," 74.

79. One instructive example reads: "Where lakhs of lives were offered up. The history of Shri RJB—a saga of eternal sacrifice." In Pandey, "Modes of History Writing: New Hindu History of Ayodhya," 1524, 1527.

80. "Hindu history poses as an epic tale, of the 'eternal' contest between good and evil, Rama and Ravana, the gods versus the demons." Pandey, "The Appeal of Hindu History," 386.

81. Eric Hobsbawm, "Inventing Traditions," in Eric Hobsbawm and Terence Ranger, eds., *The Invention of Tradition* (Cambridge 1983), 1f., 10.

82. This is applicable to most of the neo-Hindu reform movements such as the Arya Samaj, which, after all, had to legitimize itself vis-à-vis the orthodox as well.

83. K. N. Panikkar, "Search for Alternatives: Meaning of the Past in Colonial India," in Università di Roma "la Sapienza," *The East and the Meaning of History* (Rome 1994), 263–277, here 264f.

84. For the "denial of coevalness" see Johannes Fabian, *Time and the Other. How Anthropology Makes Its Object* (New York 1983). For the "otherization of actions," according to which it is the historically meaningful that is attributed preferably to non-Indian factors, see Pandey, "The Prose of Otherness," 214. For the Oriental as nonhistorical discourse see also van der Veer, "The Foreign Hand," 25, 29.

85. For the distinction between the "problematic" and the "thematic" in the emergence of Indian nationalism (following Anwar Abdel Malek, "Orientalism in Crisis," *Diogenes* 44 (1963): 102–140); see also Partha Chatterjee, *Nationalist Thought and the Colonial World: A Derivative Discourse?* (London 1986), 36ff.; Chatterjee, *The Nation and Its Fragments: Colonial and Postcolonial Histories* (Princeton 1993). The essentialist reading of the East-West confrontation is enhanced also by the neo-Hinduist inclusivism of Vivekananda and others, according to which Hinduism implies anything non-Hindu. It can serve as a counter-strategy for the dehistoricizing colonial ideology of history, inasmuch as all future developments are anticipated here. Inclusivism, by which neo-Hinduism distinguishes itself from the missionary nature of Christianity and the aggressiveness of Islam as a Hindu form of universalism, dissolves historicity as much as alterity and does not substantiate either historical progress or tolerance. See Paul Hacker, "Religiöse Toleranz und Intoleranz im Hinduismus," in Lambert Schmithausen, ed., *Kleine Schriften* (Wiesbaden 1978),

376–388; Wilhelm Halbfass, *India and Europe: An Essay in Philosophical Understanding* (New York 1988 [Basel 1981]), 403–418.

86. Cf. Ashis Nandy, *Traditions, Tyranny, and Utopias* (Delhi 1992), 11ff.

87. Nandy, "An anti-secularist manifesto," 14.

88. Cf. Jörn Rüsen, "Geschichte und Utopie," in idem, *Historische Orientierung: Über die Arbeit des Geschichtsbewußtseins, sich in der Zeit zurechtzufinden* (Cologne 1994), 48–67; and idem, *Lebendige Geschichte: Grundzüge einer Historik III: Formen und Funktionen historischen Wissens* (Göttingen 1989), 121ff.

89. Nandy describes the "hierarchy of citizens," introduced by the secularists, in which the "hierarchy of nations" is reflected. Nandy, "An Anti-secularist Manifesto," 15–16. For the "blinded eye of history" see Ashis Nandy, Zia Sardar, and Merryl Wyn Davies, *Barbaric Others: Manifesto on Western Racism* (London 1993).

90. Pandey, "Partition and the Politics of History," 15.

91. The investigations initiated by Ranajit Guha over the struggles of the subalterns, who do not figure in the perspectives of the colonial or the nationalist elite, can be seen as examples of the attempt to recover such openness of past futures: *Subaltern Studies: Writings on South Asian History and Society* (New Delhi 1982ff.)

92. Cf. Pandey, "The Prose of Otherness," 214.

93. Nandy et al., *Barbaric Others,* 94.

94. Pandey, "The Appeal of Hindu History," 387. The "internal colonialism" of the historicizer that, according to Nandy, "uses the fact of the external threat to legitimize and perpetuate itself" (*The Intimate Enemy: Loss and Recovery of Self Under Colonialism* [Delhi 1983], xii), can also appear in the form of dehistoricization. Cf. instead Chakrabarty's demand "to write over the given and privileged narratives of citizenship other narratives of human connections that draw sustenance from dreamed-up pasts and futures where collectivities are defined neither by the rituals of citizenship nor by the nightmare of 'tradition' that 'modernity' creates." Chakrabarty, "Postcoloniality," 23.

95. The form of coexistence and mutual recognition suggested here comes close to what has been described by K. Raghavendra Rao in defence of Gandhian secularism: "In a society where 'religion' is an everyday mode, not a sociological dramatisation, secularism cannot mean a structural separation between the religious and the secular, but an arrangement in which all group identities, contingently 'religious', enjoy a fair deal." Quoted in Thomas Pantham, *Political Theories and Social Reconstruction: A Critical Survey of the Literature on India* (New Delhi 1995), 145.

CHAPTER 13

The Search for Scholarly Identity: Renaming the Field of History in Late Nineteenth-Century Japan

MASAYUKI SATO

Historia and *Rekishi*

"Nifonno Cotoba to Historia uo narai xiran to fossuru Fito no tameni Xeuani yauaragetatu Feiqeno Monogatari." ("The Tale of the Heike, made simple in order to help people who want to learn the language and history of Japan.").

This caption graces the title page of the "Fabian edition" of the *Heike Monogatari,* published by the Jesuit press at Amakusa, Kyushu, in 1592. The entire passage is in Japanese, with the sole exception of the word "historia," for which the Latin term is used, and in the original the entire passage—indeed the entire transcription of the *Heike*—is rendered in romanized Japanese. Could it be that there was no appropriate term in the Japanese language of the late sixteenth century corresponding to the Latin word "historia?"

In his *Dictionarium Latino-Lusitanicm ac Iaponicum,* published at Amakusa the following year, Ambrosius Calepinus offers five Japanese words as apposite renderings for the word "historia": *yurai, coji, raireki, denqi, yengui.*

Today we would unhesitatingly use the term *rekishi* for historia, and *rekishika* for *historicus,* but to judge by the Jesuit press, the best evidence of the Japanese lexicon we have for the turn of the seventeenth century, the term *rekishi* did not exist at that time. If that is the case, then when did this word that we now use so comfortably and naturally first come to be used, and what was its meaning at that time? New terms are often coined to accommodate new concepts, and come into being against a background of new ideas and cultural innovation. In the discussion that follows, I shall examine the history of the

Notes for this section begin on page 210.

intellectual transformation implied by the emergence of this new word, *rekishi,* in Japanese discourse.[1]

Rekishi as Chinese historiography

The word *rekishi* did not exist in China at the time of the *Chunqiu* (Spring and Autumn Annals) by Confucius (551–479? BCE), the *Shiji* (Records of the Grand Historian, 97 BCE) by Sima Qian, or the *Hanshu* (History of the Han Dynasty) by Ban Gu (32–92). The earliest documented use of the word seems to be in a biography of the *Nanqi shu* (History of the Southern Qi dynasty) by Xiao Zixian (489–537) in the late Liang period.[2] The word was rarely used in China before the Republican period of 1911–1949. Even *Peiwen yunfu* (*The Great Thesaurus of Phrases and Rhymes* [1711], 16) lacks the word in its vocabulary entry. The second documented use of the word seems to be in the title of Yuan Huang's late Ming period work *Lishi gangjian bu* (Chinese *lishi,* pronounced *rekishi* in Japanese), which is dated 1606.[3] This work collects and collates all the survey histories of China composed since Sima Guang's *Zishi tongjian* (A comprehensive mirror for aid in government, 1084). The *Lishi gangjian bu* was later brought to Japan, and after being published in a 1663 Japanese facsimile edition, remained in widespread use in Japan until the Meiji period (1868–1912).

One wonders if this might not have been the catalyst that affected the Japanese language, for the term *rekishi* began to be used from time to time as of the Genroku period (1688–1704). Starting with Kose Onsen's *Honchô rekishi hyôchu* (1690), some twenty or more works with *rekishi* in their titles were published during the Edo period (1600–1868), including Okamoto Tôyô, *Rekishi shôkan* (1781), Matsuzaki Sukeyuki, *Rekishi-chô* (1799), and Kawaseki Iko, *Honchô rekishi yôryaku* (1813).

The word *rekishi* is, though only rarely, found in some books by Hayashi Gahô, Hayashi Dokkôsai, Kaibara Ekiken, Ogyû Sorai, Kondô Morishige, Yamagata Bantô, and Satô Shin'en,[4] among others, and in essays by authors including Ôta Kinjyô, Nishida Naokai, and Hirose Kyokushô in the Tokugawa period.

The word *rekishi,* therefore, has clearly been in use for over 300 years. However, it is important to note that the significance of the word as used in the examples above was substantially different from the meaning of the word as we commonly use it today. First of all, the term *shi* (Chinese *shi*) originally denoted a Chinese court official historian, one whose duties were to keep the court's records. In later ages, the term also came to denote the work-product of the record-keeper, that is, books of historical record, or what we would today call histories. Burton Watson captures this sense handily in rendering the title of Sima Qian's *Shiji* as *Records of the Grand Historian,* in which the significance of the two elements of the title is clearly conveyed.[5]

The term *reki* (Chinese *lì*) seems, from the contents of the works noted above, to be best rendered in its sense of "passage through time," or "successive," in the sense of the "histories of the successive dynasties." Thus, in the parlance of the age, to "learn *rekishi*" was not to explore the facts of the past per se, but rather to become conversant with the historical books of the successive dynasties, and to make them thoroughly and completely one's own. That is to say, in the discourse of the neo-Confucian idiom, the histories were the fountainhead of all knowledge, and particularly, they were the storehouse of moral norms. This attitude towards the books of history is eloquently expressed in a passage from the "Taiheisaku" of Ogyû Sorai (1666–1728): "Xunzi says that scholarship is the Way to give wings to one's hearing, and extend one's vision afar. If we read this to mean, 'causing the ears to fly, and lengthening the vision of one's eyes,' it means that one can remain seated and yet hear of things thousands miles away, or see things thousands of years in the past, and this is the Way of learning."[6]

Having given this conceptual definition of learning, Sorai continues later:

> Nothing excels scholarship in producing men of talent. The path into scholarship is a knowledge of one's letters, and for this one should employ the study of the successive histories (*rekishi*). ... For they contain the facts of the successive dynasties, the Way for governing the country, the facts about the [great] military campaigns, and the goings-on of the world at peace, as well as the accomplishments of loyal ministers and dutiful officials. Rather than [merely] hearing about the Principle [governing the world], nothing will move one like observing the effects [of these actions and events through the reading of history].[7]

In the words of Confucius, "All the empty words I want to write down are not as clear and startling as seeing [their meaning] in action,"[8] which is a reiteration of the position that actions speak louder than words. This position is clearly stated in a letter Sorai wrote at about the same time the "Taiheisaku" was published, and included in a collection of Sorai's letters from the years around 1720:[9]

> Xunzi writes that all in all, scholarship is the Way to extend one's hearing and lengthen the range of one's sight [lit. "to give wings to one's ears and lengthen one's eyes"]. When we hear about other countries while remaining in this country, it is as if our ears had grown wings and we had become able to fly there and when, though we are born in the world of today, we are able to feel as if we are observing with today's eyes things that occurred several thousand years ago, this is lengthening the range of one's sight. Therefore, since it is the extending of our hearing and our vision that we call scholarship, scholarship reaches its pinnacle in the [reading of] the successive histories. He who is not versed in things both ancient and modern, both Japanese and Chinese, but only looks out from the confines of the customs of this country and this age, he is truly [of vision as limited as] a frog in a well.[10]

Here we already have, in a somewhat simple sense, the beginnings of the historicist vision, but it is important to note that Sorai's first purpose is not merely the reconstruction of the facts of the past. Furthermore, while it may seem on the first reading that the word "*rekishi*" could be taken in the sense of "research on the past," Sorai in fact is referring only to the "accumulated historical works of past generations," as is confirmed by the following passage: "I hear that Your Majesty [the shôgun] has recently been looking at the successive histories [*rekishi*], which I regard as an excellent idea. I hear Your Majesty is reading the *Tongjian gangmu* [in which case may I recommend also the] *Zishi tongjian*."[11]

It may seem that the word *rekishi* could be taken in the sense of "accumulated Chinese historical works of past generations," but the word was also applied to the Japanese historical works, as is seen in his *Keishishi yôran* (Outlines of classics, philosophy and history); "thus our country possesses *Rikkokusi* [six national histories], histories [*rekishi*] of Japan."[12]

Rekishi as a translation of "history"

Currently we use the term *rekishi* to mean either: (1) The events of the past, and the development of those events (*res gestae*), or (2) books about those events and developments (*historia rerum gestarum*); these are the two meanings the term is held to have.[13] It is extremely rare to find this term used in documents predating the Meiji Restoration of 1868, when Japan made both political and cultural commitments to the importation of Western technology and culture. In those few pre-Meiji instances when the term appears, it proves on examination to have been employed in a quite different sense from its meaning in common usage today. This naturally raises the question of when and how the term came to develop into the familiar word we use today, and to take on its common modern meanings. To anticipate myself, let me state my conclusion that the term seems to have taken on this meaning around 1873, and that this was the result of both the massive importance of Western culture from late Tokugawa times into the early Meiji period, and of the 1872 promulgation the modern system of public school education.

The years from the late Tokugawa period to the early Meiji period witnessed great changes in the nature of knowledge in Japan. Specifically, the culture that was seen as the appropriate object of study—as what one should study—changed from being Chinese to that of the West. Furthermore, in the flow of modern scholarship imported from the West at that time, historical inquiry changed from taking Chinese textual criticism in the style of the Qing dynasty as its norm, to modeling itself on European-style positivism. This took place in earnest beginning in the late 1880s, but forerunners of this movement were active in the late Tokugawa period, translating and introducing Western works of history as a way of gaining information about the people and customs of the West.

What is of particular interest here is the Japanese-language terms that those translations used when translating *geschiedenis* (Dutch), *history* (English), *histoire*

(French), and *Geschichte* (German) into Japanese.[14] But for the most part the books they were translating were histories of someplace, so the translators were able to make do with terms like *shi, ki, kiryaku, shiryaku, shimatsu, shunjû,* or the like. The compound *rekishi* apparently had a concrete existence of its own, and consequently it was not initially used for the conceptual category of "history," a rather more abstract expression. Indeed, the word *rekishi* was hardly used at all until scholars turned their interest to the conceptual processes of history. Therefore, let us first turn to an examination of the Japanese terminology used by the European language—Japanese language dictionaries compiled in the last decades of the Tokugawa period and the first years of the Meiji era, to see what words were used to render these European terms in Japanese.

It is logical to begin with dictionaries of Dutch, as these were the first European-language lexicons to appear in Japan. Iiizumi Jô's *Warana bunten jibiki* (Dictionary of Dutch letters), published in 1856, rendered *geschiedenis* with the word *rekishi,* but earlier dictionaries did not. The 1848 dictionary *Yakken* (The key to translation) offers *kyûwa* (old tales), *kiroku* (records), and *setsuwa* (legends), while in Murakami Yoshishige's 1854 *Sango binran* (Handy guide to the three languages), the Dutch word *geschiedschrijver* is rendered by the Chinese character *shih* (the *shi* of *rekishi,* which has the ancient meaning of "record-keeper"), glossed with the reading *kirokukaki* (writer of records). In Katsuragawa Hoshû's 1854 *Waran ji'i* (Glossary of Dutch), which appeared the following year, 1855, he rendered both history and *geschiedenis* as *engi,* a term whose meanings range from "the history of," or "the origins of," to "legends," and thence to "omens," "portents," and "luck"; they are bound together by the concept of "origins," and in the premodern Japanese tradition the *engi* genre was dominated by more or less fabulistic tales of the histories or origins of shrines and temples. For *geschiedschrijver,* Katsuragawa Hoshû also used the phrase "*engi o tsuzuru hito*" (a person who compiles an *engi*). It should be clear, then, that there was no consensus by the mid 1850s on the term *rekishi,* and that it was not yet the common term for what we now call "history."

From the mid 1850s on, however, the main focus of interest in foreign languages in Japan shifted from Dutch to English, reflecting the intellectual impact of the appearance of Commodore M. C. Perry, and the forced entry of Japan into the Western-style comity of nations in the face of pressure that came primarily from England and America. Hori Tatsunosuke's 1862 *Eiwa taiyaku jisho* (English-Japanese dictionary) translated "history" with the two words *rekishi* and *kiroku,* and for historic, historically, historiographer, and historiography, he offers *rekishi, rekishi no gotoku* (as in history; like history), *rekishi o kaku hito* (a person who writes history), and *rekishi o kaku koto* (the writing of history), respectively.

In J. C. Hepburn's *Waei gorin shûsei* (Japanese-English dictionary), published in 1867, *rekishi* is listed along with *raireki, koji, yurai,* and *engi*. But *rekishi* does not seem to have been in very wide use at that time, as is illustrated by Petitjean's revised and expanded edition of *Calepinus's Dictionarium,* published in 1870 under the title *Lexicon*, in which Petitjean feels constrained to use the tortuous explanation, "*mucaxi arixi cotouo qiroquxi, coji raireqiuo caqi xirusu mono*" (one who records what happened in the past, or writes the history [*raireki*] of ancient matters) to render *historicus.* While the title *Ei-ka ji'i* (English-Chinese dictionary) may suggest that it is somewhat different from the other works we have examined, in this 1869 work the compiler, Yanagisawa Nobuo, renders history as *shisho* (history book), and *shiki* (historical record), and for historiographer, he offers *shikan* (official record keeper). The *Wayaku Eijisho* (Dictionary of English translated into Japanese), also published in 1869, translated historian as "*kirokusha*" (recorder), but for other relevant terms uses phrases including the term *rekishi:* history as "*rekishi*" or "*kiroku,*" historic as "*rekishi no,*" historically as "*rekishi no gotoku,*" historiographer as "*rekishi o kakuhito,*" historiography as "*rekishi o kaku koto.*"

From this point onward, most dictionaries applied the gloss *rekishi* for history, and *rekishika* for historian. And in Ishibashi Seihô's 1872 *Kaitei zôho Eigo sen* (Revised expanded English dictionary), there is an entry *rekishika,* which he renders as historian; by contrast, there was no entry for *rekishika* in the original 1861 edition of the same dictionary. It is likely that it was during this decade that the term *rekishi* developed to the point that it came to the attention of lexicographers.

This inference is buttressed by examining dictionaries of French. Murakami Hidetoshi (a.k.a. Yoshishige) published the first edition of *Futsugo meiyô* (Essential clarifications of French) in 1864. At first, he rendered *histoire* as "*shi; kiroku,*" *historien* as "*shi o kaku hito,*" *historiographe* as "*shokô no shi*" (the scribes of the feudal lords), and *historique* as "*shi no,*" but in the 1871 *Kankyo Futsugo jiten* (Officially sanctioned dictionary of French), all these terms are rendered with phrases including *rekishi*: *histoire* has become *rekishi/shôsetsu* (the term that today means novel); *historien* is *rekishika*; *historiographe* has become "*rekishi o kaku hito,*" and *historique*, "*rekishi no.*"

Beginning around the year 1872 dictionaries of German, too, started to use the word *rekishi* to translate the class of words relating to history. In the *Haiwa Shûchin jisho* (The first pocket dictionaries) (1872), *Geschichtsbuch* is rendered as "*rekishihon*" (history book) and *Geschichtskunde* as "*rekishigaku,*" while the following year Matsuda Tametsune, Senoguchi Takamasa, and Muramatsu Tsuneharu also used the term *rekishi* as the preferred gloss for German words incorporating *Geschichte* in their *Dokuwa jiten* (German-Japanese dictionary, 1873).

Judging from the way it was used in Japanese dictionaries of foreign languages, therefore, it seems safe to conclude that by 1872 the term *rekishi* had come to be the accepted Japanese equivalent for words in European languages corresponding to the English word "history." It is a relatively straightforward matter to determine when the term achieved general acceptance, but rather more difficult to say why *rekishi* came to be the agreed-upon translation for history. The suggestion has been made that the single-character term *shi* was simply too weak a word, and that *rekishi* was more euphonious, but it seems more likely that the choice was dictated by the desire to emphasize the fact that this was a new concept taken from the European intellectual tradition, as distinct from the Chinese style of historical writing and inquiry, which had constituted the Japanese historiographical tradition until then. Perhaps these lexicographers reflected the desire of historians to express, through their use of this rarely employed word, the timbre of freshness associated with this new, imported European culture.[15]

Rekishi as a school curriculum

The word *rekishi* itself seems to have staked its claim as the Japanese equivalent for history in the final years of the Tokugawa period and the first years of Meiji, but at that time its use was still limited to a very small circle of people. What spread knowledge and familiarity with the new term among large numbers of people was the modern education system being constructed under the aegis of the Meiji state.

The modern Ministry of Education, founded in 1871, set about the task of creating a modern system of education. The following year, the ministry promulgated its "Education Regulations." In the Supplement to Ministry of Education Ordinance No. 13, there was a proliferation of detailed regulations for each subject in the curriculum, and *rekishi* was included as one of those subjects. This subject was called *shigaku* in the elementary and middle school curricula, but in the curricula of schools of foreign languages, law schools, and legal preparatory schools the subject was called *rekishi*.

This term *shigaku* has often been used since ancient times in China, the oldest known use being in the *Jinshu* (History of the Jin Dynasty, 644/48). In the biography of Dan Le in that work a man is said to have been appointed to the ritual post of *shixue* (pronounced *shigaku* in Japanese). In the *Songshi* (History of the Song Dynasty, 1345), a passage in the biography of Yang Yi states that "Emperor Zhen Zong has praised his profound knowledge on *shixue*."[16] The word *shigaku* seems to be more familiar word than *rekishi* in that period.

One month after promulgation of the ministry of Education's Supplement to Ordinance No. 13, on 8 September 1872, the ministry promulgated an "Extraordinary Ministry of Education Ordinance" with even more detailed "Rules for Elementary Education." The new regulations established a "lecture course in historical studies [*shigaku*] as part of the curriculum through all the grades of the higher middle schools. The texts were prescribed for each grade level: *Survey of the Royal Dynasties*, *The Shorter History of Japan*, *The Shorter World History*, *The History of the Five Continents*.

But on 19 May 1873 the ministry promulgated ministry of Education Ordinance No. 76, which revised the title of the lecture course in historical studies from a course in *shigaku* to a course in *rekishi*. The only thing that was changed, however, was the name of the course; the list of prescribed texts was no different from what it had been before the revision.

This change in course titles effected between 1872 and 1873 was also implemented in the *Rules for Middle School Instruction*. The curriculum for each grade in the Supplemental Ministry of Education Proclamation of 8 September 1872 specified a course in *shigaku,* while the following year the ministry issued a supplemental proclamation on 23 April 1873 that revised the *Abbreviated Rules for Intermediate Education:* at every grade level in the lower middle schools, the course in *shigaku* had been renamed a course in *rekishi*. Similarly, the Ministry of Education's Proclamation No. 61, dated 3 May 1873, and Proclamation No. 74, dated 18 May 1873, *Curricular Regulations for Schools of Foreign Languages,* consistently used the term *rekishi* as the title of that course in the curriculum dealing with history. The ministry's first annual report, issued that same year, also uses the course title *rekishi* in discussing the curriculum of the Kaisei Gakkô College, and the *Regulations for Education in Normal Schools,* issued in June of 1873, mandate a course in "*Hongoku rekishi*" (The history of this country).

In the early years of Meiji, there was some confusion between the terms *shigaku* and *rekishi,* a confusion that continued for a few years. But at least, insofar as can be told from examining the formal curriculum of public school education, educators had initiated the switch from *shigaku* to *rekishi* as the terminology of choice around 1872/73. As a consequence of the firm establishment of the term *rekishi* as a course title in the elementary and middle school curriculum, the term was commonly used thenceforth in the title of school textbooks. It is for these reasons that down to the current day the word *rekishi* has become so familiar that it rolls comfortably off the tongues of very young children. The word *shigaku* is still used in the names of some historical journals, or to name a university department of history. It is, in a sense, a convenient word for avoiding the ambiguity of the two meanings of the word history, *res gestae* and *historia rerum gestarum,* both designated nowadays by the word *rekishi*. Meanwhile, the cultural transfiguration whereby the word *rekishi,* since it was introduced as a Japanese rendering of history, has also accepted the original ambiguity requires further examination.

Chinese re-importation of the term *rekishi*

In China, from the days of Yuan Huang in the early seventeenth century until the end of the Ch'ing dynasty (1644–1911), this word *rekishi* was hardly ever used.[17] With the advent of the Republican period (1911–1949), the word came to be widely used in China as well, but it seems fairly certain that it was first used in China as a result of reverse cultural importation from Japan, where it had already come into widespread use. This can be said not only about the word itself, but about changing conceptions of the content and practice of history as well.[18] One excellent example is Liang Quichao's 1922 opus, *Zhongguo lishi yanjiufa* (Research methods in Chinese history). Influenced by Ernst Bernheim and other Western European historians, this is one of the earliest attempts to set forth a methodology for the study of Chinese history based on modern Western historical concepts. But it is clear from the examination of the catalogue of Liang's personal library that he obtained most of this knowledge through Japanese mediation. And of course, in those works, and Liang's, the word *rekishi* loses its traditional meaning of the "histories of the successive Chinese dynasties," and is promoted in its new meaning of "the events of the past."[19] The word *li-shih* [*rekishi*] seems to have become a regular part of the Chinese vocabulary at this time, too.

The difficulties of understanding the concept of *rekishi*

The word *rekishi* was revived—or, it would be better to say, born—along with the modernization of Japan. Part of the background of the term's first modern use in Japan was the wave of modernization or Westernization that swept over Japan in the late nineteenth century. The issues involved do not stop with the adoption of a new word.[20]

The word *rekishi* became a fixed part of Japanese vocabulary in the early Meiji period. But did the Japanese people have an accurate understanding of the objects and practices denoted by the word at that time? In short, did the success in achieving a translation of the term necessarily resolve the fundamental problem of translating the concept behind it? This question must be considered in the context of the larger process of the importance of Western culture taking place at that time.

In the early years of Meiji, Western culture was introduced into a whole variety of fields: the humanities, engineering, natural sciences, social sciences, etc. In each case, the major problem in introducing these new fields was the question of how to translate terminology expressing the various concepts of each field into appropriate Japanese-language terminology. In the case of historical sciences, however, the situation was somewhat different from that of the natural sciences, for the traditional study of history had existed in Japan since ancient times. Moreover, the study of history had occupied an exceptionally important position in the body of Japanese tradition. In such circumstances, it was even more difficult to select words of sufficient freshness to express the newness of the field, and yet to correspond to the European terms for history.

The problem lay with the second meaning of the term, the concept of history as the study of past events. What were the methodologies it employed? How did these differ from traditional Japanese historical scholarship, and what were its special characteristics?

Mitsukuri Genpo says in the introduction to his 1856's Japanese translation of J. Bosscha's *Schets der Algemeene Geschiedenis en van die des Vaderlands* (1838): "*Geschiedenis* or history is that which records causalities of big [past] events as they really happened with such information as places and dates, without embellishment or gaudy style. And, we already have *Geschiedenis* of Netherlands, or of France. Thus, we will name *Algemeene Geschiedenis* [it should be translated into Japanese as *Zenshi* or *Sôshi*: all history; total history] that which records big [past] events which happened in every nations and peoples with the above style."[21] Evidently Genpo noticed that the concept of *Geschiedenis* or history could not adequately be expressed with the existing words that meant historiography in Japanese.

The search for answers to questions such as these began in earnest in 1875. That year the Historical Office established by the Council of State began planning for the continuation of the *Dai Nihon shi* (History of Great Japan) project conducted by the historical staff of the Mito school, originally projecting to write a formal history on the Chinese model.[22] On consideration of the style chosen for the Mito project, however, they decided that, rather than follow the *Dai Nihon shi's* organization, in which each Japanese imperial reign constituted a chapter of narrative, they would adopt a strictly chronological style for the central narrative, to which they would add "Kiji honmatsu tai" (Chronological narratives of events with their roots and branches) and commentaries, employing the historical format of the West.[23] However, in the light of the fact that at the time, "We have no personnel amongst the students of Western books who are trained in historical specialties, and we are completely without [historical] works in translation, in which we might find suitable examples of that style,"[24] in 1878 the Historical Office appointed Suyematsu Kencho (1855–1920), then posted to the Japanese Legation in London, to investigate European methods of historical survey. This marks the first beginnings of attempts in earnest to adopt the historical methods of Western Europe.

At the end of his stint in London, Suyematsu prepared an "Instructions" in English, and after negotiating with the Hungarian-born English historian G. G. Zerffi, requested of Zerffi that he prepare a textbook on the theory, philosophy, methodology, and history of European historical studies. Suyematsu's "Instructions" to Zerffi comprise twelve queries, including questions on the importance of historical research, the problem of historical narration, a request for a list of exemplary marks of history, on what might be the proper object for historical study, and on the judgment of the historian.

In response, Zerffi composed *The Science of History* (1879), which was immediately sent to Japan and translated into Japanese. Yet for a variety of reasons, it was never published.[25] Its influence on the Japanese historical community was not great, yet ironically, the specific questions put by Suyematsu in his "Instructions" to Zerffi show how unknown the concepts of European historiography were in Japan.

However, not until the arrival in Japan of Ludwig Reiss (1861–1928), in 1887, would European historical concepts take firm root among Japanese historians. Reiss, a student of Leopold von Ranke, was most active in Japan as a professor at the Imperial University in Tokyo, where he guided many young Japanese historians through their introduction to the theories and methods of modern European positivist historical scholarship.[26] With this, a conception of the field worthy of the name *rekishi* finally took root in Japan. It had been a quarter-century since the term *rekishi* was first used as a translation for "history."

Conceptual translation

I wrote at the outset that new words are born against a backdrop of new ideas and cultures. In the last analysis it is a question of how to translate and assimilate the concepts that these new terms represent. But it must not be forgotten that the traditional elements of the intellectual heritage also play a major rule in the acceptance of new concepts. At least in regard to the immediate issue, the acceptance and establishment of this new conception of *rekishi* was predicated on the historiographical tradition that was already part of Japanese culture, which carried with it the latent potential for acceptance of these new ideas.[27]

And so, inevitably, the content of what was transmitted to Japan had itself to be transformed in the process. But, if the traditional scholarship proves of no use in the acceptance and reintegration of new scholarship, then it will certainly be impossible for the culture to translate the concepts on which the new scholarship is based.

It is imperative that we keep this reciprocal relationship of indigenous and imported scholarly traditions in mind at all times. The clear understanding of this point is the starting point for the study of modern Japanese historical scholarship, and simultaneously, its ultimate destination.

Notes

1. On East Asian historiographical traditions, see my "Historiography and Historical Thought: East Asia," *International Encyclopedia of the Social and Behavioral Sciences,* vol. 10 (London 2001), 6776–6782.

2. Kobayashi Noboru, "Edojidai ni okeru shinpo no shisô (The idea of progress in early modern Japan)" in idem, *Chûgoku Nihon ni okeru rekishikan to in'itsushisô* (The idea of history and hermitism in China and Japan) (Tokyo 1983), 178.

3. Kanô Kôkichi, "Rekishi no gainen (Concept of *rekishi*)," (orig. pub. 1921) in Abe Yoshishige, ed., *Kanô Kokichi ibun shû* (Tokyo 1958), 115.

4. Sato Masayuki, *Rekishi ninshiki no jikû* (Historiographical time and space) (Tokyo 2004), 4–7.

5. Sima Qian, *Records of the Grand Historian,* 2 vols., transl. Burton Watson (New York 1961).

6. Ogyû Sorai, "Taiheisaku," in *Ogyû Sorai* (*Nihon Shisô Taikei*), vol. 36 (Tokyo 1973), 453.

7. Ibid., 485.

8. *Shijii,* chap. 70. On Confucius as a historian, see my "Kong-zi" in Lucian Boia, ed., *Great Historians from Antiquity to 1800* (New York 1989), 62–66.

9. It is not certain when Sorai composed the "Taiheisaku." See Maruyama Masao, "'Taiheisaku' kô," in Sorai, *Ogyû Sorai,* 787–829.

10. Ogyû Sorai, "Sorai sensei tômon sho," in *Kinsei bungaku ronshû* (*Nihon kotenbungaku taikei*), vol. 94 (Tokyo 1973), 187–188.

11. Ibid., 187.

12. Shimada Kenji, ed., *Ogyû Sorai zenshû* (Tokyo 1973), 526.

13. W. H. Walsh, *An Introduction to Philosophy of History* (London 1951), 119.

14. *Geschiedenis* and *Geschichte* on the one hand, and history and histoire on the other, have mutually independent etymological origins. However, since they are used synonymously within their respective linguistic fields, I have taken them together here. See Dudenredaktion, ed., *Duden: Etymologie* (Mannheim 1963), 21, 267. Also see G. Scholtz, "Geschichte," and H. W. Bartsch, "Geschichte/Histoire," in Joachim Ritter, ed., *Historisches Wörterbuch der Philosophie,* vol. 3 (Stuttgart 1974), 344–399.

15. On the other hand, Kanô Kôkichi has written, "Chinese prose is not content to express words in single characters, and one might feel that the two character term was a result of the [Chinese prose style]." Kôkichi, "Rekishi no gainen," 115. Yanagita Kunio has pointed out in his "Rekishi kyôiku ni tuite" (On the teaching of history) that "The words which start from the r series consonants leave quite a fresh impression upon a person's ear." In *The Authentic Complete Works of Kunio Yanagita,* vol. 24 (Tokyo 1963), 95–106.

16. Other examples will be found in *Peiwen yunfu* (The Great Thesaurus of Phrases and Rhymes) (1711), vol. 9. It is important to note that this enormous Chinese dictionary of quotations lacks *lishi* as the vocabulary entry, but not *shixue.*

17. Only one example is found in Zhang Xuecheng. See Noboru, "Edojidai ni okeru shinpo no shisô, 179.

18. Reverse cultural importation from Japan of scientific terms that could be seen in the Humanities brought China such important words as society, science, and philosophy, among others. See Yanabu Akira, *Honyakugo seiritsu jijyô* (Tokyo 1982).

19. For Liang's attitude toward Western historiography, see my "Rekishigakka to shiteno Tanaka Suiichirô (Tanaka Suiichirô as a Historiologus)," *Kindai nihon kenkyû* 7 (1991): 63–88.

20. See my "Historiographical Encounters: The Chinese and Western Traditions in Turn-of-the-Century Japan," *Storia della Storiografia* 19 (1991): 13–21.

21. See Ôkubo Riken, "Bakumatsu no seiyô tûshi," in Kaikoku hyakunen bunka jigyôkai, ed., *Sakokujidai nihonjin no kaigaichishiki* (Tokyo 1953), 441.

22. See my "Tokugawa Mitsukuni," in Boia, ed., *Great Historians from Antiquity to 1800,* 292–293.

23. See my "Shigeno Yasutsugu," in L. Boia, ed., *Great Historians of the Modern Age* (New York 1991), 448–449.

24. *A Written Report of the Historical Office,* 7 February 1878.

25. Jiro Numata, "Shigeno Yasutsugu and the Modern Tokyo Tradition of Historical Writing," in E. G. Beasley and E. G. Pullleyblank, eds., *Historians of China and Japan* (Oxford 1961), 264–287.

26. Kanai Madoka, *Oyatoi gaikokujin* (Tokyo 1976), 118–126.

27. See my "Japanese Historians," in Boia, ed., *Great Historians of the Modern Age,* 437–460.

CHAPTER 14

History and Cultural Identity: The Case of Japan

SHINGO SHIMADA

In a schoolbook published in 1920 by the Ministry of Education for teaching Japanese at elementary schools in Japan, one lesson bearing the title "Letters from the Isle of Truk" clearly depicts a certain attitude:[1]

> It has been three months since I came to this island, so that now I know how life functions here. Regardless of whether it is winter or spring back home, here on the island it is always like the summer we know at home ... Surprisingly enough life is quite pleasant here. As the islands nearby are all governed by our country, there are many fellow countrymen, and I do not feel lonely ... To be truthful, *the natives are not very civilized yet,* but they are peaceful and friendly towards us. The children speak Japanese fairly well, as our country has set up schools.[2]

What this passage depicts is the colonial attitude that was conveyed to the children of the mother country. Based on approaching the development of civilization from a historical aspect, one's own culture is compared to the culture of the "others" on a linear historical scale, thus justifying the superiority of one's own culture and the legitimacy of colonizing other countries. On the path to a better future, "introducing civilization" linked the colonizers to those undergoing colonization. Underlying this text is the idea of the Occidental modern spirit that represents the goal of civilization. The school as an institution, which Japan itself had adopted from the Occident only a few decades earlier, is where the first step toward becoming civilized is taken.

Worthy of note is, in particular, that a non-European society first adopted this Eurocentric approach and then, after a relatively short time, had absorbed it to such a degree that this ambivalent relationship was hardly further reflected. How does such cultural disparity arise within the relatively compact East Asian-

Notes for this section begin on page 222.

Pacific sphere? My text examines this question and the process of establishing Japan as a nation-state, whereby the concept of history as understood in the Western world was adopted and contributed to cultural identity. This chapter deals therefore with the period from 1868 to about 1910, during which the foundation for the construction of cultural identity was laid. I will focus, furthermore, on the translation, i.e., the transformation process, of a concept of time rooted in the Western world and not based on the concept of time as traditionally understood within the Japanese culture. I should like to use the introduction of the curriculum vitae to demonstrate this. By doing so I will try to show how this particular concept of time was not only accepted by the elite class but also by the common people. My thesis is that it was through this process alone that the consciousness of belonging to the Japanese nation became possible. The standardization and institutionalization of the curriculum vitae made it possible to present the "imagined community" (Anderson), which permits a new relationship towards otherness.

First, I will discuss how Japanese scholars, prior to the adoption of the Western history model, regarded the Chinese civilization as a foreign culture. I will continue then with the rise of a national identity based on the gradual introduction and acceptance of the curriculum vitae. My final topic will cover disparity, and how a new relationship evolved between Japan and Europe, China, and Korea.

The concept of progress and cultural difference

One major distinction of recorded history in Japan, before Western civilization and the Western approach towards history was adopted, seems to me to be the absence of the Western philosophy of progress. Although recording history was one of the main occupations of Japanese scholars at that time, the idea of a "universal history," which implies comparing societies and cultures along chronological lines, had obviously never occurred to them. By no means may we assume that the social processes underway in Japan were not understood. The expression *hirakeru,* which dates back to this time, stands for the process of opening. It was used to describe, for example, how a village blossomed into a prosperous city. It is important to emphasize the difference between this expression and that of progress. *Hirakeru* stands rather for a subjective impression for which there is no universally valid equivalent as is claimed for progress: *hirakeru* does not stand for the concept of mankind on his way to a better future. What is also missing is the concept of active participation on the part of man. Furthermore, *hirakeru* does not imply any qualitative-ethical improvement of the nature of man. The opposite is more likely the case, for change just as well opens the door to man's ruin.[3]

What the expression *hirakeru* does enable us to do is compare different circumstances according to a time scale. Motoori Norinaga (1730–1801), a critic of neo-Confucianism, wrote in 1789 that culture in China "opened" earlier than it did in Japan (see above). This very statement demonstrates that Chinese culture was viewed as belonging to the past. To the intellectuals of the time, whose main occupation was the study of classical Confucian texts, Chinese culture was already past history.

With this background in mind the Japanese elite of the Edo period, who favored the studies of Confucianism, were faced with a dilemma. Japan, seen from a universal Chinese-ethnocentric perspective ("Middle Kingdom"), was considered to be a barbarian country, all the more so as it met the criteria of "barbarism," such as government by warlords and not by scholars, the men's hair-style, and the rituals of daily life that did not follow Confucius. In view of this, intellectuals were forced to take up different stands. If one accepted the Chinese perspective, as the orthodox followers of Confucius did, then one's own culture had to be classified as "barbarian." In contrast to this, a "national" school of thought (*kokugaku*) developed, which proclaimed one's own universality.[4] The period of time marking this Chinese-Japanese relationship is ambivalent. To me it seems it was then that awareness arose that the importance of the Chinese civilization actually lay in the far past. Ito Jinsai (1627–1705) stressed, for example, the need to interpret classical Confucian texts as belonging to the past, and in doing so expressed his distrust of contemporary interpretations of Confucian texts in China. In the eighteenth century this distrust was eventually underlined by the adoption of knowledge from Europe: doctors performing an autopsy compared anatomical charts from Holland with what they saw and discovered that the description of the human body in Chinese literature did not correspond to the reality.[5]

This demonstrates Japan's intellectual spirit in the eighteenth century and the growing attitude that the superiority of the Chinese civilization belonged to the past. At the same time no contemporary system of measuring time was available. The Chinese past could be used to compare and view one's own reality. However, comparing the Chinese culture with one's own culture in the present like an arrow through time was too great a novelty to find acceptance. The result was that neither did Japan absorb the Chinese perspective of universality nor did it develop its own universal standpoint based on a historical approach. This ambivalent relationship to China continued up to the middle of the nineteenth century and entailed the budding desire to eventually take a definitely superior stand.

Japanese intellectuals understandably were then all too willing to adopt the Western idea of progress to define the relationship between the Chinese civilization and Japanese culture. With the opening of the country and the adoption of European civilization Japan had, according to the new time scale, overtaken China. Japan was well on its way to becoming "civilized," whereas China was approaching a "standstill."[6] The Japanese elite class of that period could thus not only determine Japan's position within the capitalistic world order but also legitimize its own class station within the country. Its members were "more" civilized than the remaining population, and they assumed it was their legitimate task to educate the "common" people. They argued, too, that they, in contrast to the Shogunate who had tried to uphold the old "feudal" structure, were the pioneers who had paved the way for "civilization."

The idea of progress and its implied meanings of civilization and enlightenment and the "idea of history" (*Geschichtsbewußtsein*), were important components in constructing a cosmology of a newly created Japanese nation.[7] Tenno's station became the nucleus of the new Japan. On the one hand, he took on the role of a modern enlightened monarch; on the other, the official "history" of Japan now centered on his 10,000-year-old genealogy. It is interesting to note that in the construction of this new cosmology, modern, rational means were applied along with conventions dating back to the "closed doors" era and traditional "mystical" rites, thus cementing the origin of Japan in the eighth-century mythologies "Kojiki" and "Nihonshoki," which were duly incorporated in the newly institutionalized school system and taught in history and ethics classes.

The adoption of historical time in Japan

Integral components of the new cosmology mentioned above included the introduction of the Gregorian calendar and the Western form of time measurement in 1872.[8] This marked the beginning of a new era. For the very first time a standard time regulating system was set up for the entire country. The importance of this step becomes evident in the changes that followed. Time was to be regulated in all of the provinces and even in the remotest cities "in the same manner." It was not easy, though, for this system to find acceptance throughout the country; in rural areas, in particular, there seemed to be no need to comply, since for practical purposes farmers still followed the seasons as prescribed under the old solar-lunar calendar. Different cosmologies collided here. The state endeavored to impose the new national cosmology on its subjects in the name of enlightenment, whereas in rural areas village life continued to be regulated according to traditional self-supporting time units. The call to maintain the old calendar was voiced by many peasants who revolted in this period.[9] However, if the villages were to be integrated into the state apparatus, the old calendar had to be dropped.

The process of integration was carried out, to be sure, by restructuring the villages into administrative units, but its primary strategies were the construction of a school system and the general acceptance of national/public holidays whose celebration was to be experienced simultaneously throughout the nation. Up to this point in Japan's history each village had had its own time regulation system for organizing everyday life; now, the restructuring and integration of villages into the public network required the coordination of time and the calendar for the nation as a whole. As modern institutions like the school, police, and administration were set up, this standard regulation of time applying to the whole country reached the villages, and local village units were integrated into the network, unifying public institutions. Various villages were reorganized, according to quantitative criteria, into administrative units, and corresponding to this, for the very first time the public territory was divided up into identical units.

Between 1873 and 1885 several school laws were passed to consolidate and expand the institution of school. Over the years a pyramid-like and centrally organized administrative unit for schools and villages was set up with its apex in the capital city Tokyo.

The restructuring process designed to adapt the provinces to the needs of a modern nation-state, unveiled a growing tendency to replace the initially proclaimed liberal ideal of education for the benefit of the individual with one grounded in nationalism. An example of how the regulation of time was linked to a public symbol is a regulation passed in 1883 requiring all teachers to declare their loyalty to the nation in writing on four newly created public holidays that celebrated Tenno as a public symbol. The elementary school law passed in 1886 was an attempt to standardize the forms of elementary education and achieve a high degree of uniformity between the capital and the provinces. Above all, official events and ceremonies became standardized in all schools.

In 1890 a revised edition of the law on elementary schools was issued in which the educational aim was clearly defined and laid down. National holidays were to be devoted to ceremonies and were obligatory for all schools. The ritualization of national holidays, in which the school played a central role, was thus enforced. The elementary school was, moreover, not only an institution of learning but a symbolic center for each local unit, the site of regularly performed political rituals that pupils, teachers, town councilors, and parents had to attend. From this moment the national holidays no longer stood for time free of the duties of school and work but for time used for ritual celebrations.

It became a custom to present schools with a portrait of Tenno. The photographed portrait of Tenno that could be acquired, to begin with, in 1874 through the Nara prefecture, was later available from all the prefectures. Later on it was distributed to all of the schools as the influence it exercised was great and inquiries were enormous. The portrait was treated with the same respect one paid to Tenno himself and was kept in a building erected especially for that purpose.[10] The reception of the portrait at each school was organized as a ceremony to which each community was invited. Thus, not only was rural religious life transformed into a national ritual, but a prescribed ritual was performed throughout the nation in every community at the exactly same time in exactly the same way. On every national holiday the same ritual was performed at the same time and before the same portrait; nationwide, this commonly shared experience strengthened spiritual bonds between the people and Tenno.[11]

During this phase teachers were faced with the question of which teaching materials were to be used to teach the children. Various books were translated from Western languages. For classes in ethics (*shûshin*) ten different translations of Wayland's *Elements of Moral Science,* for example, were published between 1873 and 1881. M. Willson's *The First Reader of the School and Family Series* (*Shôgaku dokuhon*) was translated for elementary schools in 1873 as well and was widely used.[12] In our context, however, the translation of Samuel Smiles's *Self-Help* (1859) by Nakamura Masanao (1832–1892), which was published in 1870 under the title *Seigoku risshi hen* and became a bestseller at the time, proved to be highly significant.[13] This collection of biographies of successful persons of European and North American societies demonstrates how success was related to personal effort.[14] Furthermore, from then on most new readers and books on moral science were based on *Self-Help*. In *Self-Help* the lives of Edward Jenner (1749–1823), James Watt (1735–1819), Benjamin Franklin (1706–1790), and David Livingstone (1781–1873) serve as examples. Benjamin Franklin's life deserves particular attention, as his autobiography took on a paradigmatic character for the Japanese society of the Meiji era.

Parallel to the introduction of the Western model of a personal course of life, the structure for leading a modern way of life was set up. Acceptance of this modern way of life was symbolized in the new meaning applied to "the watch." From 1900 on it became a custom in newly founded state universities to award highly talented graduates with a silver pocket watch.[15] The pocket watch thus came to mean something special and became a status symbol of the elite. At the same time this shows how the concept of an ideal course of life was laid down whereby the lives of the men mentioned above served as models: the silver pocket watch stood for the possibility of following a successful course that, contrary to the hierarchical structured society, was open to everyone. The pocket watch stood for applying a method to one's own course, a method that in everyone's imagination guaranteed success for every citizen. To be sure, the gulf between the intellectual elite, who enjoyed concrete opportunities to achieve this status symbol, and the common people was great. What counts here, however, is the establishment of a public education system that was attended as an institution, where personal experience in forming one's own life was gathered. Even though it may have remained a vision, reaching out for this status symbol was a worthy goal. The new education system, organized in a pyramid-like structure, introduced several models of a personal record, depending on when one finished school. In view of this one can understand the growing importance of school education for the common people.

It could mean moving up a class as well as relinquishing the role prescribed by one's family. The models for a personal course of life offered new possibilities for second- and third-born sons. Meanwhile, the translation of the term *katei* for the English expression "home" created a model course for women.

This expression gave birth to the ideal concept of family life, which was modeled after the European concept in which the husband was head of the family. In 1892 the newly launched magazine *Katei zasshi* (Family magazine) expressed its disfavor of the patriarchal form of the family and praised marriage based on mutual love.[16] To analyze this process the sociologist Kenji Sato referred to a bestseller published between 1903 and 1904.[17] In this novella by Murai Gensai, *Shoku dôraku* (The enjoyment of dining), everything is centered around family meals. The intention of the novella is families to certain forms of eating and etiquette. It is very instructive as it reveals the ideal concept of the family as expressed in *katei*. Moreover, it is interesting to note that it is addressed to women who are expected to feel responsible for family meals and cooking. The book depicts everyday household chores and propagates the necessity of preparing traditional homemade meals, for which the expression *katei ryôri* is introduced, a topic of great importance in the discourse on role of the today's family.

What strikes us is the new form of living together in the nuclear family, where the housewife has become responsible for the household (*kaji*). Turn-of-the century magazines and novellas written mainly for women characterized the ideal family as follows: (1.) monogamous; (2.) based on mutual love; (3.) consisting of father and mother with their legal and unmarried children (nuclear family); (4.) the husband employed; (5.) the wife a housewife.[18] In this context the expression *katei* became popular in colloquial speech. We need to keep in mind that from the beginning the expression was used to define a family living in harmony. The more widely *katei* was used, the more it tended to differentiate between gender, as it obviously occurred mainly in magazines written for women.[19] It goes without saying that this concept of family represented a translation of the idealized European family. In this respect a deep gulf remained between concept and reality, which in turn was discussed in women's magazines and reached its peak when women were seriously pressed to marry Europeans.[20] In spite of this discrepancy, the image of the housewife as a role model became established within the first decade of this century, so that the historian Masanao Kanô designated the period between 1919 and 1970 as the "age of the housewife."[21]

On the other hand, with the codification of the Civil Law in 1898 the "family" underwent a process of homogenization. A particular form of family life, i.e., patriarchal as denoted in the Japanese word *ie,* was declared binding for the whole population.[22] This process in turn prepared the ground for one of the most important principles of the Japanese nation, that of a nation based on the family system. The term "family nation" is derived from this concept, with Tenno standing at the head of the "family" as the father of the nation. The idea of the family played a central role in generating ethnic homogeneity of the population.

If biographies of famous Western personalities served as models for taking one's life in one's own hands, publishing Tenno's genealogy served to convey the course of the nation's history. Its origin lay in mythology, in the myth of "Nihonshoki." It is evident that the origin of the nation was sought along a clearly linear path. By officially proclaiming the birth of the nation, the Japanese people were constituted as a historical subject that remained the same through its whole history. Continuity was considered to be guaranteed in "Tenno's 10,000 year-old reign," which in addition was held to explain Japan's superiority over all other nations. The official proclamation of Japan's origin is interlocked with the acceptance of the Western idea of progress. With the adoption of the social theories taught by August Comte, John Stewart Mill, and above all, Herbert Spencer, it became possible to outline Japan as a society embedded in a clearly defined nation-state that, its historical origin manifest, could now be understood. This concept of a society backed by its own history thus enabled this very society to participate in shaping its future as it progressed toward a better, more civilized way of life that held the promise of happiness for everyone.[23] Individual biographies were intertwined in this common destiny, and each individual could make his personal contribution to the nation's history. In terms of Benedict Anderson's view, the concept of nation as a imagined community had now become real and presented a perspective for the future.[24]

Consequently, adopting time in a historical sense made it possible to position "others" in this newly created world order. It turned out then that the Western world was much closer to the goal of the development of civilization; other Asian societies like China and Korea in comparison lay far behind Japan. Meiji-Tenno played a dual role: on the one hand, he stood for continuity dating back to the mythical origin of his birth; on the other, he stood for the enlightened ruler who was to act as a motor in the civilization process. Each role positioned him at the opposite ends of historical time: "He is the past as well as the present, He is temporal as well as eternal." To be sure, Tenno's body represented past history, but at the same time it became also timeless. Furthermore, his body represented the nation as a geographical territory with the earth elevated to the realm of holiness.[25] These two aspects could be taught in moral science and history classes. Taking one's life in one's own hand enabled everyone to make use of the possibilities this newly founded society offered and, during the standardized rituals performed on national holidays, to experience the spiritual bonds with the nation/body. Officially proclaiming the origin of Japan to lie in the temporal past enabled each member to accept society as a "natural" one within which all people were able to take steps to shape their own way of life.

The outbreak of wars crystallized and accelerated this process of constituting history and tradition. The Chinese-Japanese War (1894–95) and the Russo-Japanese War (1904–05) serve to illustrate the central character of the school in spreading national awareness. The outbreak of war was publicly announced in the schoolyard, and starting with the Russo-Japanese War many schools were used as military training camps or temporarily as barracks. Victories were celebrated at school with the pupils, so that the school, before radio and television had come into use, functioned as a center for communication.

Genuine interest on the part of the general public was first sparked by both wars; prior to them it would have been completely inconceivable. Each personal background became intertwined with the making of the nation's history, as each person could relate his or her personal destiny to the destiny of the nation. Local tradition, ancestral roots, and family ties no longer fulfilled a primary function in decision-making processes, as it was now possible for each individual to make a personal decision to serve the cause of the nation. During this period the concept of equal "citizens" belonging to a nation gradually spread. Each person could contribute his or her share to "making history," whereby the passion supporting war is, under the circumstances, more easily understood. In an elementary school-reader published in 1894 the story is told of the heroic death of a soldier fighting a battle in the Chinese-Japanese War, which up to 1941 was to leave its mark on many generations. Although he has been seriously wounded, the soldier blows the signal to attack. He is pictured as lying on the battlefield with the instrument at his mouth. This man, who from 1904 on appeared in the state schoolbook on moral science as Kiguchi Kohei, was stylized into a national hero. He happened to come from family of farmers, so it was easy for a large part of the male population liable to conscription to identify themselves with this hero.[26]

Biography and historical time

Reflecting on what has been said up to this point, the degree to which "history" and "biography" were interlocked between 1868 and 1910 is evident. The concept of progress plays a central part in both ideas, at the same time it binds them together. The individual biography makes sense only when it can add its contribution to the progress of one's own society.

If the codification of the Civil Law Book laid down the eldest son's duty to become head of the household, meaning it was his responsibility to provide food and children for the family, then shaping the model of taking-one's-life-in-one's-own-hand permitted the other sons, according to their personal qualifications, to climb the career ladder they were destined for. The standard role women were to fill was family-oriented, with great importance placed on marriage and the birth of the children. One can conclude then that in this phase a social structure was constructed that, depending on one's gender, was conceivable as a more or less "modern" approach to shaping one's life and was practiced to a large extent. Coinciding with this, one's course of life became institutionalized as one progressed through the hierarchy of the school system. Introduced along with this was a scale by which one's performance was assessed, a factor just as important for understanding the situation today. Surely this model was applied rigorously to only a minority, but it is of importance as it had already begun to act as a model for everyone. The rise of the "middle class" can be attributed to it, and people of low income who had had the opportunity to further their education moved into the "middle class" from 1910 on. What became plausible was the idea that, based on homogenization and institutionalization of one's own personal biography, a homogenous population would form the backbone of the Japanese nation.[27]

It is important to remember that the concept of the nation with a homogeneous population became a starting point of the Japanese social sciences. The adoption of the term "society," upon which social sciences were constituted, also included the concept of the "national history" (*Gesellschaftsgeschichte*), which was derived partly from the idea of progress as propagated during the Age of Enlightenment and triumphant in the evolutionism of the nineteenth century.[28] The society was regarded as a closed unit of a nation, from within which the next stage of civilization evolved. The stage was thus prepared for the ensuing discourse on colonialism, which reached academic circles just one year before the annexation of Korea in 1910. In view of the criteria set by progress, the principle of applying a chronological system to the systematic appropriation of the past was elevated to a political platform.[29]

This situation was demonstrated by the economist Tokuzô Fukuda (1874–1930) who applied the system to determine Japan's historical standpoint in relation to Europe and Korea. Fukuda, who studied in Germany between 1889 and 1901, refers in his treatises to Karl Bücher's theory of development, which Bücher had published in his main work *Die Entstehung der Volkswirtschaft* (1893). Basing his opinion on Bücher's concept of development, Fukuda stated that Japan, like Europe, was on its way to "well-regulated evolution."[30] He maintained that Korea, where he had spent several weeks in 1902, was a negative example. Whereas the economies of Japan and Europe had developed from a feudal system with class allegiances to prosperous cities and eventually a national economy, Korea's economy had remained at the level of clan society with clan allegiances. The conclusion was that Japan's advanced culture empowered her to civilize the Korean population. This approach, even if it was not intended to justify colonization, was adopted by most social scientists and absorbed to such a degree that they failed to question its validity.

The general public surely took little note of this orientalistic discourse on colonialism, but the idea of a personal biography was one of the examples used to convey the new notion of time to each individual. Whether a man be a soldier of low rank or a farmer sent to cultivate the land, for example in Manchuria, he was well able to perceive the obvious gap between his culture and that of the population being colonized. The groundwork for this could be found in the structure of the education system, from which the feeling stemmed that all citizens shared spiritual bonds.

Finally, let me point out that the disparity in the regulation of time as presented here has to be considered within the overall degree of tension inherent in interaction between the civilizations of the Occident and the Orient. The countries of East Asia in turn adopted this time system, but this process is generally known to have been overshadowed as by the aspiration of the Japanese elite to legitimize imperialistic moves and a theory backing them. The concept of a "Great Asia" is a counterdesign to Occidental imperialism and entails the necessity to free Asia of all European powers. Culture as the Japanese understood it constituted itself at two levels, as the culture singular to the Japanese nation and a culture within the cultures of the Orient. No contradiction is involved here, as the singularity of East Asian culture is a prerequisite for defending it against advances from the Occident. The idea of a "Great Asia" as a counterdesign first developed after Japan had taken the lead and adopted the Western form of progress. Japan's understanding of itself as a nation became established between a form of occidentalism, in which the Occident was first idealized to the extreme and later negated, and a form of orientalism, in which "others" were considered to be inferior. History as the Occident understood it served as the tool for categorizing and appropriating the various cultures Japan encountered.

Notes

1. The Isle of Truk belongs to the Caroline Archipelago in Micronesia. It was under German administration from 1899 to 1919 and came under Japanese administration in 1920.

2. Quoted in Minato Kawamura, "Taishu orientarizumu to azia ninshiki" (Mass-cultural orientalism and knowledge of Asia), in Shinoo Oe, ed., *Iwanami Koza: Kindai nihon to shokuminchi,* vol. 7: *Bunka no nakano shokuminchi* (Iwanami Seminar: Modern Japan and the Colonies, vol. 7: Colonies and Culture) (Tokyo 1993), 107–136, here 111f.; italics mine.

3. Cf. Hiroshi Watanabe, "'Shinpo' to 'chûka': Nihon no baai" ("Progress" and "China:" The case of Japan), in Yûzô Mizoguchi, ed., *Ajia kara kangseru,* vol. 5: *Kindaikazô* (Thinking from Asia, vol. 5: Images of Modernization) (Tokyo 1994), 133–176, here 154f.

4. Watanabe, "'Shinpo' to 'chûka' - Nihon no baai," 161f.

5. Genpaku Sugita, *Rangaku kotohajime* (The beginnings of Dutch siences) (Tokyo 1959).

6. The intellectuals of postwar Japanese society adopted this categorization; cf. Shingo Shimada, "Zwischen Orientalismus und Okzidentalismus: Maruyama Masaos Nihonseiji Shisôshi kenkyû (Studien zur Geschichte der politischen Ideen in Japan) wieder gelesen," in Detlef Thofern, ed., *Rationalität im Diskurs: Rudolf Wolfgang Müller zum 60. Geburtstag* (Marburg 1994), 260–270.

7. I take over the expression "cosmology" from Yoshio Yasumaru, *Kindai tennô zô no keisei* (Construction of the modern images of Tenno) (Tokyo 1992), 88ff.

8. Shingo Shimada, "Überlegungen zu gesellschaftlichen Zeitlichkeitsregelungen in Japan," in Joachim Matthes, ed., *Zwischen den Kulturen? Die Sozialwissenschaften vor dem Problem des Kulturvergleichs* (Göttingen 1992), 375–392.

9. See Yasumaru, *Kindai tennô zô no keisei,* 215f.

10. It isn't difficult to find "magical" elements in this phenomenon whereby the portrait was treated as Tenno himself.

11. Cf. Nobuyoshi Yamamoto and Toshihiko Konno, *Kindai kyôiku no tennô sei ideorogî: Meijiki gakkô gyôji no kôsatsu* (The Tenno-ideology in the modern education system: Considerations to school events in the early Meiji era) (Tokyo 1973).

12. Keigo Nakamura, *Kyôkasho monogatari: Kokka to kyôkasho to minshû* (The History of teaching books: the state, books and nation) (Tokyo 1970), 47–70.

13. This book is even called the "Bible of the Meiji era" (see Nakamura, *Kyôkasho monogatari,* 60.

14. Ibid., 58f.

15. These watches were "made in USA," which was understood as a sign of high quality. The watches were conferred upon to the best graduates by Tenno himself.

16. Chizuko Ueno, *Kindai kazoku no seiritsu to shûen* (Origin and end of the modern family) (Tokyo 1994), 106f.

17. Kenji Satô, *Meiji kokka to katei ideorogî* (the Meiji state and the ideology of family), in Chizuko Ueno, ed., *Shirîzu: henbôsuru kazoku 1: kazoku no shakashi,* (Series: Changing family, vol. 1: Social history of Family) (Tokyo 1991), 78–96.

18. Ueno, *Kindai kazoku no seiritsu to shûen,* 107.

19. Between 1900 and 1910 more than ten magazines were founded with titles containing the term *katei.* Yû Ishida, *"'Ie' oyobi katei no seiji kinô"* (The political function of "ie" and family), in Masao Fukushima, ed. *Kazoku - seisaku to hô,* vol. 1: *Sôron* (Family: Policy and law, vol. 1: General) (Tokyo 1975), 311–359, here 332; cf. Kiyomi Morioka, *Gendai kazoku hendô ron* (Consideration of the changing family today) (Tokyo 1993), 85.

20. Ishida, *"'Ie' oyobi katei no seiji kinô"*, 111.

21. Masanao Kano, *Fujin, josei, onna: joseishi no toi* (Lady, woman, female: Questions for feministic history) (Tokyo 1989), 25f.

22. See Shingo Shimada, *Grenzgänge – Fremdgänge: Japan und Europa im Kulturvergleich* (Frankfurt and New York 1994), 134–140.

23. See Friedrich H. Tenbruck, "Gesellschaftsgeschichte oder Weltgeschichte?" *Kölner Zeitschrift für Soziologie und Sozialpsychologie* 41, no. 3 (1989): 417–439.

24. See Niklas Luhmann, "Die Zukunft kann nicht beginnen: Temporalstrukturen der modernen Gesellschaft," in Peter Sloterdijk, ed., *Vor der Jahrtausendwende: Berichte zur Lage der Zukunft,* vol. 1 (Frankfurt/M. 1990), 119–150.

25. We have to see this aspect in connection to Ernst H. Kantorowicz's *The King's Two Bodies* (Princeton 1957).

26. See Nakamura, *Kyôkasho monogatari - Kokka to kyôkasho to minshû,* 175f.

27. See Shimada, *Grenzgänge – Fremdgänge: Japan und Europa im Kulturvergleich.*

28. Tenbruck, "Gesellschaftsgeschichte oder Weltgeschichte?" 13.

29. San-Jun Kang, "Shakai kagakusha no shokuminchi ninshiki: Shokumin seisakugaku to orientarizumu," in Y. Yamanouchi, ed., *Iwanami koza: Shakaigaku no hohoh,* vol. 3, (Tokyo 1993), 101–130.

30. Tokuzo Fukuda, *Keizaigaku zenshu dai* (Tokyo 1925).

Bibliography

A Written Report of the Historical Office. 7 February 1878.
Abu l-Faḍl Ibrāhīm, Muhammad, ed. *Tārīh ar-rusul wal-mulūk.* 8 vols. Cairo 1960ff.
Ado-Khrawa, Ibrahim. *The Jihad in Kano.* Kano 1989.
Akhila Bharatiya Samskriti-Jnana-Pariksha (All-India Test on Cultural Knowledge). Kurukshetra n.d.
Akira, Yanabu. *Honyakugo seiritsu jijyo.* Tokyo 1982.
Al-Afghânî, J. *Al-A'mâl al-kâmila* (Complete works). Ed. M. 'Umâra. Cairo n.d.
Al-'Arwî, A. *Mafhûm at-târîkh* (The concept of history). Vol. 2. Casablanca 1992.
Al-Azmeh, Aziz. *Ibn Khaldûn in Modern Scholarship.* London 1981.
——. *Ibn Khaldun: An Essay in Reinterpretation.* London 1982.
——. "Histoire et narrative dans l'historiographie arabe." *Annales, Economies, Sociétés, Civilisations* 41 (1986): 411–431.
——. "Ifsâh al-Istishrâq." In Al-Azmeh, *Al-Turâth bayn as-Sultân wa't-târîkh.* Beirut and Casablanca 1990.
——. "Barbarians in Arab Eyes." *Past and Present* 134 (1992): 3–18.
——. *Islams and Modernities.* London 1993.
——. "Islamic Studies and the European Imagination." In Al-Azmeh, *Islams and Modernities.*
——. "Utopia and Islamic Political Thought." In Al-Azmeh, *Islams and Modernities.*
——. "Chronophagous Discourse." In D. Tracy and B. Reynolds, eds., *Religion and Practical Reason.* Albany 1994.
——. *Al-Kilâba al-târîkhîya wa'l-ma'rifa al-târîkhîya* (Historical writing and historical knowledge). 2nd ed., Beirut 1995.
——. *Reconstituting Islam.* Washington, DC., 1996.
Alekseenko, E. A. "Predstavlenija ketov o mire." In I. S. Vdovin, ed., *Priroda i celovek v religioznych predstavlenijach narodov Sibiri i severa.* Leningrad 1976.
An-Na'im, Abdullahi Ahmed. *Toward an Islamic Reformation: Civil Liberties, Human Rights and International Law.* Syracuse 1990.
Appoldt, Günter. *Zeit- und Lebenszeitkonzepte in China: Eine Untersuchung lebenszeitbezogener Vorstellungen und Handlungsstrategien im vormodernen China anhand einiger ausgewählter autobiographischer Texte.* Frankfurt/ M. 1991.
Assmann, Jan. *Das kulturelle Gedächtnis: Schrift, Erinnerung, und politische Identität in frühen Hochkulturen.* Munich 1992.
Auerbach, Erich. "Figura." In Auerbach, *Scenes from the Drama of European Literature.* Manchester 1984.
Augé, Marc. *Le sens des autres.* Paris 1994.
Azami, Muhammad Mustafa. *On Schachet's Origins of Muhammadan Jurisprudence.* New York 1985.
Bachelard, Gaston. *Dialectique de la durée.* Paris 1950.

Balakrishnan, Gopal. "The National Imagination." *New Left Review* 211 (1995): 59–61.

Ball, James Dyer. *Things Chinese: Being Notes on Various Subjects Connected with China.* London 1892.

Banerjee, R. D., ed. "Andhau Inscriptions of the Time of Rudradāman." *Epigraphia Indica* 16 (1921): 19–25.

Barnard, F. M., ed., and transl. *J.G. Herder on Social and Political Culture.* Cambridge 1969.

Bartsch, H. W. "Geschichte/Histoire." In Joachim Ritter, ed., *Historisches Wörterbuch der Philosophie.* Vol. 3. Stuttgart 1974.

Bary, William Theodore de. *The Message of the Mind in Neo-Confucianism.* New York 1988.

——, and S. Hay, eds. *Sources of Indian Tradition.* Vol. 2. 2nd ed., New Delhi 1991.

Basham, Arthur L. *The Wonder That Was India.* New York 1959.

Bauer, Wolfgang. *Chinas Vergangenheit als Trauma und Vorbild.* Stuttgart 1968.

Baumann, Hermann. "Mythos in ethnologischer Sicht," part 1. *Studium Generale* 12, no. 1 (1959): 1–17.

Bayly, Christopher A. "The Pre-History of 'Communalism'? Religious Conflict in India, 1700–1860." *Modern Asian Studies* 19 (1985): 177–203.

Bedini, Silvio A. *The Trail of Time: Time Measurement with Incense in East Asia. Shih-chien ti tsu-chi.* Cambridge 1994.

Beidelman, T. O. "Kaguru Time Reckoning: An Aspect of the Cosmology of an East African People." *Southwestern Journal of Anthropology* 19 (1963): 9–20.

Bender, John, and David E. Wellbery, eds. *Chronotypes: The Construction of Time.* Stanford 1991.

Benjamin, Walter. "Theses on History." In Benjamin, *Illuminations.* Ed. Hannah Arendt. New York 1985.

Benn, Gottfried., *Briefe: Briefe an F. W. Oelze 1932–1945.* Ed. Harald Steinhagen and Jürgen Schröder. Frankfurt/ M. 1979.

——. *Essays und Reden in der Fassung der Erstdrucke.* Ed. B. Hillebrand. Frankfurt/ M. 1989.

Bergmann, Werner. *Die Zeitstrukturen sozialer Systeme: Eine systemtheoretische Analyse.* Berlin 1981.

Berkemer, Georg. *Little Kingdoms in Kalinga.* Stuttgart 1993.

——. "The 'Centre out There' as State Archive: The Temple of Siṃhācalam." In H. Bakker, ed., *The Sacred Centre as the Focus of Political Interest.* Groningen 1992.

——. "No Heroes in Kalinga? On Death in Kalinga Inscriptions." In Elisabeth Schömbucher and Claus-Peter Zoller, eds., *Ways of Dying: Death and Its Meanings in South Asia.* New Delhi 1999.

Bernštam, T. A. "Budni i prazdniki: Povedenie vzroslych v russkoj krest'janskoj srede (XIX-nacalo XX v.)." In Albert K. Bajburin, ed., *Etniceskie stereotipy povedenija.* Leningrad 1985.

Biardeau, Madelaine. *Hinduism: The Anthropology of a Civilization.* Transl. Richard Nice. Delhi 1989.

Bielenstein, Hans. "Is There a Chinese Dynastic Cycle?" *Bulletin of the Museum of Far Eastern Antiquities* 50 (1978): 1–23.

Blanco, Walter, and Jennifer Tolbert Roberts, eds. *Herodotus: The Histories.* New York 1992.

Bloch, Marc. "Mémoire collective, tradition, et coutume: A propos d'un livre récent." *Revue de Synthèse Historique* 40 (N.S., 14) (1925): 73–83.

——. "A Contribution towards a Comparative History of European Societies." In Bloch, *Land and Work in Medieval Europe: Selected Papers.* Transl. J. E. Anderson. London 1966.

Bodde, Derk. "Cyclical and Linear Time." In Bodde, *Chinese Thought, Society, and Science: The Intellectual and Social Background of Science and Technology in Pre-modern China.* Honolulu 1991.

Bogoras, Waldemar. "Ideas of Space and Time in the Conception of Primitive Religion." *American Anthropologist* 27, no. 2 (1925): 205–266.

Bollnow, Otto Friedrich. *Existenzphilosophie.* Stuttgart 1969.

Boyer, Pascal. *Tradition as Truth and Communication: A Cognitive Description of Traditional Discourse.* Cambridge 1990.

Braudel, Fernand. *L'Identité de la France.* 2 vols. Paris 1986.

Bray, Francesca. *Science and Civilisation in China.* Vol. 6: *Biology and Biological Technology.* Part II: Agriculture, directed by Joseph Needham. Cambridge 1984.

Brémond, Claude, Jacques Le Goff, and Jean-Claude Schmitt, *L' Exemplum.* Turnhout 1982.

Büchner, Georg. *Dichtungen*. Ed. H. Poschmann. Frankfurt/ M. 1992.
——. *Complete Plays, Lenz and Other Writings*. London 1993.
Buck, David D., ed. "Dimensions of Ethnic and Cultural Nationalism in Asia: A Symposium." *Journal of Asian Studies* 53, no. 1 (1994): 3–131.
Bull, W. E. *Time, Tense, and the Verb*. Berkeley and Los Angeles 1968.
Burke, Edmund. *Reflections on the Revolution in France*. Ed. C. Cruise O'Brien. Harmondsworth 1969.
Burke, Peter. "Overture: The New History, Its Past and Its Future." In Peter Burke, ed., *New Perspectives on Historical Writing*. Cambridge 1992.
Carr, David. *Time, Narrative, and History*. Bloomington 1986.
Cassirer, Ernst. *The Philosophy of Symbolic Forms*. Transl. R. Manheim. Vol. 2. New Haven and London 1955.
Çelebi, Katib. *The Balance of Truth*. Transl. Geoffrey L. Lewis. London 1957.
Chakrabarty, Dipesh. "History as Critique and Critique(s) of History." *Economic and Political Weekly* (14 September 1991): 2162–2166.
——. "Postcoloniality and the Artifice of History: Who Speaks for 'Indian' Pasts?" *Representations* 37 (1992): 1–26.
——. "The Difference-Deferral of Colonial Modernity." In David Arnold and David Hardiman, eds., *Subaltern Studies*. Vol. 8. Delhi 1994.
Chan, Hok-lam. *Legitimation in Imperial China: Discussions under the Jurchen-Chin Dynasty (1115–1234)*. Seattle and London 1984.
Chand, Devi. *Atharvaveda: Sanskrit Text*. Delhi 1980.
Chand, Tara. *History of the Freedom Movement in India*. New Delhi 1961.
——. *Influence of Islam on Indian Culture*. Allahabad 1963.
Chandra, Bipan. *Communalism in Modern India*. New Delhi 1984.
——. "Nationalist Historians' Interpretations of the Indian National Movement." In Sabyasachi Bhattacharya and Romila Thapar, eds., *Situating Indian History: For Sarvepalli Gopal*. New Delhi 1986.
Chandra, Sudhir. *The Oppressive Present: Literature and Social Consciousness in Colonial India*. Delhi 1993.
Charnay, Jean-Paul. "Temps Sociaux et interprétation historique en Islam." *Studia Islamica* 28 (1968): 5–27.
Chatterjee, Partha. *Nationalist Thought and the Colonial World: A Derivative Discourse*. London 1986.
——. *The Nation and Its Fragments: Colonial and Post-Colonial Histories*. Princeton 1993.
——. "History and the Nationalization of Hinduism." *Social Research* 59, no. 1 (1992): 111–149.
——. "Claims on the Past: The Genealogy of Modern Historiography in Bengal." *Subaltern Studies* 8 (1994): 1–49.
——. "The Prose of Otherness." *Subaltern Studies* 8 (1994): 188–221.
——. "Secularism and Toleration." *Economic and Political Weekly* 29 (9 July 1994): 1768–1777.
Chattopadhyaya, B. *A Khila Bharatiya Samskriti-Inana-pariksa* (All-India text on cultural knowledge). Part 7. Kurukshetia n.d.
Chaudhary, Vijay Chandra Prasad. *Secularism Versus Communalism: An Anatomy of the National Debate on Five Controversial History Books*. Patna 1977.
Chen, Robert Shanmu. *A Comparative Study of Chinese and Western Cyclic Myths*. New York 1992.
Cheng, Anne. "La 'Maison des Han': Avénement et fin de l'histoire." In *Extrême-Orient / Extrême-Occident*. Vol. 9: La référence à l'histoire. Paris 1986.
Ch'ien, Ssu-ma. *Records of the Grand Historian*. 2 vols. Transl. Burton Watson. New York 1961.
Childe, V. G. *Man Makes Himself*. New York 1957.
——. *What Happened in History*. Harmondsworth 1957.
——. *Social Evolution*. London 1963.
Clark, Ronald W. *Albert Einstein: Leben und Werk. Eine Biographie*. Esslingen 1974.
Clark, S. "The Annales Historians." In Q. Skinner, ed., *The Return of Grand Theory in Human Sciences*. Cambridge, MA, 1986.

Clarke, Peter B. *West Africa and Islam.* London 1982.

Cleveland, William. *The Making of an Arab Nationalist: Ottomanism and Arabism in the Thought of Sati al-Husri.* Princeton 1971.

Clifford, James, and George E. Marcus, eds. *Writing Culture: The Poetics and Politics of Ethnography.* Delhi 1990.

Cohn, Bernard S. *An Anthropologist Amongst the Historians and Other Essays.* Delhi 1987.

——. "Notes on the History of the Study of Indian Society and Culture." In Cohn, *An Anthropologist.*

——. "The Past of an Indian Village." In Cohn, *An Anthropologist.*

Collingwood, R. G. *The Idea of History.* Oxford 1946.

Colpe, Carsten. "Die Zeit in drei asiatischen Hochkulturen." In A. Peisl and A. Mohler, eds., *Die Zeit: Schriften der Carl-Friedrich-von Siemens Stiftung 6.* Munich and Vienna 1983.

Crapanzano, Vincent. "The Hermes Dilemma: The Making of Subversion in Ethnographic Description." In James Clifford and George E. Marcus, eds., *Writing Culture: The Poetics and Politics of Ethnography.* Delhi 1990.

Cunnison, Ian. *History on the Luapula: An essay on the Historical Notions of a Central African Tribe.* Cape Town 1951.

Dalmia, Vasudha. *The Nationalization of Hindu Traditions: Bharatendu Harishchandra and Nineteenth-century Banaras.* Delhi 1997.

Deng, Francis M. *War of Visions: Conflict of Identities in the Sudan.* Washington, D.C., 1995.

Deppert, Wolfgang. *Zeit: Die Begründung des Zeitbegriffs, seine notwendige Spaltung und der ganzheitliche Charakter seiner Teile.* Stuttgart 1989.

Dharampal-Frick, Gita. "Zwischen 'Säkularismus' und 'Kommunalismus': Identitätsprobleme in der indischen Politik und Gesellschaft seit der Unabhängigkeit." In Wolfgang Reinhard, ed., *Die fundamentalistische Revolution: Partikularistische Bewegungen der Gegenwart und ihr Umgang mit der Geschichte.* Freiburg 1995.

Dicks, Anthony. "Treaty, Grant, Usage or Sufferance? Some Legal Aspects of the Status of Hongkong." *China Quarterly* 95 (1983): 429–454.

Dillon, Mary, and Thomas Abercrombie. "The Destroying Christ: An Aymara Myth of Conquest." In Jonathan D. Hill, ed., *Rethinking History and Myth: Indigenous South American Perspectives on the Past.* Urbana 1988.

Dirks, Nicholas B. "Colonial Histories and Native Informants: Biography of an Archive." In Carol Breckenridge and Peter van der Veer, eds., *Orientalism and the Postcolonial Predicament.* Philadelphia 1993.

Dudenredaktion, ed. *Duden: Etymologie.* Mannheim 1963.

Durkheim, Émile. *The Elementary Forms of the Religious Life.* New York 1968.

Dux, Günter. *Die Zeit in der Geschichte: Ihre Entwicklungslogik vom Mythos zur Weltzeit.* Frankfurt/ M. 1989.

——. *Historisch-genetische Theorie der Kultur: Instabile Welten: Zur prozessualen Logik im kulturellen Wandel.* Weilerswist 2000.

Eberhard, Wolfram. "Beiträge zur kosmologischen Spekulation Chinas in der Hanzeit." In Eberhard, *Sternkunde und Weltbild im alten China (gesammelte Aufsätze).* Taipei 1970.

Elias, Norbert. *Über die Zeit.* Frankfurt/ M. 1984.

Elliot, Henry M. "Original Preface." In Elliot, *The History of India As Told By Its Own Historians.* Calcutta 1867.

Engineer, Asghar A. *Communalism in India: A Historical and Empirical Study.* New Delhi 1995.

Evans-Pritchard, Edward E. *The Nuer: A Description of the Modes of Livelihood and Political Institutions of a Nilotic People.* Oxford 1940.

——. "Nuer Time-Reckoning." *Africa* 12 (1939): 189–216.

Fabian, Johannes. *Time and the Other: How Anthropology Makes Its Object.* New York 1983.

Falk, Harry. *Schrift in alten Indien.* Tübingen 1993.

Fettweis, E. "Orientierung und Messung in Raum und Zeit bei Naturvölkern." *Studium Generale* 11, no. 1 (1958): 1–12.

Finley, M. I. "Thucydides the Moralist." In Finley, *Aspects of Antiquity: Discoveries and Controversies.* Harmondsworth 1972.

Fleet, John F., ed. "An Inscription at Devageri." *Epigraphia Indica* 11 (1911): 1–7.

Forte, Antonino. *Mingtang and Buddhist Utopias in the History of the Astronomical Clock: The Tower, Statue and Armillary Sphere Constructed by Empress Wu.* Rome and Paris 1988.

Fortes, Meyer. "Bewußtsein." In Raymond William Firth, ed., *Institutionen in primitiven Gesellschaften.* Frankfurt/ M. 1967.

——. "Time and Social Structure: An Ashanti Case Study." Orig. pub. 1949. In Eugene A. Hammel and William S. Simmons, eds., *Man Makes Sense: A Reader in Modern Cultural Anthropology.* Boston 1970.

Fraser, Julius Thomas, Nathaniel Lawrence, and Francis C. Haber, eds. *Time, Science, and Society in China and the West.* (The Study of Time V). Amherst 1986.

Friedland, Roger, and Deidre Boden, eds. *NowHere: Space, Time, and Modernity.* Berkeley 1994.

Friedrich, Michael. "Tradition und Intuition: Zur Vorgeschichte der Schule von Chu Hsi." In Helwig Schmidt-Glintzer, ed., *Lebenswelt und Weltanschauung im frühneuzeitlichen China.* Stuttgart 1990.

Fuchs, Eckhardt, and Benedikt Stuchtey, eds. *Across Cultural Borders: Historiography in Global Perspective.* Lanham 2002.

Fukuda, Tokuzo. *Keizaigaku zenshu dai.* Tokyo 1925.

Fung, Yu-lan. *A History of Chinese Philosophy.* 2 vols. Trans Derk Bodde. Princeton 1952/53.

Fürer-Haimendorf, Christoph von. *The Chenchus.* London 1943.

Fussmann, Gerard. "L' Inscription de Rabatak et l' Origine de l' Ere Śaka." *Journal Asiatique* 286, no. 2 (1998): 571–651.

Fuzhi, Wang. *Siwen lu waipian.* 1st. ed. 1956. Beijing 1983.

Gardet, Louis. "Muslim Views on Time and History: An Essay in Cultural Typology." In UNESCO, ed., *Cultures and Time.* Paris 1976.

Geertz, Clifford. *Dichte Beschreibung: Beiträge zum Verstehen kultureller Systeme.* Frankfurt/ M. 1989.

Gell, Alfred., *The Anthropology of Time: Cultural Constructions of Temporal Maps and Images.* Oxford 1992.

Gellner, Ernest. *Reason and Culture.* Oxford 1988.

Ghoshal, U. N. *Beginnings of Indian Historiography and Other Essays.* Calcutta 1944.

Giddens, Anthony. *The Consequences of Modernity.* Stanford 1990.

Giesen, B., and K. Junge. "Der Mythos des Universalismus." In H. Berding, ed., *Mythos und Nation* (Studien zur Entwicklung des Kollektiven Bewusstseins in der Neuzeit, 3). Frankfurt/ M. 1996.

Gipper, Helmut. "Die Kluft zwischen muttersprachlichem und physikalischem Weltbild." *Physikalische Blätter* 12.3 (1956): 97–105.

Golwalkar, M. S. *We or Our Nationhood Defined.* Nagpur 1939.

Goodman, Lenn Evan. "Time in Islam." *Asian Philosophy* 2 (1992): 3–19.

Goody, Jack, and I. Watt. "The Consequences of Literacy." *Comparative Studies in Society and History* 5, no. 3 (1963): 304–345.

Gopal, Sarvepalli, ed., *Anatomy of a Confrontation: The Babri Masjid - Ramjanmabhumi Issue.* Delhi 1991.

——. *The Political Abuse of History.* New Delhi n.d.

——. "The Political Abuse of History." In Vinay Chandra Mishra and Parmanand Singh, eds., *Ram Janmabhoomi Babri Masjid: Historical Documents, Legal Opinions and Judgements.* Delhi n.d. [1991].

Gottlob, Michael, ed., *Historical Thinking in South Asia: A Handbook of Sources from Colonial Times to the Present.* New Delhi 2003.

Grabbe, Christian Dietrich. "Napoleon oder die hundert Tage." In Hans-Georg Werner, ed., *Werke in zwei Bänden.* Berlin 1987.

Granet, Marcel. *Das chinesische Denken: Inhalt, Form, Charakter.* Orig. pub. 1934. Munich 1963.

Greimas, Algirdas Julien. *Sémiotique et Sciences Sociales.* Paris 1976.

Großklaus, Götz. *Medien-Zeit, Medien-Raum: Zum Wandel raumzeitlicher Wahrnehmung in der Moderne.* Frankfurt/ M. 1995.

Guha, Ranajit. *Subaltern Studies: Writings on South Asian History and Society.* Delhi 1982ff.

——. "On some Aspects of the Historiography of Colonial India." In Guha, ed., *Subaltern Studies. Vol. I: Writings on South Asian History and Society.* Delhi 1982.

——. "The Prose of Counter-Insurgency." In Guha, ed., *Subaltern Studies.* Vol. 1. Delhi 1982.

——. *Elementary Aspects of Peasant Insurgency in Colonial India.* Delhi 1983.

——. *An Indian Historiography of India: A Nineteenth Century Agenda and Its Implications.* Calcutta 1988.

Gumplowicz, Ludwig. *Geschichte der Staatstheorien.* Innsbruck 1905.

——. "Un sociologiste arabe du XIVe siècle." In Gumplowicz, *Aperçus Sociologiques.* New York 1963 [1898].

Habermas, Jürgen. *Der Philosophische Diskurs der Moderne.* Frankfurt/ M. 1985.

Habib, Mohammad. "Presidential Address." *Proceedings of the 10th Session of the Indian History Congress.* Bombay 1947.

Hacker, Paul. "Religiöse Toleranz und Intoleranz im Hinduismus." In Lambert Schmithausen, ed., *Kleine Schriften.* Wiesbaden 1978.

Halbfass, Wilhelm. *India and Europe: An Essay in Philosophical Understanding.* Orig. pub. Basel 1981. New York 1988.

Halbwachs, Maurice. *Les cadres sociaux de la mémoire.* Paris 1925.

Hanafî, Hasan. *Dirâsât falsafiyya* (Philosophical studies). Cairo 1988.

Hansen, Kathryn. *Grounds for Play: The Nautanki Theatre of North India.* Delhi 1992.

Harbsmeier, Christoph. "Some Notions of Time and of History in China and in the West: With a Digression on the Anthropology of Writing." In Huang and Zürcher, eds., *Time and Space.*

Hartman, Charles. *Han Yü and the T'ang Search for Unity.* Princeton 1986.

Hartog, François. *Le miroir d'Hèrodote.* Paris 1980. Engl.: *The Mirror of Herodotus: The Representation of the Other in the Writing of History.* Transl. Janet Lloyd. Berkeley, Los Angeles, London 1988.

Havelock, Eric. *Preface to Plato.* Oxford 1963.

Heehs, Peter. "Myth, History, and Theory." *History and Theory* 33 (1994):1–19.

——. "Indian Communalism: A Survey of Historical and Social-Scientific Approaches." In Heehs, *Nationalism, Terrorism, Communalism: Essays in Modern Indian History.* Delhi 1998.

Heesterman, Jan C. "Ein geteiltes Haus. Die hinduistisch-muslimische Beziehung." In Shmuel N. Eisenstadt, ed., *Kulturen der Achsenzeit.* Vol. 2. Frankfurt/ M. 1992.

Hegel, Georg Wilhelm Friedrich. *Lectures on the Philosophy of History.* Transl. J. Sibree. London 1890.

——. *Reason in History.* Transl. Robert S. Hartmann. Indianapolis 1953.

——. *Werke.* Ed. E. Moldenhauer and K. M. Michel. 20 vols. Frankfurt/ M. 1969–1986.

——. *Vorlesungen über die Philosophie der Geschichte. Werke.* Vol. 12. Ed. E. Moldenhauer and K. M. Michel. Frankfurt/ M. 1970.

——. *Phänomemologie des Geistes. Werke.* Vol. 3. Ed. E. Moldenhauer and K. M. Michel. Frankfurt/ M. 1970. Engl.: *Phenomenology of Spirit.* Transl. A. V. Millar. Oxford 1977.

——. *Vorlesungen über die Philosophie der Geschichte: Sämtliche Werke.* Vol. 11. Ed. Eduard Gans and Karl Hegel. Stuttgart 1971.

——. *Lectures on the Philosophy of History.* Transl. E. S. Haldane and Frances H. Simson. Lincoln 1995.

Herder, Johann Gottfried von. *Reflections on the Philosophy of the History of Mankind.* Abridged by F. E. Manuel. Chicago and London 1968.

Hill, Jonathan D., ed. *Rethinking History and Myth: Indigenous South American Perspectives on the Past.* Urbana 1988.

Hiskett, H. *The Development of Islam in West Africa.* New York 1984.

Hobsbawm, Eric. "Inventing Traditions." In Eric Hobsbawm and Terence Ranger, eds., *The Invention of Tradition.* Cambridge 1983.

Hodgson, Marshall G. S. *Rethinking World History.* Cambridge 1993.

Hoskins, Janet. *The Play of Time: Kodi Perspectives on Calendars, History and Exchange.* Berkeley 1993.

Huang, Chun-chieh. "Historical Thinking in Classical Confucianism: Historical Argumentation from the Three Dynasties." In Huang and Zürcher, eds., *Time and Space.*

—–, and Erik Zürcher, ed. *Time and Space in Chinese Culture.* Leiden, New York, Cologne 1995.

Hudson, Charles. "Folk History and Ethnohistory." *Ethnohistory* 13 (1966): 52–70.

Hultzsch Erich, ed. *Corpus Inscriptionum Indicarum.* Vol. 1: *Inscriptions of Asoka,* new edition. Oxford 1925.

Humphreys, Stephen. "Islamic Historiography." In Joseph R. Strayer, ed., *Dictionary of the Middle Ages.* Vol. 6. New York 1985.

Huntington, Samuel P. "The Clash of Civilizations?" *Foreign Affairs* 72, no. 3 (Summer 1993): 22–49.

Husserl, Edmund. *Vorlesungen zur Phänomenologie des inneren Zeitbewußtseins.* Ed. Martin Heidegger. 2nd ed. Tübingen 1980.

Hutton, John Henry. *The Sema Nagas.* London 1921.

Ishida, Yû. "'Ie' oyobi katei no seiji kinô" (The political function of 'ie' and family). In Masao Fukushima, ed., *Kazoku - seisaku to hô.* Vol. 1: *Sôron* (Family: Policy and law. Vol. 1: General). Tokyo 1975.

Isichei, Elizabeth. *History of West Africa Since 1800.* New York 1977.

Iyer, Raghavan, ed. *The Moral and Political Writings of Mahatma Gandhi.* Vol. 1. Oxford 1986.

Jaffrelot, Christophe. *The Hindu Nationalist Movement and Indian Politics 1925 to the 1990s: Strategies of Identity-Building, Implantation and Mobilisation.* New Delhi 1996.

Jäger, Werner. *Paideia: Die Formung des griechischen Menschen.* Vol. 1. Berlin 1936.

James, Wendy. *'Kwanim Pa: The Making of the Uduk People.* Oxford 1979.

Jami, Catherine. "Western Devices for Measuring Time and Space: Clocks and Euclidian Geometry in Late Ming and Ch'ing China." In Huang and Zürcher, eds., *Time and Space.*

Jayaswal, Kashi Prasad. *Hindu Polity: A Constitutional History of India in Hindu Times.* Patna 1924.

—–, and R. D. Banerji, eds. "The Hāthīgumphā Inscription of Khāravela." *Epigraphia Indica* 20 (1929): 72–89.

Jiro, Numata. "Shigeno Yasutsugu and the Modern Tokyo Tradition of Historical Writing." In E. G. Beasley and E. G. Pullleyblank, eds., *Historians of China and Japan.* Oxford 1961.

Jones, Schuyler. "Zur Religion des Waigaltales." In Karl Jettmar, ed., *Die Religionen des Hindukusch.* Stuttgart 1975.

Jones, William. "On the Hindus." Orig. pub. 1786. In Peter J. Marshall, ed., *The British Discovery of Hinduism in the Eighteenth Century.* Cambridge 1970.

Jung, Carl Gustav. "Synchronizität als ein Prinzip akausaler Zusammenhänge." In Carl Gustav Jung and Wolfgang Pauli, eds., *Naturerklärung und Psyche.* Zurich 1952.

Kaba, Lansine. *The Wahhabiyya: Islamic Reform and Politics in French West Africa.* Evanston, IL, 1974.

Kabir, Humayun. *The Indian Heritage.* Bombay 1946.

Kakar, Sudhir. *The Colors of Violence: Cultural Identities, Religion, and Conflict.* Chicago 1996.

Kang, SanJun. "Shakai kagakusha no shokuminchi ninshiki: Shokumin seisakugaku to orientarizumu." In Y. Yamanouchi, ed., *Iwanami koza: Shakaigaku no hoho.* Vol. 3. Tokyo 1993.

Kano, Masanao. *Fujin, josei, onna: joseishi no toi* (Lady, woman, female: Questions for feministic history). Tokyo 1989.

Kant, S. *The Hāthīgumphā Inscription of Khāravela and the Bhabru Edict of Aśoka: A Critical Study.* Delhi 1971.

Kantorowicz, Ernst H. *The King's Two Bodies.* Princeton 1957.

Kaviraj, Sudipta. *The Unhappy Consciousness: Bankimchandra Chattopadhyay and the Formation of Nationalist Discourse in India.* Delhi 1995.

—–. "Crisis of the Nation-State in India." *Political Studies* 42 (1994): 115–129.

—–. "Kolonialismus, Moderne und politische Kultur: Die Krise Indiens." In Joachim Matthes, ed., *Zwischen den Kulturen?,* 219–238.

Kawamura, Minato. "Taishu orientarizumu to azia ninshiki" (Mass-cultural orientalism and knowledge of Asia). In Shinoo Oe, ed., *Iwanami Koza: Kindai nihon to shokuminchi.* Vol. 7: *Bunka no nakano shokuminchi* (Iwanami Seminar: Modern Japan and the Colonies. Vol. 7: Colonies and Culture). Tokyo 1993.

Keddie, Nikki. *An Islamic Response to Imperialism: Political and Religious Writings of Sayyid Jamal al-Din "Al-Afghânî."* Berkeley and Los Angeles 1968.

Kenji, Shimada, ed. *Ogyu Sorai zenshu*. Tokyo 1973.

Kerr, Malcolm. *Islamic Reform: The Political and Legal Theories of Muhammad Ábduh and Rashid Rida*. Berkeley and Los Angeles 1966.

Khalidi, Tarif. *Arabic Historical Thought in the Classical Period*. Cambridge 1995.

Kielhorn, F. "Dibbida Plates of Arjuna of the Matsya Familiy; Saka-Samvat 1191." *Epigraphia Indica* 5 (1898): 106–112.

——, ed., "Aihole Inscription of Pulakesin II." *Epigraphia Indica* 6 (1900): 1–12.

King, David A. "Science in the Service of Religion: The Case of Islam." *Impact of Science on Society* (UNESCO newspaper) 40, no. 159 (1990): 245–261.

Ki-Zerbo, J. "General Introduction." In J. Ki-Zerbo, ed., *General History of Africa*. Vol. 1. Berkeley 1981.

Klimkeit, Hans-Joachim. *Der politische Hinduismus: Indische Denker zwischen religiöser Reform und politischem Erwachen*. Wiesbaden 1981.

Kluge, Alexander. *Schlachtbeschreibung: Der organisatorische Aufbau eines Unglücks*. Munich 1978.

——. *Der Angriff der Gegenwart auf die übrige Zeit: Das Drehbuch zum Film*. Frankfurt/ M. 1985.

——. "Das Lesen des Textes wirklicher Verhältnisse." In Hans Dieter Müller. *Der Kopf in der Schlinge: Entscheidungen im Vorkrieg. Mit einem Nachwort von Alexander Kluge*. Frankfurt/ M. 1985.

Kokichi, Kano. "Rekishi no gainen (Concept of rekishi)." Orig. pub. 1921. In Abe Yoshishige, ed., *Kano Kokichi ibun shu*. Tokyo 1958.

Konow, Sten, ed. "Narsapatam Plates of Vajrahastadeva III.; Saka Samvat 967." *Epigraphia Indica* 11 (1911): 147–153.

Kramer, Fritz W. *Der rote Fez*. Frankfurt/ M. 1987.

——. "Über Zeit, Genealogie und solidarische Beziehung." In F. Kramer and Christian Sigrist, eds., *Genealogie und Solidarität*. (Gesellschaften ohne Staat 2). Frankfurt/ M. 1978.

——, and Gertraud Marx. *Zeitmarken: Die Feste von Dimodonko* (Sudanesische Marginalien 7). Munich 1993.

Kubin, Wolfgang. "Zeitbewußtsein und Subjektivität: Zum Problem der Epochenschwelle in China und dem Abendland." In Hans-Jürgen Gawoll and Christoph Jamme, eds., *Idealismus mit Folgen: Die Epochenschwelle um 1800 in Kunst und Geisteswissenschaften. Festschrift zum 65. Geburtstag von Otto Pöggeler*. Munich 1994.

Kulke, H. "Gibt es ein indisches Mittelalter?" *Saeculum* 33 (1982): 221–239.

——. *Kings and Cults: State Formation and Legitimation in India and Southeast Asia*. New Delhi 1993.

——. "Reflections on the Sources of the Temple Chronicles of the Mādalā Pāñji of Puri." In Kulke, *Kings and Cults*.

——. "The Chronicles and Temple Records of the Mādalā Pāñji of Puri. A Reassessment of the Evidence." In Kulke, *Kings and Cults*.

——, and D. Rothermund. *Geschichte Indiens*. Stuttgart 1982. Engl.: *A History of India*. London and Sydney 1986.

Kumar, Krishna. "Hindu Revivalism and Education in North-Central India." *Social Scientist* 18, no. 10 (1990): 4–26.

Kunio, Yanagita. "Rekishi kyoiku ni tuite" (On the teaching of history). In *The Authentic Complete Works of Yanagita Kunio*. Vol. 24. Tokyo 1963.

Kuntz, Andreas. "Erinnerungsgegenstände: Ein Diskussionsbeitrag zur volkskundlichen Erforschung rezenter Sachkultur." *Ethnologia Europaea* 20 (1990): 61–80.

Lapidus, Ira Marvin. *A History of Islamic Societies*. Cambridge 1988.

Laroui, Abdallah. *L'Idéologie arabe contemporaine*. Paris 1967.

——. "The Arabs and Cultural Anthropology." In Laroui, *The Crisis of the Arab Intellectual*. Berkeley and Los Angeles 1976.

Last, Murray. *The Sokoto Caliphate*. New York 1967.

Lau, Ulrich. *Quellenstudien zur Landvergabe und Bodenübertragung in der westlichen Zhou-Dynastie (1045?–771 v. Chr.).* Nettetal 1999.

Leach, Edmund R. "Zwei Aufsätze über die symbolische Darstellung der Zeit." In Wilhelm E. Mühlmann and Ernst W. Müller, eds., *Kulturanthropologie.* Cologne 1966.

Legge, James. *The Chinese Classics.* 5 vols. 2nd ed. Oxford 1893–95 [Hongkong 1960].

Leroi-Gourhan, André. *Hand und Wort: Die Evolution von Technik, Sprache und Kunst.* Frankfurt/ M. 1980.

Lévi-Strauss, Claude. *Race et histoire.* Paris 1948.

——. *Strukturale Anthropologie.* Frankfurt/ M. 1967.

——. *"Primitive" und "Zivilisierte."* Zurich 1972.

——. *Das wilde Denken.* Frankfurt/ M. 1973 [9th ed., Frankfurt/ M. 1994].

Li, Wai-yee. "The Idea of Authority in the Shih chi (Records of the Historian)." *Harvard Journal of Asiatic Studies* 54 (1994): 345–405.

Lienhardt, Godfrey. *Social Anthropology.* London 1966.

Lin, Li-chen. "The Notions of Time and Position in the Book of Change and Their Development." In Huang and Zürcher, eds., *Time and Space.*

Lin, Yutang. *Weisheit des lächelnden Lebens: Das Geheimnis erfüllten Daseins.* (German translation 1960.) Reinbek bei Hamburg 1988. American orig. The Importance of Living (1936).

Liu, James J. Y. "Time, Space, and Self in Chinese Poetry." *Chinese Literature, Essays, Articles, Reviews* 1, no. 2 (1979): 137–156.

Liu, Shu-hsien. "Time and Temporality: The Chinese Perspective." *Philosophy East and West* 24 (1974): 145–153.

Loewe, Michael. *Crisis and Conflict in Han China (104 BC to AD 9).* London 1974.

——."The Almanacs (*jih-shu*) from Shui-hu-ti: A Preliminary Survey." In Loewe, *Divination, Mythology and Monarchy in Han China.* Cambridge 1994.

——. "Water, Earth and Fire: The Symbols of the Han Dynasty." In Loewe, *Divination, Mythology and Monarchy in Han China.*

Löwith, Karl. *Meaning in History.* Chicago 1955.

Lotman, Juri M. "On the Metalanguage of a Typological Description of Culture." *Semiotica* 14, no. 2 (1975): 97ff.

Lovejoy, Arthur, and George Boas, eds. *Primitivism and Related Ideas in Antiquity.* Baltimore 1935.

Ludden, David. "Orientalist Empiricism: Transformation of Colonial Knowledge." In Carol Breckenridge and Peter van der Veer, eds., *Orientalism and the Postcolonial Predicament.* Philadelphia 1993.

Luhmann, Niklas. *Soziale Systeme: Grundriß einer allgemeinen Theorie.* Frankfurt/ M. 1984.

——. "Die Zukunft kann nicht beginnen: Temporalstrukturen der modernen Gesellschaft." In Peter Sloterdijk, ed., *Vor der Jahrtausendwende: Berichte zur Lage der Zukunft.* Vol. 1. Frankfurt/ M. 1990.

——. "Gleichzeitigkeit und Synchronisation." In Luhmann, *Soziologische Aufklärung 5: Konstruktivistische Perspektiven.* Opladen 1990.

Mach, Ernst. *Erkenntnis und Irrtum: Skizzen zur Psychologie der Forschung.* Darmstadt 1968.

Madan, T. N. "Secularism in Its Place." *Journal of Asian Studies* 46 (1987): 747–759.

Madoka, Kanai. *Oyatoi gaikokujin.* Tokyo 1976.

MahÁbhÁrata. Critical Edition, 19 vols. Poona 1933–1966.

Mainzer, Klaus. *Zeit: Von der Urzeit zur Computerzeit.* Munich 1995.

Majumdar, Romesh Chandra. *Ancient Indian Colonies in the Far East.* Vol. 1: *Champa.* Lahore 1927. Vol. 2: *Suvarnadvipa.* Parts 1–2. Dacca 1937–1938.

——. *Hindu Colonies in the Far East.* Calcutta 1944.

——. *History of the Freedom Movement in India.* 3 vols. Calcutta 1961–1962.

Malek, Anwar Abdel. "Orientalism in Crisis." *Diogenes* 44 (1963): 102–140.

Matthes, Joachim, ed. *Zwischen den Kulturen? Die Sozialwissenschaften vor dem Problem des Kulturvergleichs* (Soziale Welt, Supplement, vol. 8). Göttingen 1992.

——. "The Operation called 'Vergleichen'." In Matthes, ed., *Zwischen den Kulturen?*

Maxwell, Allen R. "Kadayan Ideas of Time." *Sarawak Museum Journal* 27, no. 48 (1979): 1–14.

Mbiti, John S. *Afrikanische Religion und Weltanschauung.* Berlin 1974.

Meier, Christian. "Die Entstehung der Historie." In Reinhart Koselleck and Wolf-Dieter Stempel, eds., *Geschichte: Ereignis und Erzählung* (Poetik und Hermeneutik V). Munich 1973.

Meran, Josef. *Theorien in der Geschichtswissenschaft: Die Diskussion über die Wissenschaftlichkeit der Geschichte.* Göttingen 1985.

Mercier, Paul. "Anthropologie sociale et culturelle." In Jean Poirier, ed., *Ethnologie générale* (*Encyclopédie de la Pléiade,* Vol. 24). Paris 1968.

Meskill, John, ed. *The Pattern of Chinese History: Cycles, Development or Stagnation?* Boston 1965.

Middleton, John. *Lugbara Religion: Ritual and Authority among an East African People.* London 1960.

Mill, James. *The History of British India.* London 1817.

Ministry of Human Resource Development. *National Policy on Education: 1986* (*With Modifications Undertaken in 1992*) (typescript). New Delhi n.d.

Mirashi, Vasudev V. "The Riddle of the Mewāsā Stone Inscription of the Western Kṣatrapas." *Journal of the Oriental Institute, Baroda* 28, no. 1 (1978): 56–62.

——. "Daulatpur Inscription of the Reign of Caṇḍana: Year 6." *Journal of the Oriental Institute, Baroda* 28, no. 2 (1978): 34–37.

Mitra, Subrata K., and R. Alison Lewis, eds. *Subnational Movements in South Asia.* Oxford 1996.

Mittag, Achim. "Randnotizen zu Zheng Qiao (1104–1162) und seinem Tongzhi." *Chinablätter* 19 (1994): 62–88.

——. "Zeitkonzepte in China." In Klaus E. Müller and Jörn Rüsen, eds., *Historische Sinnbildung: Struktur, Funktion und Repräsentation des Geschichtsbewußseins.* Reinbek 1997.

Mollier, Christine. *Une Apocalypse taoiste du Ve siècle: Le livre des incantations divines des grottes abyssales.* Paris 1990.

Mookerji, Radha Kumud. *The Fundamental Unity of India.* London 1914.

Morioka, Kiyomi. *Gendai kazoku hendô ron* (Consideration of the changing family today). Tokyo 1993.

Mühlmann, Wilhelm E. *Geschichte der Anthropologie.* Frankfurt/ M. 1968.

——. and Ernst W. Müller, ed. *Kulturanthropologie.* Cologne 1966

Mukherjee, Meenakshi. *Realism and Reality: The Novel and Society in India.* Delhi 1985.

Müller, Harro. *Geschichte zwischen Kairos und Katastrophe.* Frankfurt/ M. 1988.

——. *Giftpfeile: Zu Theorie und Literatur der Moderne.* Bielefeld 1994.

——. "Subjekt und Geschichte: Reflexionen zu Grabbes Naploeon-Drama'." In Müller, *Giftpfeile.*

——. "Theater als Geschichte—Geschichte als Theater: Büchners Dantons Tod." In Müller, *Giftpfeile.*

——. "Identität, Paradoxie, Differenz: An Essay on Conceptions of Time in the Literature of Modernity." In Jürgen Fohrmann and Harro Müller, eds., *Systemtheorie der Literatur.* Munich 1996.

Müller, Klaus E. "Epistemologische Grenzfälle: 'Höhere' Erkenntnis in traditionellen Gesellschaften." *Zeitschrift für Parapsychologie und Grenzgebiete der Psychologie* 32, no. 3/4 (1990): 137–152.

——. "Reguläre Anomalien im Schnittbereich zweier Welten." *Zeitschrift für Parapsychologie und Grenzgebiete der Psychologie* 34, no. 1/2 (1992): 33–50.

——. "'Prähistorisches' Geschichtsbewußtsein: Versuch einer ethnologischen Strukturbestimmung." *ZiF-Mitteilungen* 3 (1995): 3–17.

Müller, Werner. "Raum und Zeit in Sprachen und Kalendern Nordamerikas und Alteuropas." *Anthropos* 68, no. 1/2 (1973): 157–180.

Munn, Nancy D. "The Cultural Anthropology of Time: A Critical Essay." *Annual Review of Anthropology* 21 (1992): 93–123.

Murty, M .L. K., and Gunter D. Sontheimer. "Prehistoric Background to Pastoralism in the Southern Deccan in the Light of Oral Traditions and Cults of Some Pastoral Communities." *Anthropos* 75 (1980): 163–184.

Musil, Robert. *Der Mann ohne Eigenschaften.* Hamburg 1952. Engl.: *The Man without Qualities.* Transl. Sophie Wilkins and Burton Pike. London 1995.

——. *Aus den Tagebüchern.* Frankfurt/ M. 1981.
——. *Diaries 1899–1941.* Ed. Mark Mirsky. Transl. Philip Payne. New York 1998.
Mylius, K. *Geschichte der altindischen Literatur.* Bern 1988.
Nakamura, Keigo. *Kyôkasho monogatari: Kokka to kyôkasho to minshû* (The history of teaching books: The state, books and nation). Tokyo 1970.
Nandy, Ashis. *The Intimate Enemy: Loss and Recovery of Self Under Colonialism.* Delhi 1983.
——. "An anti-secularist manifesto." *Seminar* 314 (Oct. 1985): 14–24.
——. "The Politics of Secularism." In Veena Das, ed., *Mirrors of Violence: Communities, Riots, and Survivors in South Asia.* Delhi 1990.
——. *Traditions, Tyranny, and Utopias.* Delhi 1992.
——. *The Illegitimacy of Nationalism: Rabindranath Tagore and the Politics of Self.* Delhi 1994.
——. "History's Forgotten Doubles." *History and Theory* 34, no. 2 (1995): 44–66.
——. *Creating a Nationality: The Ramjanmabhumi Movement and Fear of the Self.* Delhi 1995.
——, Zia Sardar, and Merryl Wyn Davies. *Barbaric Others: Manifesto on Western Racism.* London 1993.
Nassehi, Armin. *Die Zeit der Gesellschaft.* Opladen 1993.
National Steering Committee. *National Steering Committee on Textbook Evaluation: Recommendations and Report.* Parts 1 and 2 (typescript). New Delhi n.d.
Needham, Joseph. *Science and Civilisation in China.* Vol. 2: *History of Scientific Thought.* Cambridge 1956.
——. *Science and Civilisation in China.* Vol. 3: *Mathematics and the Sciences of the Heavens and the Earth.* Cambridge 1959.
——. *Science and Civilisation in China.* Vol. 4: *Physics and Physical Technology.* Part II: *Mechanical Engineering,* with the collaboration of Wang Ling. Cambridge 1965.
——. "Time and Eastern Man." In Needham, *The Grand Titration.* London 1969.
——. "Der Zeitbegriff im Orient." In Needham, *Wissenschaftlicher Universalismus: Über Bedeutung und Besonderheit der chinesischen Wissenschaft.* Frankfurt/ M. 1979.
——. "The Missing Link in Horological History." In Needham, *Clerks and Craftsmen in China and the West.* Cambridge 1979.
——, Wang Ling, and Derek J. Price de Solla. *Heavenly Clockwork: The Great Astronomical Clocks of Medieval China.* Cambridge 1960.
Negt, Oskar, and Alexander Kluge. *Geschichte und Eigensinn.* Frankfurt/ M. 1985.
——. *Maßverhältnisse des Politischen: 15 Vorschläge zum Unterscheidungsvermögen.* Frankfurt/ M. 1992.
Nehru, Jawaharlal. *The Discovery of India.* Calcutta and New York 1946.
——. *Glimpses of World History.* Orig. pub. 1934–1935. New Delhi 1982.
Ng, On-cho. "Mid-Ch'ing New Text (*chin-wen*) Classical Learning and its Han Provenance: The Dynamics of a Tradition of Ideas." *East Asian History* 8 (1994): 1–32.
Nietzsche, Friedrich. *Thus Spoke Zarathustra.* Transl. R. J. Hollingdale. London 1961.
——. *Werke.* Ed. K. Schlechta. 3 vols. Munich 1966.
——. *On the Genealogy of Morality.* Ed. Keith Ansell-Pearson. Transl. Carol Diethe. Cambridge, MA, 1994.
——. *Human, All Too Human.* Transl. R. J. Hollingdale. Cambridge, MA, 1996.
——. "David Strauss, the Confessor and the Writer." In Nietzsche, *Untimeley Meditations.* Ed. Daniel Breazeale. Transl. R. J. Hollingdale. Cambridge, MA, 1997.
——. "On the Uses and Disadvantages of History for Life." In Nietzsche, *Untimely Meditations.* Ed. Daniel Breazeale. Transl. R. J. Hollingdale. Cambridge, MA, 1997.
——. *The Birth of Tragedy and Other Writings.* Ed. Raymond Geuss and ed., transl. Ronald Speirs. Cambridge, MA, 1999.
Noboru, Kobayashi. "Edojidai ni okeru shinpo no shiso (The idea of progress in early modern Japan)." In Noboru, *Chugoku Nihon ni okeru rekishikan to in'itsushiso* (The idea of history and hermitism in China and Japan). Tokyo 1983.
Nock, Arthur D. *Conversion.* Oxford 1933 [1961].
O'Flaherty, Wendy Doniger. *Dreams, Illusion and Other Realties.* Chicago and London 1984.

Oak, P. N. "Institute for Rewriting Indian History." In Oak, *Some Blunders of Indian Historical Research*. New Delhi 1966.
Odouev, Stepan. *Par les sentiers de Zarathoustra*. Transl. C. Emery. Moscow 1980.
Oppenheimer, Franz. *System der Soziologie*. Vol. 2. 2nd ed., Stuttgart 1964.
Osborne, Peter. *The Politics of Time: Modernity and Avant-Garde*. London 1995.
Pakdaman, Homa. *Djamal-Ed-Din Assad Abadi dit Afghânî*. Paris 1969.
Pandey, Gyanendra. *The Construction of Communalism in Colonial North India*. Delhi 1990.
——. "Which of us are Hindus?" In Pandey, ed., *Hindus and Others: The Question of Identity in India Today*. New Delhi 1993.
——. "Modes of History Writing: New Hindu History of Ayodhya." *Economic and Political Weekly* 29 (18 June 1994): 1523–1528.
——. "Nationalism, Communalism, and the Struggle over History." In Mehdi Arslan and Janaki Rajan, eds., *Communalism in India: Challenge and Response*. New Delhi 1994.
——. "The Prose of Otherness." *Subaltern Studies* 8 (1994): 188–221.
——. "The Appeal of Hindu History." In Vasudha Dalmia and Heinrich von Stietencron, eds., *Representing Hinduism: The Construction of Religious Traditions and National Identity*. New Delhi 1995.
——. "Partition and the Politics of History." In Madhusree Dutta, Flavia Agnes, and Neera Adarkar, eds., *The Nation, the State: Indian Identity*. Calcutta 1996.
Panikkar, K. N., ed. *Communalism in India: History, Politics and Culture*. New Delhi 1991.
——. "Search for Alternatives: Meaning of the Past in Colonial India." In AA. VV. Università di Roma "la Sapienza," ed., *The East and the Meaning of History*. Rome 1994.
Pantham, Thomas. *Political Theories and Social Reconstruction: A Critical Survey of the Literature on India*. New Delhi 1995.
Pargiter, F. E. *Ancient Indian Historical Tradition*. Delhi 1972.
Parla, Taha. *The Social and Political Thought of Ziya Gökalp*. Leiden 1986.
Parry, Nevill Edward. *The Lakhers*. London 1932.
Parthasarathy, Malini. "Soft Hindutva and Nationalism." *Seminar* 411 (Nov. 1993): 19–22.
Pathak, V. S. *Ancient Historians of India*. Bombay 1966.
Pillai, Swamikannu L. D. *Indian Chronology (Solar, Lunar and Planetary): A Practical Guide to the Interpretation and Verification of Tithis, Nakshatras, Horoscopes and Other Indian Time-Records, B.C. 1 to A.D. 2000*. Madras 1911 [repr. Delhi 1982].
Pomian, Krzyzstof. *L'ordre du temps*. Paris 1984.
Prakash, Gyan. "Writing Post-Orientalist Histories of the Third World: Indian Historiography Is Good to Think." In Nicholas B. Dirks, ed., *Colonialism and Culture*. Ann Arbor 1992.
Prasad, Beni. *The Hindu-Muslim Question*. Allahabad 1941.
Prasad, H. S., ed. *Time in Indian Philosophy*. Delhi 1992.
Qureshi, I. H. *The Muslim Community of the Indo-Pak Subcontinent (610–1947): A Brief Historical Analysis*. Delhi 1985.
Rancière, Jacques. *The Names of History: On the Poetics of Knowledge*. Transl. H. Melehy. Minneapolis 1994.
Rao, V. Naryana. "Purāṇa as Brahminic Ideology." In W. Doniger, ed., *Purāṇa Perennis*. Albany 1993.
Ray, H. C. "Was State-Socialism Known in Ancient India? A Study in Kautiliya´s Arthasastra." In University of Calcutta, ed., *Sir Asutosh Mookerjee Silver Jubilee Volumes*. Vol. 3, part I. Calcutta 1922.
——. "Economic Policy and Functions of the Kautilyan State." *Journal of the Departments of Letters* 13 (1926): 1–30.
Redfield, Robert. "Die 'Folk'-Gesellschaft." In Wilhelm Ernst Mühlmann and Ernst W. Müller, eds., *Kulturanthropologie*, Cologne 1966.
——. *The Primitive World and Its Transformations*. London 1968.
Reetz, Dietrich. "In Search of the Collective Self: How Ethnic Group Concepts Were Cast through Conflict in Colonial India." *Modern Asian Studies* 31 (1997): 285–315.

Reeve, Mary-Elizabeth. "Cauchu Uras: Lowland Quichua Histories of the Amazon Rubber Boom." In Hill, *Rethinking History and Myth.*

Reischauer, Edwin O. "The Dynastic Cycle." In Meskill, ed., *The Pattern of Chinese History.*

Revel, Jacques, ed., *Jeux d' échelle: La micro-analyse á l'experience.* Paris 1996.

Ricoeur, Paul. *Time and Narrative.* 3 vols. Chicago 1984–1988.

——. *Zeit und Erzählung.* Vol. 1: *Zeit und historische Erzählung.* Munich 1988.

Riehl, Wilhelm Heinrich. *Die Naturgeschichte des Volkes als Grundlage einer deutschen Social-Politik.* 4 vols. Stuttgart 1862–1869.

Riese, Berthold. "Zeitstrukturen in Mesoamerika." In Klaus E. Müller and Jörn Rüsen, eds., *Historische Sinnbildung: Problemstellungen, Zeitkonzepte, Wahrnehmungshorizonte, Darstellungsstrategien.* Reinbek bei Hamburg 1997.

Riken, Okubo. "Bakumatsu no seiyo tushi." In Kaikoku hyakunen bunka jigyokai, ed., *Sakoku-jidai nihonjin no kaigaichishiki.* Kangensha 1953.

Roetz, Heiner. *Chinesische Ethik der Achsenzeit: Eine Rekonstruktion unter dem Aspekt des Durchbruchs zu konventionellem Denken.* Frankfurt/ M. 1992.

——. *Konfuzius.* Munich 1995.

Rosenthal, Franz. *History of Muslim Historiography.* Leiden 1968.

Rudolph, L. I., and S. H. Rudolph. "Rethinking Secularism: Genesis and Implications of the Textbook Controversy, 1977–1979." *Pacific Affairs* 56 (1983): 15–37.

Rüsen, Jörn. *Historische Vernunft: Grundzüge einer Historik I: Die Grundlagen der Geschichtswissenschaft.* Göttingen 1983.

——. *Rekonstruktion der Vergangenheit: Grundzüge einer Historik II: Die Prinzipien der historischen Forschung.* Göttingen 1986.

——. "Historical Narration: Foundations, Types, Reason." *History and Theory* 26 (1987): 87–97.

——. *Lebendige Geschichte: Grundzüge einer Historik III: Formen und Funktionen historischen Wissens.* Göttingen 1989.

——. *Zeit und Sinn: Strategien historischen Denkens.* Frankfurt/ M. 1990.

——. *Studies in Metahistory.* Pretoria 1993.

——. "Geschichte und Utopie." In Rüsen, *Historische Orientierung: Über die Arbeit des Geschichtsbewußtseins, sich in der Zeit zurechtzufinden.* Cologne 1994.

——. "Some Theoretical Approaches to Intercultural Comparison of Historiography." *History and Theory*, Theme Issue: *Chinese Historiography in Comparative Perspective* 35 (1996): 5–22.

——. *Zerbrechende Zeit: Studien über den Sinn der Geschichte.* Cologne 2001.

——. *History: Narration – Interpretation – Orientation.* New York 2005.

——. "Historische Methode und religiöser Sinn-Vorüberlegungen zu einer Dialektik der Rationalisierung des historischen Denkens in der Moderne." In Wolfgang Küttler, Jörn Rüsen, and Ernst Schulin, eds., *Geschichtsdiskurs 2: Anfänge modernen historischen Denkens.* Frankfurt/ M. 1994.

Rüsen, Jörn, Achim Mittag and Helwig Schmidt-Glintzer, eds. *Collective Identity- Experiences of Crisis-Traumata: New Approaches to Chinese Historiography and Historical Culture in a Comparative Perspective.* (Forthcoming).

Sacks, Jonathan. *The Dignity of Difference: How to Avoid the Clash of Civilizations.* London 2003.

Sahlins, Marshall. *Culture and Practical Reason.* Chicago 1976.

——. *Islands of History.* Chicago 1985.

——. *How "Natives" Think, about Captain Cook, for Example.* Chicago 1995.

Sarkar, H. B. *Cultural Relations between India and Southeast Asian Countries.* Delhi 1985.

Sastri, Hirananda, ed. "Nalanda Stone Inscription of the Reign of Yasovarmmadeva." *Epigraphia Indica* 20 (1929): 37–46.

Satô, Kenji. *Meiji kokka to katei ideorogî* (The Meijistate and the ideology of family). In Chizuko Ueno, ed., *Shirîzu: henbôsuru kazoku 1: kazoku no shakashi,* (Series: Changing family. Vol. 1: Social history of family). Tokyo 1991.

Sato, Masayuki. *Rekishi ninshiki no jiku* (Historiographical time and space). Tokyo 2004.

——. "Kong-zi." In Lucian Boia, ed., *Great Historians from Antiquity to 1800.* New York 1989.

——. "Tokugawa Mitsukuni." In Boia, ed., *Great Historians from Antiquity to 1800.*

——. "Japanese historians." In L. Boia, ed., *Great Historians of the Modern Age.* New York 1991.

——. "Shigeno Yasutsugu." In Boia, ed., *Great Historians of the Modern Age.*

——. "Historiographical Encounters: The Chinese and Western Traditions in Turn-of-the-Century Japan." *Storia de la Storiografia* 19 (1991): 13–21.

——. "Rekishigakka to shiteno Tanaka Suiichiro (Tanaka Suiichiro as a historiologus)." *Kindai nihon kenkyu* 7 (1991): 63–88.

——. "Historiography and Historical Thought: East Asia." *International Encyclopedia of the Social and Behavioral Sciences.* Vol. 10. Amsterdam 2001.

Savarkar, Vinayak Damodar. *The Indian War of Independence 1857.* Orig. pub. 1909. Bombay 1947.

——. *Hindutva: Who is a Hindu?* Orig. pub. 1923. 6th ed., Delhi 1989.

Saxena, N. C. "Historiography of Communalism in India." In Mushirul Hasan, ed., *Communal and Pan-Islamic Trends in Colonial India.* Delhi 1981.

Schacht, Joseph. *The Origins of Muhammadan Jurisprudence.* Oxford 1950.

Scheftelowitz, J. *Die Zeit als Schicksalsgottheit in der Indischen und Iranischen Religion (Kāla und Zruvan).* Stuttgart 1929.

Scheibe, Erhard. "Die Einheit der Zeit." In Erich Fries, ed., *Festschrift für Joseph Klein zum 70. Geburtstag.* Göttingen 1967.

Schelling, Friedrich Wilhelm. *Philosophie der Mythologie.* Berlin 1856.

Schipper, Kristofer, and Hsiu-huei Wang. "Progressive and Regressive Time Cycles in Taoist Ritual." In Thomas J. Fraser, N. Lawrence, and F. C. Haber, eds., *Time, Science, and Society in China and the West.* Amherst 1986.

Schirokauer, Conrad. "Chu Hsi's Sense of History." In Schirokauer and Robert P. Hymes, eds., *Ordering the World: Approaches to State and Society in Sung Dynasty China.* Berkeley 1993.

——. "Chu Hsis Einstellung zur Geschichte." In Schmidt-Glintzer, ed., *Lebenswelt und Weltanschauung.*

Schlee, Günther. *Das Glaubens- und Sozialsystem der Rendille, Kamelnomaden Nord-Kenias.* Berlin 1979.

Schmidt-Glintzer, Helwig. "Traditionalismus und Geschichtsschreibung in China: Zur Maxime 'shu erh pu-tso.'" *Saeculum* 28 (1977): 42–52.

——. "Zeitbewußtsein im älteren China." In M. Horvat, ed., *Das Phänomen Zeit.* Vienna 1984.

——, ed. *Lebenswelt und Weltanschauung im frühneuzeitlichen China.* Stuttgart 1990.

Schneider, Ulrich. *Die großen Felsenedikte Ashokas.* Wiesbaden 1978.

Schnellenbach, C. *Geschichte als 'Gegengeschichte'? Historiographie in Kalhaṇas Rājataranginī.* Marburg 1996.

Scholtz, G. "Geschichte." In Joachim Ritter, ed., *Historisches Wörterbuch der Philosophie.* Vol. 3. Stuttgart 1974.

Schönherr, Ulrich. *Das Unendliche Altern der Moderne: Untersuchungen zum Romantrilologie Gert Jonkes.* Vienna 1994.

Sen-Gupta, Nares Chandra. *The Abbey of Bliss.* Calcutta 1906.

Settar, S., and G. D. Sontheimer. *Memorial Stones.* Dharwad and Heidelberg 1982.

Shah, A. M. "Social Anthropology and the Study of Historical Societies." *Economic and Political Weekly,* Special Number (July 1959).

——. "Towards a Sociological Understanding of Ancient India." *Contributions to Indian Sociology* 20.1 (1986).

Shalash 'A., ed., *Al-A'mâl al-majhûla* (Unknown works). London 1987.

Sharabi, Hisham. "The Neopatriarchal Discourse: Language and Discourse in Contemporary Arab Society." In George N. Atiyeh and Ibrahim M. Oweiss, eds., *Arab Civilization: Studies in Honor of Constantine K. Zurayk.* New York and Albany 1988.

Sharâra, Waddâh. *Hawl ba'd mushkilât ad-dawla fi'l-fikr wa'l-mujtama' al-'arabiyyayn* (On some problems concerning the state in modern Arab thought and society). Beirut 1981.

Shastri, H. Krishna. "Kuniyur Plates of Venkata II." *Epigraphia Indica* 3 (1894): 236–258.

Shaughnessy, Edward L. *Sources of Western Zhou History: Inscribed Bronze Vessels.* Berkeley, Los Angeles, Oxford 1991.

Shih, Vincent Yu-chung. *The Literary Mind and the Carving of Dragons.* Hong Kong 1983.

Shimada, Shingo. *Grenzgänge – Fremdgänge: Japan und Europa im Kulturvergleich.* Frankfurt and New York 1994.

——. "Überlegungen zu gesellschaftlichen Zeitlichkeitsregelungen in Japan." In Joachim Matthes, ed., *Zwischen den Kulturen?*

——. "Zwischen Orientalismus und Okzidentalismus: Maruyama Masaos Nihonseiji Shisôshi kenkyû (Studien zur Geschichte der politischen Ideen in Japan) wieder gelesen." In Detlef Thofern, ed., *Rationalität im Diskurs: Rudolf Wolfgang Müller zum 60. Geburtstag.* Marburg 1994.

Shukowskaja, Natalija L. *Kategorien und Symbolik in der traditionellen Kultur der Mongolen* [Studia Eurasia 4]. Berlin 1996.

Siddiqi, M. H. "History-Writing in India." *History Workshop Journal* (1980): 184–191.

Siemons, Mark. "Die Welt gehört demjenigen, der nicht fühlt." *Frankfurter Allgemeine Zeitung* (17 Nov. 1994): R2.

Simpson, David. *Romanticism, Nationalism, and the Revolt against Theory.* Chicago and London 1993.

Sims-Williams, Nicholas. "Further Notes on the Bactrian Inscription of Rabatak." *Beiträge zur Iranistik* 17.1 (1998): 79–92.

Sims-Williams, Nicholas, and Joe Cribb. "A New Bactrian Inscription of Kanishka the Great." *Silk Road Art and Archaeology* 4 (1995/96): 75–142.

Sircar, Dines Chandra. *Indian Epigraphy.* Delhi 1965.

——. "Some Performers of the Asvamedha." In K. K. Pillai, A. S. Narayana Pillai, and V. I. Subrahmaniam, eds., *Professor P. Sundaram Pillai Commemoration Volume.* Tinnevelly 1957.

Sivin, Nathan. "Cosmos and Computation in Early Chinese Mathematical Astronomy." *T'oung Pao* 55 (1969): 1–73.

Smith, Jane I., and Yvonne Y. Haddad. *The Islamic Understanding of Death and Resurrection.* Albany 1981.

Smith, Robert Sydney. *Warfare and Diplomacy in Pre-colonial West Africa.* London 1989.

Smith, Woodruff D. *Politics and the Sciences of Culture in Germany, 1840–1920.* New York 1991.

Sontheimer, Gunter D. *Pastoral Deities in Western India.* Transl. Anne Feldhaus. New York 1989.

Sorai, Ogyu. "Sorai sensei tomon sho." In *Kinsei bungaku ronshu: Nihon kotenbungaku taikei.* Vol. 94. Tokyo 1973.

——. "Taiheisaku." In *Ogyu Sorai: Nihon Shiso Taikei.* Vol. 36. Tokyo 1973.

Spiegel, Gabrielle M. "History, Historicism, and the Social Logic of the Text in the Middle Ages." *Speculum* 65 (1990): 59.

Spier, Leslie. "Inventions and Human Society." In Harry L. Shapiro, ed., *Man, Culture, and Society.* London 1971.

Spivak, Gayatri Chakravorty. *In Other Worlds: Essays in Cultural Politics.* New York 1988.

——. *The Post-Colonial Critic: Interviews, Strategies, Dialogues.* New York 1990.

Stanner, W. E. H. "The Dreaming: An Australian World View." In Peter B. Hammond, ed., *Cultural and Social Anthropology: Selected Readings.* New York 1966.

Stein, Lothar. *Die Sammar-Gerba: Beduinen im Übergang vom Nomadismus zur Seßhaftigkeit.* Berlin 1967.

Stein, Marc Aurel. *Kalhaṇa's Rājataranginī or Chronicle of the Kings of Kashmir: Sanskrit text with critical notes.* Bombay 1982 (repr. Delhi 1960).

——. *Kalhaṇa's Rājataranginī, a Chronicle of the Kings of Kashmir: Translation with Introduction, Commentary, and Appendices.* Westminster 1900, [Delhi 1961].

Stevenson, Ronald C. *The Nuba of Kordofan Province.* Khartoum 1984.

Straub, Jürgen, ed. *Narration, Identity, and Historical Consciousness.* (Making Sense of History, vol. 3). New York 2005.

Sturtevant, William C. "Anthropology, History, and Ethnohistory." *Ethnohistory* 13 (1966): 1–51.

Sugita, Genpaku. *Rangaku kotohajime* (The beginnings of Dutch sciences). Tokyo 1959.

Tenbruck, Friedrich H. "Gesellschaftsgeschichte oder Weltgeschichte?" *Kölner Zeitschrift für Soziologie und Sozialpsychologie* 41, no. 3 (1989): 417–439.

Tenenti, A. *Die Grundlegung der modernen Welt* (Fischer Weltgeschichte, vol. 12). Frankfurt/ M. 1981.

Thapar, Romila. *From Lineage to State: Social Formations in the Mid-First Millennium B.C. in the Ganga Valley.* Delhi 1990.

——. *Interpreting Early India.* Delhi 1992.

——. "Historical Realities." *Seminar* 329 (1987): 42–45.

——. "Interpretations of Ancient Indian History." In Thapar, *Ancient Indian Social History: Some Interpretations.* Hyderabad 1978.

——. "Society and Historical Consciousness: The Itihasa-Purana Tradition." In S. Bhattacharyya and Romila Thapar, eds., *Situating Indian History for Sarvepalli Gopal.* Delhi 1986, New Delhi 1991.

——. "Syndicated Moksha." *Seminar* 313 (Sept. 1985): 14–22.

——. "The Ramayana Syndrome." *Seminar* 353 (Jan. 1989): 71–75.

——. "Communalism and the Historical Legacy: Some Facets." In Panikkar, ed., *Communalism in India.*

Thapar, Romila, Harbans Mukhia, and Bipan Chandra. *Communalism and the Writing of Indian History.* Orig. pub. Delhi, Ahmedabad, Bombay 1969. 2nd ed., New Delhi 1977.

Theis, Joachim. *Nach der Razzia: Ethnographie und Geschichte der Koma* (Sudanesische Marginalien 3). Munich 1995.

Thurnwald, Richard. *Psychologie des primitiven Menschen.* Munich 1922.

Tillich, Paul, ed. *Kairos.* Darmstadt 1926.

——. *Writings in the Philosophy of Religion.* Ed. John Clayton. Berlin 1987.

Tillman, Hoyt Cleveland. *Utilitarian Confucianism: Ch'en Liang's Challenge to Chu Hsi.* Cambridge, MA, and London 1982.

Tjan, Tjoe Som. *Po hu t'ung: The Comprehensive Discussions in the White Tiger Hall.* 2 vols. Leiden 1949/1952.

Todorov, Tzvetan. *Symbolisme et interpretation.* Paris 1978.

Tripathi, G. C., and H. Kulke, eds. *Kaṭakarājavaṃśāvaliḥ: A Traditional History of Orissa.* Vol. 1. Allahabad 1987.

Tripathi, K. B. *Evolution of Oriya Language and Script.* Cuttack 1962.

Troll, Christian W. *Sayyid Ahmad Khan: A Reinterpretation of Muslim Theology.* New Delhi 1978.

Turner, B. S. *Weber and Islam.* London 1974.

Ueno, Chizuko. *Kindai kazoku no seiritsu to shûen* (Origin and end of the modern family). Tokyo 1994.

Unschuld, Paul U. "Therapie der Welt und ihrer Krankheiten." In Schmidt-Glintzer, ed., *Lebenswelt und Weltanschauung.*

van der Sprenkel, Otto Berkelbach. "Chronographie et historiographie chinoises." *Mélanges de l'Institut des Hautes Études Chinoises* 2 (1960): 407–421.

van Nieuwenhuize, C. A. O. "Muslim Attitudes towards Planning." *Internationales Jahrbuch für Religionssoziologie* 5. Special Issue: *Religion, Kultur und sozialer Wandel* (1969): 89–117.

Vansina, Jan. *Oral Tradition: A Study in Historical Methodology.* London 1965.

Vasu, Nagendra Nath, ed.97 "Copper-Plate Inscription of Nṛsiṃha-dēva II of Orissa, Dated 1217 Śaka." *Journal of the Royal Asiatic Society of Bengal* (1896): 229–271.

Veer, Peter van der. *Religious Nationalism: Hindus and Muslims in India.* Delhi 1996.

——. "The Foreign Hand." In Carol Breckenridge and Peter van der Veer, eds., *Orientalism and the Postcolonial Predicament.* Philadelphia 1993.

Walsh, W. H. *An Introduction to Philosophy of History.* London 1951.

Wansbrough, John. *The Sectarian Mileu: Content and Composition of Islamic Salvation History.* Oxford 1978.

Watanabe, Hiroshi. "'Shinpo' to 'chûka': Nihon no baai" ("Progress" and "China:" The case of Japan). In Yûzô Mizoguchi, ed., *Ajia kara kangseru.* Vol. 5: *Kindaikazô* (Thinking from Asia. Vol. 5: Images of modernization). Tokyo 1994.

Weber, Max. "Die 'Objektivität' sozialwissenschaftlicher und sozialpolitischer Erkenntnis." In Weber, *Gesammelte Aufsätze zur Wissenschaftslehre.* 3rd ed., Tübingen 1968. Engl.: *The Method-*

ology of the Social Sciences. Transl. and ed., Edward A. Shils and Henry A. Finch. New York 1949.

Webster, J. B. *The Revolutionary Years: West Africa Since 1800.* London 1980.

Wendorff, Rudolf. *Zeit und Kultur: Geschichte des Zeitbewußtseins in Europa.* Opladen 1980.

Westermann, Diedrich. *Geschichte Afrikas: Staatenbildungen südlich der Sahara.* Cologne 1952.

Whitehead, Alfred North. *Abenteuer der Ideen.* Frankfurt/ M. 1971.

Wilcox, Donald J. *The Measure of Times Past: Pre-Newtonian Chronologies and the Rhetoric of Relative Time.* Chicago 1987.

Wilhelm, Hellmut. "Der Zeitbegriff im Buch der Wandlungen." In H. Wilhelm, *Sinn des I Ging.* Düsseldorf and Cologne 1972.

Wilhelm, Richard. *Dschuang Dsi: Das wahre Buch vom südlichen Blütenland.* Jena 1912.

——. *Frühling und Herbst des Lü Bu We.* Orig. pub. 1928. Düsseldorf 1971.

——. *I Ging: Das Buch der Wandlungen.* Orig. pub. 1924. Munich 1989.

Wilson, Thomas W. *Genealogy of the Way: The Construction and Uses of the Confucian Tradition in Late Imperial China.* Stanford 1995.

Wu Youru. *Wu Yuru huabao, Shanghai Magazine.* 3 vols. Shanghai 1983.

Yamamoto, Nobuyoshi, and Toshihiko Konno. *Kindai kyôiku no tennô sei ideorogî: Meijiki gakkô gyôji no kôsatsu* (The Tenno-ideology in the modern education system: Considerations to school events in the early Meiji era). Tokyo 1973.

Yang Lien-sheng. "Schedules of Work and Rest in Imperial China." *Harvard Journal of Asiatic Studies* 18 (1955): 301–325. Repr. In Yang, *Studies in Chinese Institutional History.* Cambridge, MA, 1961.

——. "The Organization of Chinese Official Historiography." In W. G. Beasley and E. G. Pulleyblank, eds., *Historians of China and Japan.* London, New York, Toronto 1961.

Yasumaru, Yoshio. *Kindai tennô zô no keisei* (Construction of the modern images of Tenno). Tokyo 1992.

Yu-wie, K'ang. *Ta T'ung Shu: Das Buch von der Großen Gemeinschaft.* Transl. Laurence G. Thompson. Ed. Wolfgang Bauer. Düsseldorf and Cologne 1974.

Zimmermann, James H. "Die Zeit in der chinesischen Geschichtsschreibung." *Saeculum* 23 (1972): 332–350.

Zuncke, Christoph. *Eschatologie und historische Wirklichkeit in der frühen chinesischen Tradition.* Unpublished M.A. thesis. Munich 1993.

Zürcher, Erik. "'In the Beginning': 17th-Century Chinese Reactions to Christian Creationism." In Huang and Zürcher, eds., *Time and Space.*

Notes on the Contributors

Abdullahi A. An-Na'im, Dr., Professor of Law at Emory University in Atlanta, Georgia, USA. His research interests are in human rights in Islamic and cross-cultural perspectives, and comparative constitutionalism. His publications include *Toward an Islamic Reformation: Civil Liberties, Human Rights and International Law* (1990) and *Human Rights in Cross-Cultural Perspectives: Quest for Consensus* (1992). He has also published more than sixty articles and book chapters on human rights, constitutionalism, Islamic law, and politics.

Aziz Al-Azmeh, Dr., is Distinguished Visiting Professor at the Central European University, Budapest, and Fellow of the Collegium Budapest (Institute for Advanced Study). His books include *Ibn Khaldun* (1982), *Arabic Thought and Islamic Societies* (1986), *Muslim Kingship: Power and the Sacred in Christian, Muslim and Pagan Polities* (1997), *Islam and Modernities* (1992), and *Historical Knowledge and Historical Culture* (1983, in Arabic).

Georg Berkemer, Dr., Assistant Professor, Department of South Asian History at the South Asia Institute, University of Heidelberg. General research topics: historiography of South Asia, epigraphy, regional history of Orissa and Andhra Pradesh. His recent publications include *Orissa Bibliographic Database* (2005, with N. Aulike and W. Ulbrich), *Sharing Sovereignty: The Little Kingdom in South Asia* (2003, ed. with M. Frenz), and articles in collections of the Orissa Research Project, journals, and other collections.

B. D. Chattopadhyaya, Dr., is retired Professor of Ancient Indian History at the Centre for Historical Studies, Jawaharlal Nehru University, Delhi. His interests center on the study of the history of South Asia before the fifteenth century and span the disciplines of economic and social history, archaeology, numismatics, epigraphy, and literature. Some of his recent publications are, as editor with D. D. Kosambi, *Combined Methods in Indology and Other Writings* (2002), *Archaeology, Texts, and Historical Issues* (2003), and as editor with Gautam Sen-gupta and Sambhu Chakrabarty, *An Annoted Archaeological Atlas of West Bengal,* vol. 1 (2005).

Michael Gottlob, Dr., Language Teacher and Human Rights Activist, living in Bergamo and Berlin. General research topics: Historical writing and political culture in modern India. His recent publications include "Geschichtspolitik in Indien," *International Textbook Research* 23 (2001) and *Historical Thinking in South Asia: A Handbook of Sources from Colonial Times to the Present* (ed., 2003).

Fritz W. Kramer, Dr., Professor of Visual Anthropology at the Academy of Fine Arts, Hamburg, Germany. Main areas of interest: anthropology, art, and religion. His publications include *The Red Fes* (1993), *Zeitmarken* (1993), and *Bikini* (2000).

Achim Mittag, Dr., Professor at the Department of Sinology, Eberhard Karl University, Tübingen, Germany, is currently working on a project of comparative historiography and compiling a source book in Chinese historiography and historical thinking. His forthcoming publications include a collection of research studies entitled *Pacing the Past: Essays on Chinese Historiography and Historical Thought.*

Harro Müller, Dr., Professor at the Department of Germanic Language, Columbia University New York, New York. General Research Topics: German Literature from the eighteenth to the twentieth centuries, and theory and methodology of literary criticism. His books include *Giftpfeile: Zu Theorie und Geschichte der Moderne* (1994); and *Systemtheorie der Literatur* (ed. with Jürgen Fehrmann, 1996). He has published numerous articles on Schiller, Kleist, Büchner, Grabbe, Döblin, H. Mann, Brecht, Kluge, Schuldt, Benjamin, Adorno, Habermas, Luhmann, de Man, and Derrida.

Klaus E. Müller, Dr. Phil., Professor Emeritus of Ethnology at the University of Frankfurt am Main. Research interests: cognitive ethnology, ethnology of behavior, group identity, historical thinking, history of ethnology. His publications include: *Geschichte der antiken Ethnographie und ethnologischen Theoriebildung,* 2 vols. (1972–1980); *Das magische Universum der Identität: Elementarformen sozialen Verhaltens* (1987); *Die fünfte Dimension: Soziale Raumzeit und Geschichtsverständnis in primordialen Kulturen* (1999); *Wortzauber: Eine Ethnologie der Eloquenz* (2001); "Perspectives on Historical Anthropology," in Jörn Rüsen, ed., *Western Historical Thinking: An Intercultural Debate* (2002); *Der sechste Sinn: ethnologische Studien zu Phänomenen der außersinnlichen Wahrnehmung* (2004); "Der Ursprung der Geschichte," in Jan Assmann and Klaus E. Müller, eds., *Der Ursprung der Geschichte: archaische Kulturen, das Alte Ägypten und das Frühe Griechenland* (2005).

Jörn Rüsen, Dr., Professor of General History and Historical Culture at the University of Witten/Herdecke (Germany) and President of the Institute for Advanced Study in the Humanities at Essen in the Scientific Center of Northrhine-Westfalia. Fields of research: theory and methodology of history, history of historiography, modern intellectual history, historical consciousness, historical learning, processes of sense-generation, strategies of intercultural comparison and communication, general issues of cultural orientation in modern societies. His many publications include *Historische Vernunft* (1983), *Zeit und Sinn* (1990), *Geschichte im Kulturprozeß* (2002), *Kann Gestern besser werden?* (2003), and *History: Narration—Interpretation—Orientation* (2005).

Masayuki Sato, Dr., is now Professor of Social Studies in the Faculty of Education and Human Sciences at Yamanashi University in Japan. He is President of the International Commission for the History and Theory of Historiography (2005–2010), and a board member of the International Society for History Didactics (2002–2008). Works in English, German, and French include "The Two Historiographical Cultures in Twentieth Century Japan," in Rolf Torstendhal, ed., *An Assessment of the 20th century Historigraphy* (2000). "The Construction and Division of Time," in Solvi Sogner, ed., *Making Sense of Global History* (2001). "Historiography and Historical Thought: East Asia" (vol. 10, 6776–6782) and "Time, Chronology and Periodization in History"(vol. 23, 15686–15692) in *International Encyclopedia of the Social and Behavioral Sciences* (2002), apart from many works in Japanese including his latest book, *Historiographical Time and Space* (2004).

D. L. Sheth, Dr., is Senior Fellow and former Director of the Centre for the Study of Developing Societies, Delhi. He was appointed Member (Social Scientist) to the National Commission for Backward Classes (1993–1996). He is associated with Human Rights Movement and served as President of the Delhi chapter of People's Union for Civil Liberties (PUCL) (1991–1993). Sheth has edited *Citizens and Parties* (1995); with Ashis Nandy, *Multiverse of Democracy* (1996); and with Gurpreet Mahajan, *Minorities, Identities and the Nation State* (1998). He is editor of the journal *Alternatives: Social Transformation and Humane Governance* and a founding member of *Lokayan: Dialogues of the People.* A political sociologist, he has made major contributions to the study of comparative politics, the politics of social issues, and grassroots movements.

Shingo Shimada, Dr., Professor at the Department of Social Anthropology at the University of Halle-Wittenberg, Germany. General research topics: theory and method of cross-cultural comparison, sociology of knowledge, modern Japanese society. His publications include *Development of Sociology in Japan* (2004, ed. with Ilja Srubar), *Erfindung Japans: Kulturelle Wechselwirkung und politische Identitätskonstruktion* (2000), "Politik zwischen Differenz und Anerkennung: Multikulturalismus und das Problem der Menschenrechte," in Friedrich Jaeger, ed., *Handbuch der Kulturwissenschaften,* vol. 3: *Themen und Tendenzen* (2004).

Otfried Weintritt, Dr. (Islamic Studies). Research topics: Arabic historiography, literature, history of Islamic civilization. Publications include: *Formen spätmittelalterlicher islamischer Geschichtsdarstellung* (1992); *As-Safadi's Biographical Dictionary' Kitab al-Wafi bil-wafayat',* part 27 (ed., 1997); *Familie im Islam* (2001); "Transport in der islamischen Welt," in Rolf Peter Sieferle, ed., *Transportgeschichte im Internationalen Vergleich* (2004); and several articles in all his fields of research.

Index of Names